Biodiversity, Temperate Ecosystems, and Global Change

NATO ASI Series

Advanced Science Institutes Series

A series presenting the results of activities sponsored by the NATO Science Committee, which aims at the dissemination of advanced scientific and technological knowledge, with a view to strengthening links between scientific communities.

The Series is published by an international board of publishers in conjunction with the NATO Scientific Affairs Division

A Life Sciences B Physics	Plenum Publishing Corporation London and New York
C Mathematical and Physical Sciences D Behavioural and Social Sciences E Applied Sciences	Kluwer Academic Publishers Dordrecht, Boston and London
F Computer and Systems Sciences G Ecological Sciences H Cell Biology I Global Environmental Change	Springer-Verlag Berlin Heidelberg New York London Paris Tokyo Hong Kong Barcelona Budapest

NATO-PCO DATABASE

The electronic index to the NATO ASI Series provides full bibliographical references (with keywords and/or abstracts) to more than 30000 contributions from international scientists published in all sections of the NATO ASI Series. Access to the NATO-PCO DATABASE compiled by the NATO Publication Coordination Office is possible in two ways:

- via online FILE 128 (NATO-PCO DATABASE) hosted by ESRIN, Via Galileo Galilei, I-00044 Frascati, Italy.

- via CD-ROM "NATO Science & Technology Disk" with user-friendly retrieval software in English, French and German (© WTV GmbH and DATAWARE Technologies Inc. 1992).

The CD-ROM can be ordered through any member of the Board of Publishers or through NATO-PCO, Overijse, Belgium.

Series I: Global Environmental Change, Vol. 20

Biodiversity, Temperate Ecosystems, and Global Change

Edited by

Timothy J. B. Boyle

Christopher E. B. Boyle

Canadian Forest Service
351 St. Joseph Blvd.
Hull, Quebec K1A 1GS
Canada

Springer-Verlag
Berlin Heidelberg New York London Paris Tokyo
Hong Kong Barcelona Budapest
Published in cooperation with NATO Scientific Affairs Division

Proceedings of the NATO Advanced Research Workshop on Biodiversity, Temperate Ecosystems and Global Change, held at Montebello, Canada, August 15–19, 1993

ISBN 3-540-57950-8 Springer-Verlag Berlin Heidelberg New York
ISBN 0-387-57950-8 Springer-Verlag New York Berlin Heidelberg

Library of Congress Cataloging-in-Publication Data
Biodiversity, temperate ecosystems, and global change / edited by Timothy J.B. Boyle, Christopher E.B. Boyle. p. cm. – (NATO ASI series. Series I, Global environmental change ; vol. 20)
"Proceedings of the NATO Advanced Research Workshop on Biodiversity, Temperate Ecosystems and Global Change, held at Montebello, Canada, August 15-19, 1993"–T. p. verso. Includes bibliographical references and index.
ISBN 0-387-57950-8 (New York: alk. paper). – ISBN 3-540-57950-8 (Berlin: alk. paper)
1. Bioclimatology–Congresses. 2. Climatic changes–Environmental aspects–Congresses. 3. Biological diversity consercation–Congresses. 4. Biotic communities–Congresses. 5. Man–Influene on nature–Congresses. I. Boyle, Timothy James Butler, 1957- . II. Boyle, Christopher E. B., 1943- . III. NATO Advanced Research Workshop on Biodiversity, Temperate Ecosystems, and Global change (1993: Montebello, Québec) IV. Series. QH543.B54 1994 574.5'222–dc20 94-14990

Printed in Germany

Typesetting: Camera ready by authors
SPIN 10101840 31/3130 - 5 4 3 2 1 0 - Printed on acid-free paper

TABLE OF CONTENTS

Biodiversity, Temperate Ecosystems, and Global Change: Introduction to the NATO Advanced Research Workshop

T.J.B. Boyle
Canadian Forest Service
351 St. Joseph Blvd.
Hull, Quebec, Canada

In the second half of the 20th century, the burgeoning global human population, combined with dramatic developments in the power of human technology, has resulted in an immense increase in the impacts of human activities on the global environment. Huge, self-contained factory ships have replaced small, local fishing vessels; in place of axes and chainsaws, feller-buncher vehicles now work 24 hours a day harvesting forests; and new varieties of agricultural crops allow clearance of land previously unsuitable for agriculture. Thankfully, technological developments have also produced more rapid, more comprehensive, and more dramatic images of the extent of human impacts. Only with the advent of satellite images has the enormous extent of deforestation and erosion become obvious, and it is only with modern computing power that Global Circulation Models can be used to make predictions of future climates.

The idea of global resources being limited and the global environment being threatened is nevertheless difficult to comprehend when human interactions inevitably occur at a very local, small scale. This is certainly one reason why the response to the threats posed to the global environment has lagged behind the pace of the damage. It took the report of the World Commission on Environment and Development ("*Our Common Future*", WCED 1987) to draw the world's attention to the magnitude of the problem. The WCED issued a call to action, and popularized the concept of sustainable development as being a suitable goal for human activities.

NATO ASI Series, Vol. I 20
Biodiversity, Temperate Ecosystems, and Global Change
Edited by T. J. B. Boyle and C. E. B. Boyle

Since the production of the WCED report, there has indeed been some action. In 1992, the United Nations Conference on Environment and Development (UNCED) was held in Rio de Janeiro, and at that conference international treaties were signed on biodiversity and climate change. Since UNCED, international discussions have begun on desertification, and the idea of an international treaty on forests is still being pursued.

The fact that biodiversity was one of the first issues to gain the attention of the world's nations reflects an appreciation that biodiversity is both affected by global change, and provides a safeguard against the effects of global change. Obviously, land degradation and extinction events are two examples of global change issues that result in losses of biodiversity. However, the very fact that life exists on earth at all is the result of biodiversity, which allows organisms, species, and communities to respond to a changing environment.

A NATO Advanced Research Workshop on the subject of *Biodiversity, Temperate Ecosystems, and Global Change*, was held at Chateau Montebello, Quebec, Canada, August 15-19 1993. The goal of the workshop was to review the current state of knowledge concerning the relationship between global change and biodiversity of temperate ecosystems, and to determine which research topics should be given priority in order to promote increased understanding of the relationship and an improved ability to conserve biodiversity under conditions of global change. Although the theme of the workshop is widely recognized to be extremely topical and important, there is a degree of ambiguity concerning each of the terms used in the title of the Advanced Research Workshop. The use of the terms, and definitions adopted for the purposes of the workshop, are discussed below.

BIODIVERSITY

Biodiversity is a concept that encompasses all levels of biological diversity. A variety of definitions have been offered (appropriately, given the subject) for the concept of biodiversity. All of them attempt to capture essentially the same ideas, with varying degrees of clarity and succinctness. In the "Global Biodiversity Strategy", the term is defined simply as:

> "*the totality of genes, species and ecosystems in a region*" (World Resources Institute *et al.* 1991).

The definition given in "Caring for the Earth" is:

> "*the variety of life in all its forms, levels and combinations. [It] includes ecosystem diversity, species diversity and genetic diversity*" (World Conservation Union *et al.* 1991).

while McNeely et al. (1990) offer a description, rather than a definition, that reads in part:

> "*Biological diversity encompasses all species of plants, animals and microorganisms and the ecosystems and ecological processes of which they are parts. It is an umbrella term for the degree of nature's variety, including both the number and frequency of ecosystems, species or genes in a given assemblage*".

Perhaps the best, and most widely accepted definition was proposed by the U.S. Government's Office of Technology Assessment (U.S. Congress, Office of Technology Assessment 1987), which stated that:

> "*Biological diversity refers to the variety and variability among living organisms and the ecological complexes in which they occur. Diversity can be defined as the number of different items and their relative frequency. For biological diversity, these items are organized at many levels, ranging from complete ecosystems to the chemical structures that are the molecular basis of heredity. Thus, the term encompasses different ecosystems, species, genes and their relative abundance.*"

This definition emphasizes the two main points concerning biodiversity: first, that it comprises all levels of biological organization, from genes through to ecosystems; and secondly, that it refers not only to numbers but also to frequencies.

However, biodiversity is a concept that causes difficulty in application. For example, measurement of biodiversity, which is essential if changes are to be detected, strictly speaking requires the assessment and integration of measures of biodiversity at all levels in the biological hierarchy. This is not realistic; what is required is a comprehensive, but practical set of indicators of biodiversity, but as yet no consensus exists on what such indicators should be.

TEMPERATE ECOSYSTEMS

Temperate ecosystems are subject to several different interpretations, but the most widely used classification system is that based on the "Köppen system" (Bailey 1989). Under this classification system, temperate ecosystems are defined as follows:

Temperate oceanic:	8 months >10°C, warmest month <22°C
Temperate continental/warm summer:	4-8 months >10°C, warmest month >22°C
Temperate continental/cool summer:	4-8 months >10°C, warmest month <22°C.

Temperate climates are distinguished from boreal climates by having four or more months warmer than 10°C, while boreal climates are distinct from polar climates in having at least one month over 10°C. However, for the purpose of the Advanced Research Workshop, boreal ecosystems were included in the definition of temperate.

GLOBAL CHANGE

Global change is variously defined, and there is no consensus on what processes should be included and the most logical grouping of issues. The US Global Change Research Program supports research under the themes of:

- Climate Modelling and Prediction,
- Global Water and Energy Cycles,
- Global Carbon Cycle,
- Ecological Systems and Population Dynamics, and
- Other Research.

The last category included, for 1992, research on stratospheric ozone, human interactions, solid earth processes, and solar influences (Committee on Earth and Environmental Sciences 1992).

The Global Change Program of the Royal Society of Canada (RSC) recognizes seven global change issues. These are:

- Climate Change,
- Energy and Resource Consumption,

- Air and Water Pollution,
- Stratospheric Ozone Depletion,
- Population Increase,
- Extinction Events, and
- Land and Soil Degradation (Royal Society of Canada 1992).

Although the two programmes mostly cover the same issues, there are obvious differences in the priority or the emphasis placed on different issues. For example, the Ecological Systems and Population Dynamics component of the US programme is accounted for by the Extinction Events and Land and Soil Degradation issues of the RSC programme, while Stratospheric Ozone is identified as an issue in the RSC programme, but is included only under "other" in the US programme. In order to provide a consistent framework for discussion of global change in the Advanced Research Workshop, the seven issues identified in the RSC programme were used to describe global change.

Not all global change issues are equally significant to temperate ecosystems, nor to all levels of the biological hierarchy. Soulé (1991) assessed the relative impacts on terrestrial biodiversity at the gene and population, species, community, and ecosystem levels of six global change issues: habitat loss, habitat fragmentation, overexploitation, exotic species, pollution, and climate change. Soulé also distinguished between impacts in "poorer" and "richer" countries, the latter obviously encompassing most of the temperate zone. He suggested that the most significant impacts in richer countries are due to exotic species, at all levels of the hierarchy; pollution, at all levels except communities; climate change, at all levels except species; and habitat fragmentation, at the genes and populations, and species levels. In contrast, in the poorer countries habitat loss, habitat fragmentation, and overexploitation were identified as most significant, their impacts concentrated mainly at the lower levels of the hierarchy. Exotic species are also significant in poorer countries at all levels, and pollution is significant at the ecosystem level. Obviously, when dealing with broad generalizations, some of these suggestions could be contested, for example the impact of climate change on species in the temperate zone might be argued to be as significant as its impact at other levels. Nevertheless, Soulé's analysis does provide a useful framework for consideration of research priorities related to global change and biodiversity.

With this background, workshop sessions were organized on five topics:

- The threats posed by global change to biodiversity in temperate ecosystems,
- Levels and spatial patterns of diversity in temperate ecosystems,
- The impact of global change on genetic diversity,
- The effects of disturbance (natural and anthropogenic) on temperate ecosystems, and
- Existing research priorities and programmes.

Many of the presentations in each session focused on forested ecosystems, for example those by Hansen *et al.* (Chapter 12), Williams (Chapter 18), and Cimon (Chapter 20), reflecting the historical dominance of forests in temperate, including boreal, landscapes. However, in the modern world, although boreal landscapes continue to be dominated by forests, in many parts of the temperate zone agricultural or agro-forestry landscapes now dominate. Presentations by Thompson (Chapter 4), Hermy (Chapter 7), and Lenz (Chapter 14), among others, examined the current status of such landscapes. Far more attention has been paid to temperate ecosystems than to aquatic ecosystems, a reflection of the fact that impacts on terrestrial ecosystems are far more apparent, and that terrestrial systems are easier to study. However, the presentation by Wolff *et al.* (Chapter 17) described the status and prospects of a marine ecosystem which, because of its international significance, has been intensively studied. Research on socio-economic factors has also received relatively little attention, but the presentations by Machlis and Forrester (Chapter 2), and Hodges and Regens (Chapter 22) provide a good assessment of the requirements for future research. Presentations by Namkoong (Chapter 8), Gregorius (Chapter 9), and Skrøppa and Johnsen (Chapter 10) addressed the impacts of global change on genetic diversity, which is the basic level of diversity from which diversity at higher levels in the hierarchy is ultimately derived. Finally, outlines of existing research agendas *e.g.* Younès (Chapter 23) and Väisäinen and Jäppinen (Chapter 19) were presented to assist in the formulation of workshop conclusions, concerning an international research agenda for Biodiversity, Temperate Ecosystems, and Global Change.

REFERENCES

Bailey RG (1989) Explanatory supplement to ecoregions map of the continents. Environmental Conservation 16: 307-309.

Committee on Earth and Environmental Sciences (1992) Our Changing Planet: The FY 1992 US Global Change Research Program. Office of Science and Technology Policy, Washington DC.

McNeely, JA, KR Miller, WV Reid, RA Mittermeier and TB Werner (1990) Conserving the world's biological diversity. World Conservation Union, Gland, Switzerland.

Office of Technology Assessment (1987) Technologies to maintain biological diversity. US Government Printing Office, Washington, DC.

Royal Society of Canada (1992) Canadian Global Change Program. Royal Society of Canada, Ottawa.

Soulé ME (1991) Conservation: Tactics for a constant crisis. Science 253: 744-757.

WCED (World Commission on Environment and Development) (1987) Our Common Future. Oxford University Press, Oxford, U.K.

World Conservation Union, United Nations Environment Programme and World-Wide Fund for Nature (1991) Caring for the Earth: A Strategy for Sustainable Living. World Conservation Union (IUCN), Gland, Switzerland.

World Resources Institute, World Conservation Union, and United Nations Environment Programme (1991) The Global Biodiversity Strategy. World Resources Institute, Washington, DC.

Chapter 1: Conservation strategies in the face of Global Change

David Neave, C. Caza, and S. Henneman

Wildlife Habitat Canada

1704 Carling Ave.

Ottawa, Ont.

Canada

Over the past eight years, Wildlife Habitat Canada (WHC) - a non-profit national foundation dedicated to habitat conservation - has adopted a landscape approach to the conservation of habitat. In so doing WHC seeks to balance acquisition of critical habitat resource lands with the stewardship of private lands, ensure adequate policies and planning, and encourage public participation and education. Projects for wetland conservation are a priority for funding. We also carry out research and education projects across Canada. Through our projects and our publications, Wildlife Habitat Canada is recognised nationally and internationally as an authority on habitat issues.

Working for a wildlife conservation foundation, we approach the subject of global change from a somewhat different perspective than many of you. The perspective we bring to this topic is that of conservationists faced with the challenge - indeed, given the statistics on decline in wildlife, the imperative - to change the way others manage land to ensure that there is room in our future landscapes for wildlife, in its broadest sense.

We would like to demonstrate to you how wildlife and wildlife habitat fit into this workshop by outlining how important wildlife is, not only as a natural resource, but as a publicly understood barometer of the health of all our landscapes and as a yardstick to measure the success of conservation strategies (or their more recent analogues, strategies for sustainable development). Seen from this perspective, wildlife and wildlife habitat provide an holistic view of the effects of global change.

NATO ASI Series, Vol. I 20
Biodiversity, Temperate Ecosystems, and Global Change
Edited by T. J. B. Boyle and C. E. B. Boyle

They are the end products of the sum of various changes that reflect social, economic and ecological forces on the landscape.

We will then show how the landscape approach to habitat conservation represents one approach to sustainable development. We will show how this approach integrates some of the lessons we have already learned from global change; we will also demonstrate, using this approach, some of the challenges and opportunities in applying strategies for sustainable development and we will illustrate by reference to a few of the many projects we are involved in across Canada. Our aim is to turn our focus from apprehension over dire predictions of global disaster, to a positive approach to the conservation of global ecosystems.

Historically, conservation was treated primarily as a site-specific or a species-specific issue. In the 1970s and 1980s, scientists began to realize that there were significant global impacts resulting from human activities. They realized that air, soil and water pollution, loss of species and genetic diversity, were no longer local phenomena.

Habitats are a product of a changing landscape to which wildlife must constantly adapt. Most, if not all of the issues of global change which you are considering this week, are having a significant impact on habitat. As the retention of biodiversity is a cornerstone of sustainable development, it follows that the status of habitat will indicate our success in achieving sustainability.

Some of the potential impacts of global change on habitat have been well described in our 1991 report "*The Status of Wildlife Habitat in Canada*". There are two major ways in which habitat has been affected by global change. The first, is obviously the conversion of land - the quantity of various habitats. Examples include:

- loss of wetlands due to agricultural drainage; and
- the loss of certain forest types due to logging activities.

The second way, is the change in quality of habitats, affected by various forms of pollution, including:

- changes in water quality from agricultural practices; and
- the acidification of lakes due to acidic precipitation.

Studies of global change have taught us about the scale of these types of impacts, the subtle synergy of contaminants, the results of cumulative impacts, and the existence of ecological thresholds.

The World Conservation Strategy (World Conservation Union *et al.* 1980) was perhaps the first attempt to organize these concerns into a strategy for global sustainability. In 1987, the World Commission on Environment and Development released their report, "*Our Common Future*" (WCED 1987). At this time, the social and economic dimensions were added to the environmental dimension and documents like "*Caring for the Earth: A Strategy for Sustainable Living*" (World Conservation Union *et al.* 1991) appeared. This report represented a step forward in our thinking about conservation strategies.

Canada was quick to jump aboard the conservation train and, more recently, the sustainable development rocket, but largely at the paper stage. That is, conservation strategies (and their more recent relatives, sustainable development strategies) have been or are being developed in many provinces across Canada and in a number of national forums. Provincial and sector strategies have endeavoured to incorporate the basic principles of the World Conservation Strategy in the development and implementation of environmental and related programmes and policies.

Unfortunately, none of these has yet to be fully implemented and it is still unclear what impacts any will have on actual management practices. This is because Canada's current approach to environmental management still does not appear to reflect many of the lessons we have learned from global change. Scientists must take some of the blame for often being unable to translate their growing concerns over our changing world into appropriate policy or programme directions for governments.

In Canada, the Federal Government's goal is to set aside 12% of our land mass in protected areas, such as parks, by the year 2000. Yet the majority of our protected areas were not selected on the basis of ecological representation or biological integrity. The biodiversity at greatest risk lies outside of protected areas, on privately held, or corporately leased, lands. In addition, a protected areas' strategy cannot protect the biodiversity <u>inside</u> protected areas from the effects of cross-boundary impacts or landscape-level impacts, certainly not in relation to the effects of global warming in respect to shifting ecosystems. As a result, in the long-term, protected areas will fail to conserve wildlife and we will not have ensured the long-term survival of our wildlife and their habitats even if we reach the highly publicized goal of 12% for protected areas in Canada.

Our other current environmental management strategy is to sustain the *status quo*, and it relies on the enforcement of minimum environmental standards "to protect the environment", in areas other than parks. These standards ignore cumulative, or system-level, impacts across the landscape and treat each new proposal for resource development as if it existed in isolation in time and space. This means we will not even be able to maintain current levels of sustainability - as reflected in habitat conditions and wildlife populations. The inevitable consequence will be a continued decline in environmental quality because many of the impacts of global change are at the system-level. It is this which is causing alarm - how can we be satisfied with a minimum standards and site-specific approach? Yet this is the basis of the much publicized new Environmental Assessment Act of the Federal Government.

We know enough to be doing a better job. We should be doing comprehensive land-use planning at scales of space and time that reflect the scales at which natural systems function. Coupled with clear goals for sustainability, we will then end up with effective sustainable conservation strategies that achieve our potential.

The current emphasis within governments on process is almost counterproductive and will remain so until the goals and objectives of sustainable development are defined, understood and accepted by society. Canadians should be calling for the formulation of goals for environmental protection, and exercises, such as the Federal Government's State of the Environment Reporting, should be measuring progress towards those goals. Goals such as complete compliance with minimum environmental standards, and a target of 12% for protected areas in Canada, will not, by themselves, move Canada closer to sustainable development. We do not need a 12% solution -- we need a 100% solution based on explicit goals for the conservation of the environment across the entire landscape and strategies to achieve these goals. We must develop a vision of the future we desire and it must be a future that this land is capable of providing.

What role can research and researchers on global change play in this? We believe that you can provide the tools to reach a vision through research such as the following:

Economic Research:

- to help us measure our progress towards sustainable development by calculating the costs of environmental degradation;
- we need economic indicators of the environment (other than growth and employment rates) that also account for the value of our renewable resources to allow us to develop a national balance sheet that accounts for the ecological capital;
- we need to focus on national budgets rather than State of the Environment reports as the major annual government commitment/policy statements. We need to be able to understand the impact of subsidies, incentives and tax breaks that are aimed at encouraging economic activity and reflect social issues rather than focusing on budgets of environmental agencies; and
- to balance incentive programmes, reflecting social concerns with the current regulatory approach. How do we take advantage of the trade settlements that will limit subsidies but allow recognition of broad social objectives such as the environment?

Social Research:

- to determine what it is people want to sustain in our landscapes;
- we must determine what our social values are; and
- we must define society's relationship to the natural environment.

Ecological Research:

- we must be able to provide accepted definitions and an understanding of terms such as landscape diversity and ecological integrity;
- we need to experiment with adaptive management strategies in forest and agricultural management which incorporate many objectives, including wildlife; and

- there is a particular need for research and experimentation at the landscape level in such areas as: the behaviour of populations in a landscape; the relationship between recolonization and sustainability in local areas within landscapes; and the effects of different patch sizes and distributions on habitat quality and persistence.

In addition, we need:

- to be able to integrate information from various disciplines onto a landscape basis;
- to develop useful ecological indicators that measure the success in achieving our goals;
- research to test the validity of indicators and compare alternative classes of indicators; and
- to use tools such as Visual Impact Analysis (V.I.S.) which can help us visualize future landscapes. With the use of historical data and trends in land use, we can paint images of alternative futures based on current patterns of land management. Tools such as V.I.S. and appropriate ecosystem land classification information can be used in forest management.

With the assistance of this type of research, what might our conservation strategies look like? Wildlife Habitat Canada offers what we call the landscape approach as one example. This is the approach WHC has adopted for habitat conservation over the last eight years and we have organized our programme and goals around it. In so doing, our programme identifies three major elements:

- **Policy reform** - ensuring governments establish clear goals that recognize environmental implications;
- **Development of stewardship initiatives** - ensuring that landowners and corporations which are currently conserving the natural resources and attributes on their land effectively are recognized before paying poorer managers for remedial measures; and

- **Protection of critical areas** - as ecological benchmarks and to protect sensitive areas from inappropriate disturbance.

What is a landscape approach?

This approach includes urban, coastal, agricultural, northern, and forested landscapes. It is based on the principles of landscape ecology and is applied with the techniques of landscape management; it is regional in scale, process-oriented, ecosystem-oriented, and includes all elements of the landscape, including protected and unprotected lands, as well as humans and their activities. It is an approach rooted in a stewardship ethic - respect for land that is reflected in planning; it provides a practical framework for management and it is based on a vision of the future landscape.

There are many initiatives which are currently underway across Canada that have adopted this approach or aspects of it, and many others which are being developed that are doing so. We have new partnerships and programmes with which we could practise sustainable development. Examples include:

- The North American Waterfowl Management Plan - an ambitious international 15 year plan to restore waterfowl populations and wetland habitats throughout North America at a cost of 1.5 billion (US) dollars;
- The National Forest Strategy - a 5 year blueprint for changes in forest management towards "sustainable forestry" developed and to be implemented by a broad partnership of business, academic, government, non-government, and native groups;
- Some of the agricultural initiatives funded by the Federal Government's Green Plan have enabled farmers' associations to develop "Our Farm Environmental Agenda" programmes and "Best Management Practices" programmes;
- In New Brunswick, legislation requires that the amount and distribution of forest habitat required to maintain specific wildlife populations in all 10 crown licences is defined similarly to timber forecasts through computer forecasting

models. Similar projects are underway in Nova Scotia, Saskatchewan, Alberta and Manitoba;

- Practical approaches for landowners, in Quebec for example, which financially encourage private woodlot owners to enhance winter habitat for deer; and
- The Montague River Watershed of Prince Edward Island, a community-based forestry project with a 100% landowner involvement, which has implemented a range of forestry and wildlife conservation initiatives including combatting soil and water erosion.

Unfortunately some of these examples still lack that most important component - this vision of the future landscape. And as a result, many of our current policies are barriers to change rather than tools for change. In this regard, Wildlife Habitat Canada has recently collaborated with key stakeholders to work toward visions of future forest and agricultural landscapes, as described in our reports entitled "*Natural Forest Landscape Management in Canada - Setting a Global Standard for Implementing Sustainable Development*", and "*Wildlife Stewardship on Farm Land*".

What might future forested landscapes look like in Canada? This question does not seem to be at the centre of Canada's current debate with other countries regarding the adoption of sustainable forest practices. Harvesting a tree from a plantation seems to be the easiest solution to many critics and policy makers. Yet the encouragement of plantations in Canada would be folly from an economic, social and particularly ecological perspective. We clearly need to maintain the ecological integrity of the immense breath of our boreal forest rather than retaining a few parks and promoting tree farms.

In drafting the vision, "*Natural Forest Landscape Management in Canada*" (Booth *et al.* 1993), a group of us attempted to paint the picture which would urge Canadians, as stewards of a globally significant resource, to maintain ecological integrity and forest industry competitiveness. This paper advocates a landscape approach based on the principle of ecological sustainability that focuses on maintaining the full range of natural forest ecosystems. In practice, the approach views the forest in its entirety moving from management of stands to the management of forest landscapes. It leads to a "lighter hand" approach with more attention to natural regeneration and longer rotations for most of the landbase. In so doing, it provides a good fit between economics and environment while providing a wider range of recreational and other community activities.

This is what our conservation strategies must include: a vision of the future which the land is capable of providing, goals for this future, and strategies to achieve these goals. Only when we meet these objectives will we be able to build the framework in which sustainable development can work.

In conclusion, the decade of the 1990's is an exciting time for conservation. There is increasing synergy between government interest and action to protect the environment, and the growing willingness and capability of the private sector and citizens to deliver programmes. In contributing to this movement, Wildlife Habitat Canada has promoted:

- a better balance between legislative control and private stewardship initiatives;
- a need to strengthen the philosophical underpinnings and use existing implementation programmes to achieve sustainable development; and
- a landscape approach to forest management that will ensure the retention of the integrity of our natural forests. We have promoted this through partnerships and by speaking out for the need of a common vision. We have also helped define the first vision of our future natural forests.

Through this workshop, we hope you will be more aggressive in seeking changes in policy and in ensuring that what you know is making a difference in what is happening on the ground. At a time when trade wars are creating questionable environmental criteria, we believe it is imperative that scientists speak out on their visions of our future landscapes - not dire prophecies of gloom, but landscape-oriented visions with appropriate management scenarios. We need very different forms of research to reach these specific goals.

REFERENCES

Booth DL, Boulter DWK, Neave DJ, Rotherham AA, Welsh DA (1993) Natural forest landscape management: A strategy for Canada. The Forestry Chronicle 69: 141-145.

WCED (World Commission on Environment and Development) (1987) Our Common Future. Oxford University Press, Oxford, U.K.

World Conservation Union, United Nations Environment Programme and World-Wide Fund for Nature (1980) World Conservation Strategy: Living Resource Conservation for Sustainable Development. World Conservation Union (IUCN), Gland, Switzerland.

World Conservation Union, United Nations Environment Programme and World-Wide Fund for Nature (1991) Caring for the Earth: A Strategy for Sustainable Living. World Conservation Union (IUCN), Gland, Switzerland.

Chapter 2: Social factors as driving forces: Towards interdisciplinary models of global change

Gary E. Machlis and Deborah J. Forester
Department of Forest Resources
University of Idaho
Moscow, Idaho 83843
USA

INTRODUCTION

Homo sapiens is not so fortunate as to have its ecological challenges be separate, distinct and amenable to simple solutions. Biodiversity loss and climate change are examples. The loss of biodiversity is a significant ecological problem, and an important component of global environmental change (Ehrenfeld, 1972; Soulé, 1986; Wilson, 1988a; Lovejoy, 1988; NRC, 1992). The major cause of biodiversity loss is human action, primarily land-use that alters habitat (Pimm and Gilpin, 1989; Freedman, 1989). Climate change is also a critical ecological problem (Peters and Lovejoy, 1992; Gable *et al.*, 1990; Peters, 1988), and anthropogenic factors such as fossil fuel use and release of trace gases are now considered significant driving forces (NRC, 1992; Goodess *et al.*, 1992; Glantz and Krenz, 1992). Further, there are important linkages between climate change and biodiversity loss (Peters and Lovejoy, 1992; Root and Schneider, 1993), and these relationships are often interactive, indirect and nonlinear.

In addition, mitigation of both challenges relies upon human institutions of some complexity and unreliability. Culture and technology empowers and confounds us as a species, and contrary to the economist's assumption, our behaviour is often far from rational. It is not given that policy and actions aimed at solving one environmental problem will help solve another.

NATO ASI Series, Vol. I 20
Biodiversity, Temperate Ecosystems, and Global Change
Edited by T. J. B. Boyle and C. E. B. Boyle

Hence, unravelling the relationship between social factors and global change is unlikely to be easy, amenable to single-cause explanations or ideologies, or simple to test and verify. The task is made difficult by the orthodoxies of scientific disciplines (often myopic), the style of modern scientific inquiry (heavily empiricist), and the reluctant and wary cooperation between biological and social scientists (Buttel and Taylor, 1992). The purposes of this paper are to 1) provide a primer on social factors influencing biodiversity loss and climate change, and 2) describe our efforts to build interdisciplinary models of biodiversity loss . As this is an interdisciplinary exercise, some preparatory remarks are in order.

The earth transformed by human action

Human impacts at the global level have historically been examined by the scientific community; three efforts in English literature are milestones. George Perkins Marsh published "*Man and Nature; or Physical Geography as Modified by Human Action*" in 1864 (Marsh, 1907). Marsh used an encyclopædic set of examples and first-hand observations to document how mankind had modified nature, often unintentionally and with adverse consequences. Almost a century later, a second major review was published, based upon an international symposium sponsored by the German Wenner-Gren Foundation. Out of the symposium (held in 1955) came an edited volume "*Man's Role in Changing the Face of the Earth*" (Thomas, 1956). It included reviews and analyses by numerous scholars, and was based primarily upon case studies. *Man's Role* demonstrated both the quickening pace of global change and the complex interaction between social and biological systems.

Half a century later, Clark University in the USA sponsored a third major inventory of human-caused global change. The resulting volume, "*The Earth as Transformed by Human Action*", was published in 1990 (Turner *et al.*, 1990). It focused on changes occurring in the past 300 years, and took advantage of the vast improvement in available data on global change and environmental conditions. The Clark-led effort found that "humankind has altered or transformed virtually every element of the biosphere" (Kates *et al.*, 1990). The authors recite a litany of impacts:

"... *an area the size of France submerged by artificial reservoirs; [in the] 15% of all plant species now estimated to be threatened with extinction; [in the] estimated threefold increase in sediment load in typical large rivers...industrial emissions alone now multiply the annual natural releases of arsenic by 3, of cadmium by 7, of mercury by 10, and of lead by 25* ..." (Kates *et al.*, 1990:6).

In addition to these major inventories, numerous other reviews (more limited in scope or data) have been prepared (for example, see Meadows *et al.*, 1972; Clark and Munn, 1986; for a summary, see Kates *et al.*, 1990). The literature repeats the theme: the earth is being transformed by human action and, in many instances and regions, the magnitude and rate of transformation are dramatically increasing.

Our knowledge has increased since Marsh and the Wenner–Gren meeting. Data on global change have improved in recent years, partly due to advanced information technology and the increased economic value such data commands. Likewise, our understanding of human factors influencing environmental change has become more insightful, explicit and useful. What has been described by the US National Research Council as research on the "human dimensions of global change" (NRC, 1992) is in reality a growing collection of interdisciplinary work that:

"... *strives to understand the interactions between human systems and environmental systems...and to understand the aspects of human systems that affect those interactions*" (NRC 1992:33).

Broadly conceived, the human dimensions of global change are independent variables (causes), intervening variables (influences), and dependent variables (consequences). For example, at local or regional scales, human population structure (such as the relative size of each birth cohort) can be a cause, influence, and/or consequence of resource depletion. This is a source of much confusion (particularly when ecologists and social scientists attempt a dialogue). In this paper, we focus on human dimensions as independent and intervening variables. The socioeconomic consequences of global change (human dimensions as the dependent variable) have been discussed elsewhere (see, for example, Goody, 1982).

Biodiversity loss and climate change as exemplars

Global change is a comprehensive and ambiguous term; numerous definitions exist (see, for example, NRC, 1992). The seven "issues" identified by the Royal Society of Canada as constituting global change are: climate change, energy and resource consumption, air and water pollution, stratospheric ozone depletion, population increase, extinction events, and land and soil degradation (Boyle, 1993). Some of these issues are causes (population increase), some are influences (energy consumption and air pollution), and some are consequences (extinction events or climate change). All may have impacts upon biodiversity; biodiversity loss has itself been identified as a critical source and consequence of global change (Wilson, 1992).

Biodiversity loss and climate change are exemplars, and can illustrate how the human dimensions of global change are important and amenable to scientific analysis. Both are ecological conditions that have recently become important international issues (witness the Montreal Protocol on chlorofluorocarbons and the UNCED Biodiversity Treaty). Both reflect the serious concerns of the scientific community (Peters and Lovejoy, 1992) as well as the general public (Dunlap, 1993). Both have emerged as fields of expanded research in the last decade and have produced empirical, quantitative databases (whether it be temperature records or species lists). Importantly, biodiversity loss and climate change each have significant human dimensions only recently being addressed (NRC, 1992; Machlis, 1992).

Using biodiversity loss and climate change as exemplars allows us to focus our analysis of the human dimensions of global change, and makes the task of developing preliminary models somewhat more tractable. Hence, in this effort, social factors are treated as causes, climate change as an intervening influence (as we shall see, one of several), and biodiversity loss as a dependent variable—a consequence of the earth transformed by human action.

Overview of the paper

After this introduction, we begin with a set of basic definitions, for interdisciplinary communication is often hampered by technical vocabulary and concepts left unclear. Biodiversity loss, climate change and social factors are defined, and we present a general and brief description of current conditions and trends. Following this, we provide a basic primer on the human dimensions of biodiversity loss and climate change, so that readers unfamiliar with the social sciences can evaluate their potential contribution.

We then present our work towards an interdisciplinary model of biodiversity loss as an example of first efforts. Climate change is explicitly considered as an intervening variable. Finally, we outline key objectives for improving our ability to model the human dimensions of global change.

DEFINITIONS AND CURRENT CONDITIONS

Climate change

Climate change refers to long-term shifts in normal climate (MacDonald, 1989). Natural forces such as orbital forcing, ocean circulation and heat storage, solar activity, and volcanism may influence climate change (Goodess *et al.*, 1992; Jäger and Barry, 1990). Increasingly, human activities such as energy use and deforestation are influencing climate, by causing the build up of gases in the atmosphere that trap radiative heat and may lead to warming of the earth's atmosphere (Abrahamson, 1989; Ciborowski, 1989; Goodess *et al.*, 1992; Hammond *et al.*, 1991; Root and Schneider, 1993; Schneider, 1989; Sedjo, 1990; Woodwell, 1989). The major greenhouse gases responsible for this effect are water vapour, carbon dioxide, methane, nitrous oxide, tropospheric ozone, and chlorofluorocarbons (Abrahamson, 1989; Ciborowski, 1989; Gable *et al.*, 1990; Glantz and Krenz, 1992; Hammond *et al.*, 1991; Turco, 1992). Questions remain about the timing and magnitude of climate change, though there is much agreement that it will occur (Graham *et al.*, 1990; Schneider, 1989; Root and Schneider, 1993).

Global warming will likely result in climatic zone shifts, altered rainfall patterns, soil moisture changes, rising sea levels, and the occurrence of extreme weather events (Abrahamson, 1989; Flavin, 1989; Gable *et al.*, 1990; Hansen, 1989; Postel, 1988; Schneider, 1989). These climatic changes are likely to cause adaptations in biological and social systems, including changes in vegetation patterns and animal distributions, biodiversity loss, interference with agricultural activities, deterioration of air quality, reduced freshwater availability, public health problems, and changes in energy demand (Cohn, 1989; Flavin, 1989; Peters, 1989; Postel, 1988; Schneider, 1989; WRI, 1990). The greatest temperature changes likely to occur with global climate change will be in the middle and higher latitudes (Graham *et al.*, 1990). The oceans, which take more time to warm than air, effectively create "committed warming" due to past greenhouse gas releases (Abrahamson, 1989).

Hence, both temporal and spatial scale considerations are keys to defining and identifying climate change (Jäger and Barry, 1990). Numerous definitions of climate change exist. McDowell *et al.* (1990), expanding on work by Clark (1985), have defined variations in weather and climate at several spatial and temporal scales. For example, major droughts and El Niño are labelled short period variations, and are located at regional scales. Global warming and cooling operate at long time scales (10^9 years) and over large areas (10^9 km^2).

The simplest models used for examining global climate change are energy or radiation balance models that equate solar energy and terrestrial radiation (MacDonald, 1989; Schneider, 1992). A hierarchy of gradually more complex models is used to study the potential responses of the atmosphere to external forcing (Schneider, 1992). The most complex are global circulation models (GCMs). These are three-dimensional computer models that work at scales of hundreds of square kilometres (MacDonald, 1989; Root and Schneider, 1993; Schneider, 1992). They attempt to represent the "interactions among the atmosphere, oceans, ice, biota, and land" (Root and Schneider, 1993).

Climate models look at "what-if" scenarios (Dickinson, 1992). Changes in the chemical composition of the atmosphere, such as the equivalent doubling of atmospheric carbon dioxide (estimated to occur in about 50 years [Schneider *et al.*,

1992]), are used as the basis for the models. The Intergovernmental Panel on Climate Change (IPCC) considers some of the predictions based on these models to be very reliable; part of their basic outline is displayed in Table 1.

Table 1. Predictions of climatic effects resulting from a doubling of atmospheric CO_2.

Highest Confidence	* warming of earth's surface and lower atmosphere * cooling of stratosphere
Very Good Confidence	* global average increase in precipitation and evaporation (the larger the warming the larger the increase) * area of sea ice and seasonal snow cover diminish
Good Confidence	* the earth's temperature will rise between 1.5°C and 4.5°C * surface warming at higher latitudes will be greater than the global average in winter and smaller in summer * the least warming and seasonal variation will be in the tropics * precipitation will increase at high latitudes throughout the year * precipitation and evaporation will increase 3-15% * soil moisture will increase in high latitudes in winter

Source: Maunder, 1992:40

Biodiversity loss

Biological diversity has numerous definitions (Huntley, 1988); most treat diversity as a qualitative state at genetic, species, ecosystem, or landscape levels (Wilson, 1988a). In the sciences of population biology and ecology, diversity is a multi-dimensional concept (MacArthur, 1985). Wilcox (1984:640) defines biodiversity as "*the variety of life forms, the ecological roles they perform, and the genetic diversity they contain.*" A definition by the U.S. Office of Technology Assessment is similar:

> "*Biological diversity refers to the variety and variability among living organisms and the ecological complexes in which they occur. Diversity can be defined as the number of different items and their relative frequency. For biological diversity, these items range from complete ecosystems to the chemical structures that are the molecular basis of heredity. Thus, the term encompasses different ecosystems, species, and genes, and their relative abundance.*" (1987:1).

There is disagreement over how to measure biological diversity (Hurlbert, 1971; Norton, 1986); current measures select different components of ecosystems for emphasis. Potential indicators include number of species, or "richness" (Scott *et al.*, 1987; Magurran, 1988), abundance and distribution of populations (Krebs, 1972; Westman, 1990), number of endangered species, centres of species-richness with high endemism (Myers, 1988), and degree of genetic variability (Allen, 1963; Ruffie, 1982). Other approaches treat ecosystem functions (Ray, 1988), interactions (Janzen, 1988), natural communities (The Nature Conservancy, 1975; Western *et al.*, 1989), successional stages (Franklin, 1988), or ecological redundancy (Walker, 1992) as key measures of diversity.

There is widespread agreement that global biodiversity is being reduced at an accelerated rate (Myers, 1980; Wilson, 1988a). There is less agreement about the actual level of biodiversity loss (Harwood, 1982; Lovejoy, 1986), compounded by the wide range of operational measures, variation between biomes, and lack of baseline knowledge about the number of species and taxa (Freedman, 1989). Even when there is a common measure (say, species-richness) and biome (tropical forests), differences in estimated rates are large. Lugo (1988) compared several estimates of species extinction in the tropics; they ranged from 15% to 50% of all species by the year 2000.

Using any of these estimates, the loss of biodiversity is an important social and biological concern. Lost species may have commodity values; an example is the rosy periwinkle, a species of plant yielding pharmaceuticals worth $100 million annually (Farnsworth, 1988). Species may have intrinsic values in a socio-cultural system (Ehrenfeld, 1976; Norton, 1987; Shiva *et al.*, 1991), and landscapes that provide habitat may have symbolic meanings (Burch, 1971; Kellert, 1986), provide an economic resource base (Brown, 1985), or serve legally mandated functions such as wilderness (Hendee, 1990). Biologically, loss of species may lead to synergistic effects on other species, altered energy flows and nutrient cycles (Freedman, 1989), reduced ecosystem services such as oxygen production and climate modification (Mueller-Dombois *et al.*, 1983; Norton, 1986), and a decline in ecosystem resilience (Margalef, 1968; Wilson, 1988a). Hence, biodiversity loss is a global concern (Soulé, 1986) and in some locations, such as the Amazonian tropical forests, a critical problem (Myers, 1988).

Social factors as driving forces

The industrial revolution, demographic explosion of *Homo sapiens* and rise of a global exchange economy represent major social factors influencing global environmental change. Social factors can be defined as social, cultural, economic and political variables that act as driving forces in affecting ecological systems. A report by the US National Research Council (1992) describes five key driving forces: 1) population growth, 2) economic growth, 3) technological change, 4) political-economic institutions, and 5) attitudes and beliefs. The mechanisms by which these forces influence natural systems at the global level are not fully understood, though a general description is possible.

Population growth increases overall resource use (including area, intensity and magnitudes), even with a stable or modest decline in per capita use levels. Human demographic trends at global scales can be presented in at least three ways: arithmetic-exponential, logarithmic-logistic, and arithmetic-logistic—all demonstrate increasing population (Whitmore *et al.*, 1990). Table 2 shows population growth rates since 1900; 1985-2020 growth rates are predicted to be highest for Africa, Asia (excluding China and Japan) and Latin America (Demeny, 1990). These regions have high biodiversity values as well.

Table 2. Average annual rates of population increase (percent). Estimates: 1900-1985; Projection: 1985-2020.

	1900-1950	1950-1985	1985-2020
World total	0.84	1.88	1.45
Africa	1.0	2.6	2.7
Asia[a]	0.8	2.1	1.4
China	0.5	1.9	0.9
Japan	1.3	1.0	0.2
Rest of Asia	1.0	2.3	1.8
Europe[a]	0.6	0.6	0.1
USSR	0.7	1.2	0.6
North America[b]	1.2	1.3	0.6
Latin America[c]	1.6	2.6	1.6
Oceania	1.6	1.9	1.2

Notes: [a] Excluding the USSR
[b] Canada and the United States
[c] South and Central America, Mexico, and the Caribbean
Source: Adapted from Demeny, 1990.

Economic growth increases resource use, by increasing the per capita demand and supply of goods and services (Simmons, 1989). Withdrawals from natural cycles are significant; Table 3 shows examples of land subsidence due to groundwater extraction. Critical to global change, outputs such as pollution by-products or waste stream materials are also increased by economic growth. Table 4 illustrates wastes generated in the production of toilet paper; even illegal economic activity (such as cocaine manufacture) can have significant impacts upon ecosystem functioning (Morales, 1989; Machlis, 1992; Goodman, 1993).

Technological change includes not only physical technology, but the cultural rules and organizational strategies needed to operate tools. An example is the "bundle" of organizational requirements (such as management skills and specialized training) needed to operate a nuclear power plant. Technological change has historically

Table 3. Examples of land subsidence due to groundwater extraction.

Locality	Maximum subsidence (m)	Area affected (km^2)
San Joaquin Valley, CA	9.00	13500
Houston-Galveston, TX	2.75	12170
Eloy-Pichacho, AZ	3.60	8700
Tokyo area, Japan	4.60	2400
Nobi plain, Japan	1.50	800
Po valley, Italy	3.00	780
London, England	0.35	450
Venice area, Italy	0.14	400
Mexico City, Mexico	8.70	225

Source: Adapted from Coates, 1983.

Table 4. Wastes generated in the production of one ton of toilet paper (standard brightness) in lbs/ton.

Gases		
	Chlorine	1.2
	Chlorine dioxide	0.6
	Sulphium dioxide	5.6
	Hydrogen sulphide	25.5
	Particulates	57.5
Liquids		
	Dissolved inorganic solids	263.0
	Dissolved organic solids	244.0
	Suspended organic solids	113.0
	Suspended inorganic solids	4.5
	BOD_5	147.0
Solids		
	Inorganic solids	82.0
	Organic solids	0.0

Source: Adapted from Kneese and Bower, 1979.

increased per capita energy use and concentrated pollution outputs (Kates *et al.*, 1990; Burch and DeLuca, 1984); its current potential is both to increase and mitigate negative impacts upon ecological systems. For example, Table 5 shows significant variations in energy efficiency and emissions from alternative electricity-generating technologies.

Table 5. Pollutants from various electricity-generating technologies.[1]

Technology	Conversion Efficiency (percent)	Emissions NO_x	SO_2	CO_2
		(grams per kilowatt hour)		
High-sulphur coal-fired steam plant (with scrubbers)	36	4.3	2.1	889
Low-sulphur coal-fired fluidized bed plant	32	0.3	1.2	975
Integrated gasification combined cycle plant (coal gasification)	38	0.2	0.3	747
Gas turbine combined cycle plant (advanced)	55	0.03	0.0	331

[1] The figures in this table are for particular plants that are representative of ones in operation or under development.

Source: Adapted from Flavin, 1992.

Political-economic institutions are crucial forces, for human use of natural systems necessarily takes place through these organizational forms. As Bennett and Dahlberg note (1990), "*anything in nature, society or culture can become a "resource" when humans incorporate it into institutions of purposeful accomplishment.*" Institutions are culturally-specific solutions to universal human concerns: public schools for education, armies for defense, courts for dispensation of justice, and so forth. Means and methods (such as type of legal system or government policies toward education) can mediate or intensify both individual behaviour and aggregate human action. For example, capitalist economic incentives can encourage deforestation (Hecht and Cockburn, 1989); democratic political structures can encourage public participation in resource decisions (Machlis, 1990).

Finally, *attitudes and beliefs* limit what is acceptable in the human use of resources and significantly influence how resources are, in fact, used. In sociological terms, beliefs reflect deeply-held values ("animals have rights") while attitudes are opinions derived from beliefs ("hunting should be banned"). Attitudes and beliefs directly influence resource consumption and use of nature; they also have indirect effects i.e., attitudes toward birth control influence population growth.

Attitudes and beliefs can be measured at international and global levels. In 1992 the Gallup International Institute conducted a survey of environmental concern in 24 countries (Dunlap, 1993). The results of the survey demonstrate worldwide concern about environmental problems (see Table 6). Significantly, there was greater concern in the less industrialized countries than more industrialized nations. Those with a "great deal" of concern varied from a high in Nigeria of 71% to a low of 5% in Poland. While some of the variation may be due to cultural differences and survey methods, studies such as this provide valuable information to track changing attitudes over time. Such shifts in public values and opinions can have significant impact upon other social factors, and ultimately upon environmental conditions.

Together, the social factors just described are critical to a realistic understanding of climate change, biodiversity loss and other global change. Models that ignore them may reflect only proximate causes, and fail to reflect both the predictability and uncertainty that social systems add to global ecosystem functioning. However, the social sciences have not intensively addressed such issues. A research review by the U.S. National Research Council (1992) notes:

> "*Although each of these driving forces is important at certain times and under certain conditions, much remains unknown about what determines their relative importance, how they affect each other, and how the driving forces in particular places combine to produce global effects.*" (NRC, 1992:3).

Table 6. Percent of respondents who are personally concerned a "great deal" or "fair amount" about environmental problems.

	Country	Great Deal	Fair Amount	Total
More Industrialized Nations	Portugal	46	44	90
	Canada	37	52	89
	USA	38	47	85
	Great Britain	28	53	81
	Norway	19	58	77
	Ireland	22	51	73
	Netherlands	17	54	71
	Japan	23	44	67
	Germany	15	48	63
	Finland	17	46	63
	Denmark	12	41	53
	Switzerland	12	30	42
Less Industrialized Nations	Philippines	55	39	94
	Nigeria	71	15	86
	Mexico	50	33	83
	Uruguay	38	44	82
	Brazil	52	28	80
	Korea (Rep)	22	58	80
	Hungary	32	47	79
	Russia	41	37	78
	India	34	43	77
	Chile	30	40	70
	Turkey	12	28	40
	Poland	5	20	25

Source: From Dunlap, 1993.

Discovery and insight will partly depend on the creative participation of the social sciences.

The social sciences described

A history and description of the social sciences is neither possible nor necessary here; a primer on the scope of the social sciences may be useful. Orthodox approaches place six disciplines in the social sciences: anthropology, economics, geography (human rather than physical), psychology, political science, and sociology. History is marginally excluded. Contrary to conventional wisdom, the social sciences are not particularly young; economics for example, long precedes the development of modern chemistry and most of the social sciences precede ecology.

These sciences have much in common: research techniques such as observation, social surveys and experiments are used in all. Boundaries between the sciences are nebulous and prone to arcane distinctions; subfields such as social psychology and economic sociology flourish in academe. New specializations emerge yearly, tracking the growth of knowledge (some of it usable knowledge) and the search for "relevance," funding, or both.

For those interested in understanding the human dimensions of global change, what may be useful is a comparison of each discipline's special focus, i.e., where the discipline has traditionally concentrated intellect and effort. A "map" of the social sciences can be described in general terms. Table 7 provides a basic outline, organizing the sciences around their key units of analysis (the scale of things they study) and the central "engine" of change (the driving forces considered most important).

Anthropology focuses primarily upon social groupings that are intensely cultural: communities, subcultural groups, and even entire cultures themselves. The driving forces are primarily cultural change, with the role of tradition being a critical interest. *Economics* (which could be split into macro and microeconomics) treats markets, industries, and economies as key units of study; the driving force of change is economic value (broadly defined). *Geography*

Table 7. A basic outline of the social sciences.

Discipline	Key Units of Analysis	Engine of Change (Driving Forces)
Anthropology	communities subcultures cultures	tradition and culture
Economics	markets industries	economic value
Geography	regions landscapes	spatial distribution
Psychology	individuals	communication
Political Science	institutions states	power
Sociology	social groups organizations communities	conflict and cohesion

(specifically human geography) treats regions, landscapes, and other spatial units (governmental, environmental, and so forth) as critical, and the spatial distribution of people, resources, and culture is seen as a significant driving force. *Psychology's* key unit is the individual, and communication of meaning (within and between individuals) is a central driving force. *Political science* focuses upon the institutions of state (at many levels); the central engine of change to many political scientists is power and its use. *Sociology* treats social groups, organizations, and communities as key units of analysis, with conflict and cohesion as central forces driving change.

Several patterns emerge. First, the social sciences overlap considerably as to their units of analysis; no clear demarcation exists between, for example, anthropology and sociology applied at the community level. Second, the social sciences have remained largely unfamiliar with global level analyses (there are exceptions, see for example Sklair, 1991). And third, the sciences reflect the complexity of human social behaviour: tradition, value, power and space are critical to understanding the human condition. If the social sciences can help to describe accurately the mechanisms that link social and biological systems, and help in predicting these interactions, they can make a significant contribution to global change research (Machlis, 1992; NRC, 1992).

INTERACTIONS

Social factors and climate change

The industrial revolution marked the beginning of large-scale human activities with the potential to alter weather and climate (MacDonald, 1989; Kates *et al.*, 1990). Since 1850 atmospheric carbon dioxide has risen about 25% due to anthropogenic releases (Schneider, 1989). A variety of human activities result in the release of greenhouse gases, including the burning of fossil fuels, deforestation and urbanization. Table 8 illustrates the relationship between human action and released gases .

Table 8. Greenhouse gases released as a result of various human activities.

Burning of fossil fuels	CO_2, CH_4, N_2O
Tropical deforestation	CO_2, CH_4, N_2O
Irrigated rice production	CH_4
Cattle production (digestion)	CH_4
Use of aerosols, cleaning solvents, and refrigerants	CFCs
Use of chemical fertilizers and other agricultural activities	N_2O
Landfills	CH_4

Sources: Glantz and Krenz, 1992; NRC, 1992.

Various human activities may influence climate change at different scales (global, regional or local), and changes at one scale may affect the others. Globally, energy use is the single most important activity (NRC, 1992; Schneider, 1989). In addition to releasing gases to the atmosphere directly, human actions such as burning wood result in a net decrease in carbon storage (Aber 1992). The chemical alterations in the earth's atmosphere due to these actions is contributing to global climate changes.

At the regional scale, climates are heavily impacted by changes in land use. Deforestation and overgrazing are examples. Removing the vegetation changes the albedo and alters the hydrologic cycle; warmer temperatures and drier conditions are the result (Jäger and Barry, 1990). At local scales, human activities that impact climates include power plants (cooling mechanisms encourage cloud formation), shelterbelts that alter wind patterns, and urban "heat islands" caused by changes in land use, air flows and heat release (Jäger and Barry, 1990). These regional and local climate changes may have cumulative effects.

As described earlier, the driving forces behind these human activities may be found in social, economic, and political systems. For example, energy use is driven by population growth and economic development (Burch and DeLuca, 1984; Schneider, 1989). The increase of methane in the environment tracks population growth curves (Glantz and Krenz, 1992). Government policies may promote deforestation or encourage urbanization. A clearer understanding of the relationships between driving forces, human actions and climate change is needed.

Social factors and biodiversity loss

Homo sapiens plays a dominant role in contemporary biodiversity loss; Wilson (1988b) suggests that for tropical rain forests, the current per-species rate and absolute loss is 1,000 to 10,000 times higher than before human intervention. Contemporary extinctions are almost entirely caused by human action, although natural forces such as demographic stochasticity, natural catastrophes, social dysfunction, and genetic deterioration contribute a "background" extinction rate (Soulé and Simberloff, 1986). Diamond (1984), echoed by Pimm and Gilpin

(1989), suggests four major causes of biodiversity loss: 1) destruction and degradation of habitat, 2) anthropogenic over-exploitation of plants and animals, 3) introduction of alien species, and 4) secondary effects of extinctions. Three of the four are directly related to the activities of *Homo sapiens*. Soulé (1991) suggests six classes of "human interference": loss of habitat; fragmentation of remaining habitat; over-exploitation; spread of exotic species; air, water and soil pollution; and climate change.

Habitat destruction is considered the most widespread anthropogenic cause of biodiversity loss (Myers, 1986; Brown, 1985; Wilson, 1988a). IUCN and UNEP (1986a & b) estimate that 65% of the original ecosystems south of the Sahara have been subject to major ecological disturbance, and 67% of natural habitat has been lost in southeast Asia. In western Europe the percentage of habitat loss is even higher (Schücking and Anderson, 1991). Habitat fragmentation may differentially increase vulnerabilities (Lovejoy, 1986; Vermeij, 1986); extinction selectivity suggests that certain habitats and species are more at risk (Terborgh and Winter, 1980; Slobodkin, 1986). Particular attention has been paid to the tropical rain forests because of their high biodiversity and accelerated rate of deforestation (Myers, 1988; WRI, 1986). Other affected tropical habitats include coral reefs (Salm, 1984) and mangrove forests (Johannes and Hatcher, 1986). Case studies of species decline due to anthropogenic habitat change also exist in the temperate zone; for example, the ivory-billed woodpecker, dusky seaside sparrow, and black-footed ferret (Freedman, 1989; Wilson, 1988a).

A decline in biodiversity has been associated with specific land-use changes such as powerline construction (Nickerson *et al.*, 1989), urbanization (Leidy and Fiedler, 1985), colonization adjacent to protected areas (Neumann and Machlis, 1989), and fragmentation of forests (Harris, 1984; Hanson *et al.*, 1990). Currently, the literature is anecdotal and descriptive rather than systematic and analytical. Comparative studies are rare, and predictive ability is especially meagre. Hence, there is a need to understand better the relation between human action and biodiversity loss.

TOWARDS AN INTERDISCIPLINARY MODEL OF BIODIVERSITY LOSS

A specific interdisciplinary model for biodiversity loss requires that 1) socio-economic indicators serve as measures for key social variables, 2) the social variables have specific impacts upon environmental variables, 3) the variables and relationships be derived from biological and social science theory, and 4) biodiversity loss be operationalized for measurement over time. To be specific, the model should be constructed for a particular locale and spatial scale; to be generalizable it should contain variables likely to be operating at other locales and scales (Turner and Meyer, 1991). In Levins' (1966) terms, this exploratory model-building trades off *precision* for *generality* and *realism.*

To better understand the specific processes and causative pathways that lead to biodiversity loss, Machlis and colleagues are attempting to develop a series of testable models (Machlis, 1992; McKendry and Machlis, 1991; Forester Tear, 1992; Machlis and Forester, in press). The modelling is explicitly interdisciplinary, and assumes that changes in human social systems are the driving forces behind most contemporary biodiversity loss. A conceptual model is shown in Figure 1. Social, economic, and political forces are considered as independent variables driving changes in the human use of resources. These uses have specific ecosystem impacts, and ecological fluxes (such as heightened ecosystem stress) lead to biodiversity loss. Feedback loops explicitly complicate the conceptual model.

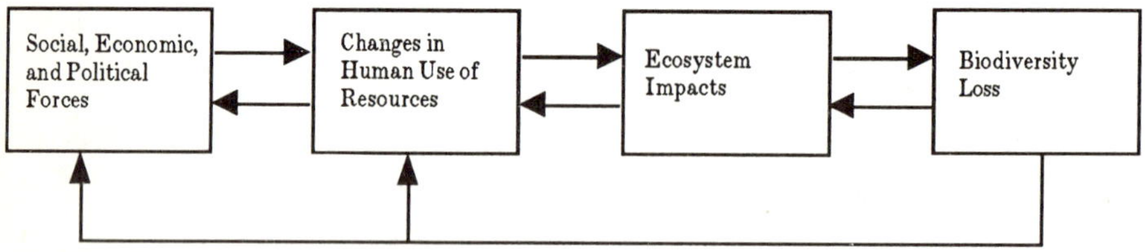

Figure 1. Conceptual model. Source: Machlis, 1992.

From the conceptual model, a preliminary version of a predictive model was developed (Figure 2). The various subsystems or component variables of the system are represented in boxes. The arrows represent causal relations, and the hypothesized nature of the relationship is identified as increasing (+), decreasing (-), or causing an unspecified change (Δ). Model variables are aggregated by the general class of activities to which they belong, such as economic exchange, politics, resource use, and industry. Definitions for each of the variables are outlined in Table 9.

The model illustrates that a variety of variables and relations are important for understanding human impacts on biodiversity. For example, population immigration (a component of demographic change) leads to an increase in forest clearing (land-use conversion) which results in habitat fragmentation. This fragmentation causes biodiversity loss. Concentration of monetary capital leads to an increase in industrial activity. Industrial activity contributes to anthropogenic climate change, which stresses ecosystems and ultimately results in biodiversity loss. The model shows that the paths by which *Homo sapiens* impact biodiversity are many, varied and complex.

Different variables may play a larger or smaller role in explaining biodiversity loss, depending on the spatial and temporal scales and on the geographic locations to which the model is applied. For example, demographic change might be an important driving force behind changes in human use of resources at the local level, whereas government policies may play a bigger role at state or national scales. Industrial activity may play a large role in ecosystem stress in industrialized countries, whereas the over-harvesting of selected species (natural resource production activity) might be more important in agricultural or primary production regimes. Global warming (anthropogenic climate change) may have a greater direct impact on temperate than on tropical ecosystems.

A literature review provided the basis for the model structure. While several of the component interactions are based on empirical evidence, others are based on descriptive case studies. As an example, several studies empirically

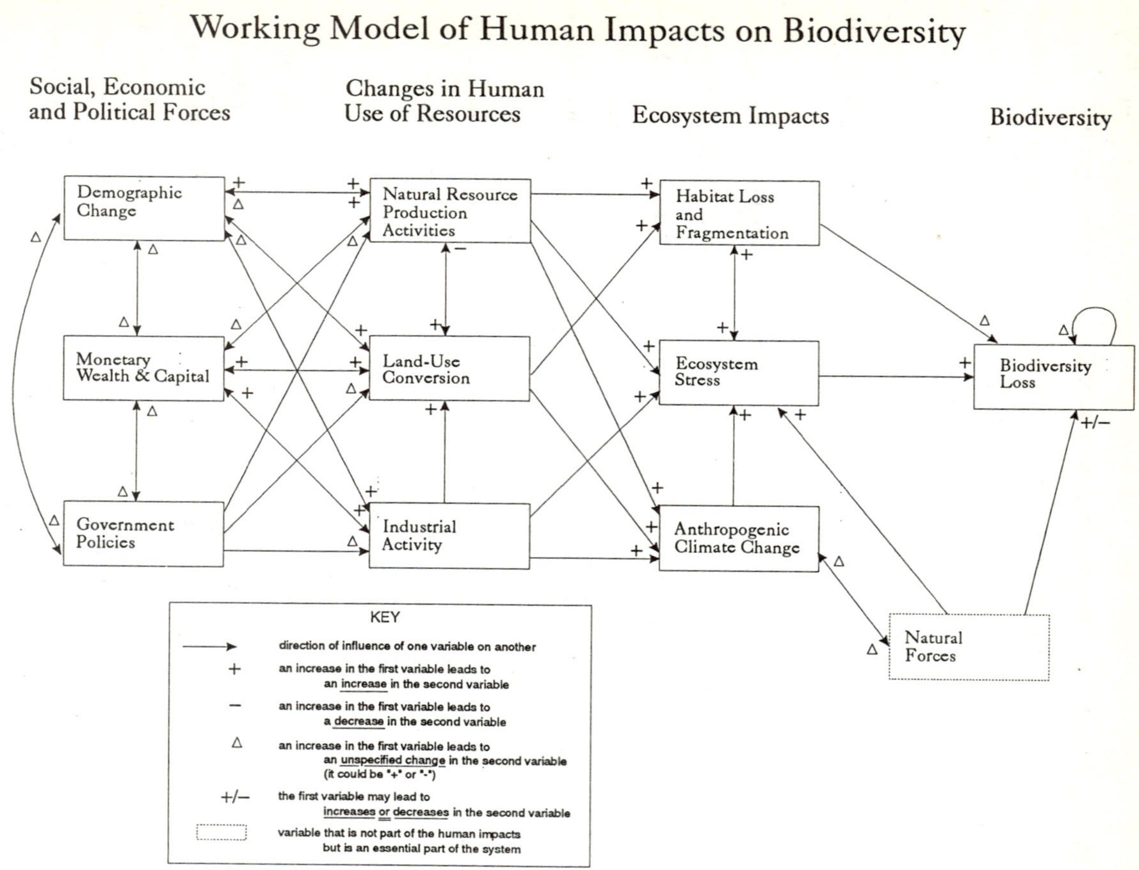

Figure 2. Working model of human impacts on biodiversity.

establish a positive link between population growth (an aspect of demographic change) and deforestation (an aspect of land use conversion) in developing countries (Allen and Barnes, 1985; Rudel 1989). Goodess *et al.* (1992) explain how population growth in Asia leads to more irrigated rice paddies and therefore increased methane releases that fuel global warming. Reh and Seitz (1990) demonstrate a link between land-use conversion (roads) that fragments habitat and reduced genetic diversity in animal populations (frogs) residing in smaller habitat patches.

The working model has several limitations, all of which must be considered in its test and revision. The aggregation level may be too coarse to adequately capture all of the important processes that ultimately influence biodiversity loss. Reciprocal relations or feedback loops may be difficult to conclusively test and validate. There is no standard way to measure biodiversity or its loss; the lack of a consistent and precise measure of the dependent variable may limit the usefulness of analyses. The model fails to take into account the potential time lag between the main system components. Various indicators could be used for each of the model variables. Although this flexibility of application was one of the considerations behind building the model, it may be difficult to compare the results of different analyses. Its usefulness to resource managers, planners, policy makers, and others remains theoretical until validated.

Nevertheless, the model has potential value. It can be properly specified, though some variables (e.g., wealth and capital) may require refinement. Its explanatory power is directed specifically at biodiversity loss and, if validated, could provide predictions of vulnerability for specific locales.. It is reasonably parsimonious, and intervening social and biological variables are explicitly considered. The model is testable given clear scope conditions, because it is constructed of a series of individual hypotheses about human/nature interactions.

Table 9. Definition of model components for Figure 2.

Social, Economic and Political Forces

Demographic change refers to changes in the size, distribution and density of human populations in a given area. Examples include population growth and migration.

Monetary wealth and capital refers to the presence of money and money available for use in producing more money. Note that wealth and capital do not refer to natural resources, labour or other human and biological resources. Examples include income and savings.

Government policies refers to laws, regulations and practices adopted or enforced by government at any of its several levels (national, regional, local). Inaction by a government also constitutes a government policy. Examples include taxes, hunting regulations and presence of (or lack of) zoning laws.

Changes in Human Use of Resources

Natural resources production activities refers to the use and exploitation of natural resources to produce goods and services which may be used directly or in industrial activity. Examples include forestry, agriculture, hunting, mining, and fishing activities.

Land-use conversion refers to the transformation of land from one use to another, usually from a less to a more developed state. Examples include the building of roads and railways where they did not exist previously, and the conversion of land from one use to another, e.g. agriculture to housing.

Industrial activity refers to practices and processes involved in the manufacture of goods by industry (which is usually the transformation of natural resources). Examples include gas emissions, waste disposal, transport,and energy use.

Ecosystem Impacts

Habitat fragmentation refers to the physical division of a habitat into smaller units. An example would be the division of one area into several due to road construction or timber harvesting. Habitat loss refers to changes in a habitat such that it can no longer support a keystone species. Examples include the outright destruction of a habitat by clearing an area or a significant change in a habitat due to cattle grazing.

Table 9 (contd.).

Ecosystem stress refers to changes in natural systems beyond a dynamic equilibrium state to compensate for a disturbance. Examples include harvesting a species beyond some threshold level, loss of a keystone species, or forest die-back due to acid rain.

Anthropogenic climate change refers to long-term variation (over 100 years) in mean values of a climatic parameter (such as temperature or precipitation) that are due to human actions. An example is greenhouse gas–induced temperature rises.

Natural forces refers to biological, geological, climatological, and meteorological activities that occur without a dominant influence from humans. This would include natural (background) extinctions, genetic adaptation and speciation, and natural climate change.

Biological Diversity

Biodiversity loss refers to a significant reduction (e.g. beyond dynamic fluctuation) in the variety and variability of genes, species or habitats. This reduction may change or impair ecological processes. Species extinctions and loss of heterozygosity are examples.

Evaluating the model's insight awaits its application. In our work, we are exploring the use of cross-national datasets and structural equation models to test the hypotheses represented in the working model (see Forester Tear, 1992; Machlis and Forester, in press).

A RESEARCH AGENDA

As we have described and attempted to model, social factors are important driving forces upon global change. Climate change and biodiversity loss are exemplars of this condition. The root causes of contemporary biodiversity loss are human actions, imbedded in social systems that are inextricably intertwined with natural systems. Hence, understanding the relationship between social factors and biodiversity loss is a prerequisite for wise and effective ecosystem management. Such understanding will largely rest on sound and careful science. While we have detailed elsewhere a strategic and broad research agenda (see Machlis, 1992;

McKendry and Machlis, 1993; Machlis and Forester, in press), a brief listing of key objectives related to modelling may be useful.

Building a synthesis of theory

There is a need to develop testable theories of global change that integrate social and natural systems. Social and biological variables must be considered as integral parts of any coherent explanation. Rather than "grand theory" attempts, theories of the middle range may prove the most tractable, and care must be taken to define each variable, causal mechanisms and scope conditions of the theory. This work can serve as a foundation for modelling activities.

The construction of explicit biodiversity models

There is a need to develop interdisciplinary models of specific global change conditions - i.e. "unpack" broad concepts and focus on distinct ecosystem perturbations as dependent variables. The construction of explicit biodiversity models (as we have begun) is a first step. Similar efforts should be (and are being) undertaken for climate change, air and water pollution and so forth (for an example, see Costanza and Daly, 1992). In some cases, global models can evolve from existing efforts at local and regional scales; in all cases, social factors must be evaluated as potential driving forces.

Sustained efforts at model-testing and revision are necessary, for first efforts are likely to be crude and imprecise. The shift from conceptual to quantitative models is necessary and inevitable. A diversity of model-testing strategies is required, and the robustness of the models in the face of data variability is a critical concern.

Improved data sources and analysis techniques

If theory on the human dimensions of global change is to advance, and useful models are to be created, vastly improved and accessible datasets are necessary. The NRC (1992) lists several concerns: 1) availability and access to existing data, 2) quality control and interpretability of data, 3) missing data for parts of the world, 4) inadequate time series for some variables, 5) lack of measures for key variables, and 6) incomparability of datasets across time and space. Each of these limit the range, quality and interdisciplinary potential of global change research. A concerted effort should be raised to gather and disseminate existing datasets, and carefully plan the acquisition, development and maintenance of new and needed data. The overall quality of human dimensions of global change research could be greatly improved, and our understanding advanced.

CONCLUSION

As a species, *Homo sapiens* lives within two grand systems. One is largely natural, driven by biophysical forces and subject to the laws of physics, energy conservation, human biology and ecosystem functioning. The other is largely sociocultural, and uniquely human: an advanced technology, unparalleled organizational skills, self-awareness and a rising spiral of wants and needs. The two systems have always been conjoined. We use the natural system to accomplish and satisfy the social system; this is partly what makes us human. What has changed is the scope and intensity of our planetary modifications. Bill McKibben, writing in his essay "*The End of Nature*" speaks simply:

> "*We have changed the most basic forces around us...We have changed the atmosphere, and that will change the weather. The temperature and rainfall are no longer to be entirely the work of some separate, uncivilizable force, but instead in part a product of our habits, our economies, our ways of life.*" (1989:47).

In a thought both "great and terrible," he warns us that the result may be the cultural extinction of the *natural*, the death of what nature has meant to human societies. For this unimaginable event, social factors *are* driving forces. A sad irony may lie in science's inability to grasp what is the most significant of global changes: our loss of understanding about our place on earth.

REFERENCES

Aber JD (1992) Terrestrial ecosystems. Pages 173-200 *in* Trenberth KE (ed) Climate system modelling. Cambridge University Press, New York

Abrahamson DE (1989) Global warming: the issue, impacts, responses. Pages 3-34 *in* Abrahamson DE (ed) The challenge of global warming. Island Press, Washington DC

Allen JC, Barnes DF (1985) The causes of deforestation in developing countries. Ann. of the Assoc. of Amer.Geog. 75(2):163-184

Allen JM (ed) (1963) The nature of biological diversity. McGraw Hill Inc., New York

Bennett JW, Dahlberg KA (1990) Institutions, social organizations, and cultural values. Pages 69-86 *in* Turner BL II, Clark WC, Kates RW, Richards JF, Mathews JT, Meyer WB (eds) The earth as transformed by human action: global and regional changes in the biosphere over the past 300 years. Cambridge University Press with Clark University, New York

Boyle TJB (1993) Personal communication, 26 May

Brown LR (ed) (1985) State of the world 1985: a Worldwatch Institute report on progress toward a sustainable society. W.W. Norton & Co., New York

Burch WR Jr. (1971) Daydreams and nightmares: a sociological essay on the American environment. W.W. Norton & Co., New York

Burch, WR, Jr., DeLuca, DR (1984) Measuring the social impact of natural resource policies. University of New Mexico Press, Albuquerque

Buttel FH, Taylor, PJ (1992) Environmental sociology and global environmental change: A critical assessment. SNR 5:211-230

Ciborowski P (1989) Sources, sinks, trends, and opportunities. Pages. 213-230 *in* Abrahamson DE (ed) The challenge of global warming. Island Press, Washington DC

Clark WC (1985) Scales of climatic impact. Clim.Cha. 7:5-27

Clark WC, Munn RE (eds) (1986) Sustainable development of the biosphere. Cambridge University Press, Cambridge

Coates DR (1983) Large-scale land subsidence. Pages 212-234 *in* Gardner R, Scoging H (eds) Mega-Geomorphology. Clarendon Press, Oxford

Cohn JP (1989) Gauging the biological effects of the greenhouse effect. BioSci. 39(3):142-146

Costanza R, Daly HE (1992) Natural capital and sustainable development. Cons. Biol. 6(1):37-46

Demeny P (1990) Population. Pages. 41-54 *in* Turner BL II, Clark WC, Kates RW, Richards JF, Mathews JT, Meyer WB (eds) The earth as transformed by human action: global and regional changes in the biosphere over the past 300 years. Cambridge University Press with Clark University, New York

Diamond J (1984) Historic extinction: a Rosetta stone for understanding prehistoric extinctions. Pages. 824-862 *in* Martin PS, Klein RG (eds) Quaternary extinctions: a prehistoric revolution. The University of Arizona Press, Tucson

Dickinson RE (1992) Changes in land use. Pages 689-701 *in* Trenberth KE (ed) Climate system modelling. Cambridge University Press, New York

Dunlap RE (1993) Global concern for the environment. Env., Tech., and Soc. 70:1-11

Ehrenfeld D (1972) Conserving life on earth. Oxford University Press, New York

Ehrenfeld D (1976) The conservation of non-resources. Amer.Sci. 64:654

Farnsworth NR (1988) Screening plants for new medicines. Pages 83-97 *in* Wilson EO (ed) Biodiversity. National Academy Press, Washington DC

Flavin C (1989) Slowing global warming: a worldwide strategy. Worldwatch Paper 91. Worldwatch Institute, Washington DC

Flavin C (1992) Building a bridge to sustainable energy. Pages 27-45 *in* Brown LR (ed) State of the world 1992: a Worldwatch institute report on progress toward a sustainable society. W.W. Norton & Co., New York

Forester Tear DJ (1992) Human impacts on biodiversity loss in developing countries: a preliminary analysis. MS Thesis University of Idaho, Moscow

Franklin JF (1988) Structural and functional diversity in temperate forests. Pages 166-175 *in* Wilson EO (ed) Biodiversity. National Academy Press, Washington DC

Freedman B (1989) Environmental ecology: the impacts of pollution and other stresses on ecosystem structure and function. Academic Press Inc., San Diego

Gable FJ, Gentile JH, Aubrey DG (1990) Global climate issues in the coastal wider Caribbean region. Env. Cons. 17(1):51-60

Glantz MH, Krenz JH (1992) Human components of the climate system. Pages 27-49 *in* Trenberth KE (ed) Climate system modelling. Cambridge University Press, New York

Goodess CM, Palutikof JP, Davies TD (1992) The nature and causes of climate change: assessing the long term future. Lewis Publishers, Boca Raton

Goodman B (1993) Drugs and people threaten diversity in Andean forests. Science 261:293

Goody, RM (1982) Global change: impacts upon habitability. A scientific basis for assessment. Institute of Technology, Pasadena CA

Graham RL, Turner MG, Dale VH (1990) How increasing CO_2 and climate change affect forests. BioSci. 40(8):575-587

Hammond AL, Rodenburg E, Moomaw WR (1991) Calculating national accountability for climate change. Envir. 33(1):11-15,33-35

Hansen JE (1989) The greenhouse effect: impacts on current global temperature and regional heat waves. Pages 35-43 *in* Abrahamson DE (ed) The challenge of global warming. Island Press, Washington DC

Hanson JS, Malanson GP, Armstrong MP (1990) Landscape fragmentation and dispersal in a model of riparian forest dynamics. Ecol. Model. 49:277-296

Harris LD (1984) The fragmented forest. The University of Chicago Press, Chicago

Harwood M (1982) Math of extinction. Audubon Magazine 84:18-21

Hecht S, Cockburn A (1989) The fate of the forest: Developers, destroyers and defenders of the Amazon. Verso, New York

Hendee JC (1990) Wilderness management. Fulcrum Press, Fort Collins

Huntley BJ (1988) Conserving and monitoring biotic diversity -some African examples. Pages 284-260 *in* Wilson EO (ed) Biodiversity. National Academy Press, Washington DC

Hurlbert SH (1971) The nonconcept of species diversity: a critique and alternative parameters. Ecology 52:577-586

International Union for the Conservation of Nature and Natural Resources and United Nations Environmental Programme (1986a) Review of the protected areas system in the Indo-Malayan realm. IUCN, Gland, Switzerland

International Union for the Conservation of Nature and United Nations Environmental Programme (1986b) Review of the protected areas system in the Afro-tropical realm. IUCN, Gland, Switzerland

Jäger J, Barry RG (1990) Climate. Pages 335-351 *in* Turner BL II, Clark WC, Kates RW, Richards JF, Mathews JT, Meyer WB (eds) The earth as transformed by human action: global and regional changes in the biosphere over the past 300 years. Cambridge University Press with Clark University, New York

Janzen DH (1988) Tropical dry forests - the most endangered major tropical forest ecosystem. Pages 130-137 *in* Wilson EO (ed) Biodiversity. National Academy Press, Washington DC

Johannes RE, Hatcher BG (1986) Shallow tropical marine environments. Pages. 371-382 *in* Soulé ME (ed) Conservation Biology. Sinauer Associates Inc., Sunderland, MA

Kates RW, Turner BL II, Clark WC (1990) The great transformation. Pages 1-17 *in* Turner BL II, Clark WC, Kates RW, Richards JF, Mathews JT, Meyer WB (eds) The earth as transformed by human action: global and regional changes in the biosphere over the past 300 years. Cambridge University Press with Clark University, New York

Kellert SR (1986) Social and perceptual factors in the preservation of animal species. Pages 50-73 *in* Norton BG (ed) The preservation of species. Princeton University Press, Princeton

Kneese AV, Bower BT (1979) Environmental quality and residuals management. Johns Hopkins University Press, Baltimore

Krebs CJ (1972) Ecology. Harper & Row Publishers Inc., New York

Leidy RA, Fiedler, PL (1985) Human disturbance and patterns of fish species diversity in the San Francisco Bay drainage, California. Biol. Cons. 33:247-267

Levins R (1966) The strategy of model building in population biology. Amer. Sci. 54:421-431

Lovejoy TE (1986) Species leave the ark one by one. Pages 13-27 *in* Norton BG (ed) The preservation of species. Princeton University Press, Princeton

Lovejoy TE (1988) Diverse considerations. Pages 421-427 *in* Wilson EO (ed) Biodiversity. National Academy Press, Washington DC

Lugo AE (1988) Estimating reductions in the diversity of tropical forest species. Pages 58-70 *in* Wilson EO (ed) Biodiversity. National Academy Press, Washington DC

MacArthur H (1985) Patterns of species diversity. Biol. Rev. 40:510-533

MacDonald G (1989) Scientific basis for the greenhouse effect. Pages 123-145 *in* Abrahamson DE (ed) The challenge of global warming. Island Press, Washington DC

Machlis GE (1990) The tension between local and national conservation groups in the democratic regime. SNR 3:267-279

Machlis GE (1992) The contribution of sociology to biodiversity research and management. Biol. Cons. 62(3):161-170.

Machlis GE, Forester DJ (In press) The relationship between socio-economic factors and the loss of biodiversity: First efforts at theoretical and quantitative models. *In* Szaro, R (ed) Biodiversity in managed landscapes: Theory and practice. Oxford University Press, Oxford

Magurran AE (1988) Ecological diversity and its measurement. Princeton University Press, Princeton

Margalef R (1968) Perspectives in ecological theory. The University of Chicago Press, Chicago

Marsh GP (1907) The earth as modified by human action (a last revision of Man and Nature). Charles Scribner's Sons, New York

Maunder WJ (compiler) (1992) Dictionary of global climate change. UCL Press, London

McDowell PF, Webb T III, Bartlein PJ (1990) Long-term environmental change. Pages 143-162 *in* Turner BL II, Clark WC, Kates RW, Richards JF, Mathews JT, Meyer WB (eds) The earth as transformed by human action: global and regional changes in the biosphere over the past 300 years. Cambridge University Press with Clark University, New York

McKendry JE, Machlis GE (1991) "The role of geography in extending biodiversity gap analysis." Paper presented at the American Association of Geographers Annual Meeting, 13 April 1991, Miami FL, and at the World Bank, 25 April 1991, Washington DC

McKendry JE, Machlis GE (1993) The role of geography in extending biodiversity gap analysis. App.Geog. 13:135-152

McKibben, B (1989) The end of nature. Doubleday, New York

Meadows DH, Meadows DL, Randers J, Behrens WW III (1972) The limits to growth. Universe Books, New York

Morales E (1989) Cocaine. The University of Arizona Press, Tucson

Mueller-Dombois D, Kartawinata K, Handley LL (1983) Conservation of species and habitats: a major responsibility in development planning. Pages 1-51 *in* Carpenter RA (ed) Natural systems for development: what planners need to know. MacMillan Publishing Company, New York

Myers N (1980) The problem of disappearing species: what can be done? Ambio 9(5):229-235

Myers N (1986) Tackling mass extinction of species: A great creative challenge. Albright Lecture, University of California, Berkeley

Myers N (1988) Tropical forests and their species--going, going,...? Pages 218-235 *in* Wilson EO (ed) Biodiversity. National Academy Press, Washington DC

National Research Council (1992) Global environmental change. National Academy Press, Washington DC

Nickerson NH, Dobberteen RA, Jarman NM (1989) Effects of power-line construction on wetland vegetation in Massachusetts, USA. Env. Mgmt. 13(4):477-483

Neumann R, Machlis GE (1989) Land-use and threats to parks in the neotropics. Env. Cons. 16(1):13-18

Norton BG (1986) On the inherent danger of undervaluing species. Pages 110-137 *in* Norton BG (ed) The preservation of species. Princeton University Press, Princeton

Norton BG (1987) Why preserve natural variety? Princeton University Press, Princeton

Office of Technology Assessment (1987) Technologies to maintain biological diversity. U.S. Government Printing Office, Washington DC

Peters RL (1989) Effects of global warming on biological diversity. Pages 82-95 *in* Abrahamson DE (ed) The challenge of global warming. Island Press, Washington DC

Peters RL, Lovejoy TE (eds) (1992) Global warming and biological diversity. Yale University Press, New Haven

Pimm SL, Gilpin ME (1989) Theoretical issues in conservation biology. Pages 287-305 *in* Roughgarden J, May RM, Levin SA (eds) Perspectives in ecological theory. Princeton University Press, Princeton

Postel S (1988) Global view of a tropical disaster. Amer.For. 94(11&12):25-29,69-71

Ray CG (1988) Ecological diversity in coastal zones and oceans. Pages 36-50 *in* Wilson EO (ed) Biodiversity. National Academy Press, Washington DC

Reh W, Seitz A (1990) The influence of land use on the genetic structure of the common frog *Rana temporaria*. Biol. Cons. 54(3):239-249

Root TL, Schneider SH (1993) Can large-scale climatic models be linked with multiscale ecological studies? Cons. Biol. 7(2):256-270

Rudel TK (1989) Population, development, and tropical deforestation: a cross-national study. Rur. Soc. 54(3):327-338

Ruffie J (1982) The population alternative: a new look at competition and the species. Pantheon Books, New York

Salm RV (1984) Marine and coastal protected areas: A guide for planners and managers. IUCN, Gland, Switzerland

Schneider SH (1989) The greenhouse effect: science and policy. Science 243(4892):771-781

Schneider SH (1992) Introduction to climate modelling. Pages 3-26 *in* Trenberth KE (ed) Climate system modelling. Cambridge University Press, New York

Schneider SH, Mearns L, Gleik PH (1992) Climate–change scenarios for impact assessment. Pages 38-55 *in* Peters RL, Lovejoy TE (eds) Global warming and biological diversity. Yale University Press, New Haven

Schücking H, Anderson P (1991) Voices unheard and unheeded. Pages 13-41 *in* Shiva V, Anderson P, Schücking H, Gray A, Lohman L, Cooper D (eds) Biodiversity: Social & ecological perspectives. Zed Books, New Jersey

Scott JM, Csuti B, Jacobi JD, Estes JE (1987) Species richness: a geographic approach to protecting future biological diversity. BioSci. 37(11):782-788

Sedjo RA (1990) The global carbon cycle: are forests missing the sink? J. of For. 88(10):33-34

Shiva V, Anderson H, Schücking H, Gray A, Lohmann L, Cooper D (1991) Biodiversity: social & ecological perspectives. Zed Books, New Jersey

Simmons IG (1989) Changing the face of the earth: culture, environment, history. Basil Blackwell, New York

Sklair L (1991) Sociology of the global system. The Johns Hopkins University Press, Baltimore

Slobodkin LB (1986) On the susceptibility of different species to extinction: Elementary instructions for owners of a world. Pages. 226-242 *in* Norton, BG (ed) The preservation of species. Princeton University Press, Princeton

Soulé ME (ed) (1986) Conservation biology: the science of scarcity and diversity. Sinauer Associates Inc., Sunderland, MA

Soulé ME (1991) Conservation: tactics for a constant crisis. Science 253:744-750

Soulé ME, Simberloff D (1986) What do genetics and ecology tell us about the design of nature reserves? Biol. Cons. 35:19-40

Terborgh J, Winter B (1980) Some causes of extinction. Pages. 119-133 *in* Soulé ME, Wilcox BA (eds) Conservation biology: An evolutionary-ecological perspective. Sinauer Associates Inc., Sunderland, MA

The Nature Conservancy (1975) The preservation of natural diversity: a survey and recommendations. The Nature Conservancy, Washington DC

Thomas WL Jr. (ed) (1956) Man's role in changing the face of the earth (vol. 1). The University of Chicago Press, Chicago

Turco RP (1992) Atmospheric chemistry. Pages 201-240 *in* Trenberth KE (ed) Climate system modelling. Cambridge University Press, New York

Turner, BL II, Clark WC, Kates, RW, Richards, JF, Mathews, JT, Meyer, WB (eds) (1990) The earth as transformed by human action: Global and regional changes in the biosphere over the past 300 years. Cambridge University Press with Clark University, New York

Turner BL II, Meyer WB (1991) Land use and land cover in global environmental change: considerations for study. ISSJ 43(4):669-679

Vermeij GJ (1986) The biology of human-caused extinction. Pages. 28-49 *in* Norton BG (ed) The preservation of species. Princeton University Press, Princeton

Walker BH (1992) Biodiversity and ecological redundancy. Cons.Biol. 6(1):18-22

Western D, Pearl MC, Pimm SL, Walker B, Atkinson I, Woodruff D.S (1989) An agenda for conservation action. Pages 304-323 *in* Western D, Pearl MC (eds) Conservation for the twenty-first century. Oxford University Press, New York

Westman W (1990) Managing for biodiversity. BioSci. 40(1):26-33

Whitmore TM, Turner BL II, Johnson DL, Kates RW, Gottschang TR (1990) Long-term population change. Pages 25-39 *in* Turner BL II, Clark WC, Kates RW, Richards JF, Mathews JT, Meyer WB (eds) The earth as transformed by human action: global and regional changes in the biosphere over the past 300 years. Cambridge University Press with Clark University, New York

Wilcox BA (1984) Concepts in conservation biology: applications to the management of biological diversity. Pages. 639-647 McNeely JA, Miller KR (eds) National parks, conservation, and development: the role of protected areas in sustaining society. Smithsonian Institution Press, Washington DC

Wilson EO (1992) The diversity of life. W.W. Norton & Co., New York

Wilson EO (ed) (1988a) Biodiversity. National Academy Press, Washington DC

Wilson EO (1988b) The current state of biological diversity. Pages 3-18 *in* Wilson EO (ed) Biodiversity. National Academy Press, Washington DC

Woodwell GM (1989) Biotic causes and effects of the disruption of the global carbon cycle. Pages 71-81 *in* Abrahamson DE (ed) The challenge of global warming. Island Press, Washington DC

World Resources Institute (1986) World Resources 1986: An assessment of the resource base that supports the global economy. Basic Books, New York

World Resources Institute (1990) World Resources 1990-91: A guide to the global environment. Oxford University Press, New York

Chapter 3: The Natural Impermanency of Most Temperate Wetlands

Richard L. Bowen
Department of Geology
Box 8152
University of Southern Mississippi
Hattiesburg, Mississippi 39406
USA

TIME SCALES AND ENVIRONMENTAL CONCERNS

The current generation of mankind is the first to be concerned with the global consequences of human interactions with ecosystems. An unbiased analysis of ecosystems requires the understanding of their individual natural histories. In the following discussion of temperate wetlands and their associated ecosystems, major emphasis is placed on discrimination between temperate wetlands, which formed as artefacts of the global chilling and warming of the Pleistocene glacial and interglacial episodes, and those other wetlands whose natural histories have little or no direct relation with Late Cenozoic climatic variability.

It is a common view that wetlands (however such are defined both legally and informally) as they now appear distributed in scale and space represent the *status quo*. Some groups have called attention to the widely extensive modifications to wetlands that existed some centuries ago. Examples of these include: large portions of the Netherlands that have been converted to other uses such as agriculture, settlement, and industry; extensive portions of San Francisco Bay that have been drained, landfilled, and sliced up for service as airports, salt extraction brine ponds, settlement, agriculture, and industry; and the Nile Delta, which has become the granary of Egypt. Innumerable examples of modifications by humans on a smaller scale could easily be compiled. Many people in North America and western Europe, including those with some influence, would wish that, by legal and political persuasion, many such

NATO ASI Series, Vol. I 20
Biodiversity, Temperate Ecosystems, and Global Change
Edited by T. J. B. Boyle and C. E. B. Boyle

converted wetland regions should forcefully be reverted to their former character. They seek what can be termed a *status quo ante*. Both in the case of the portion of the public which would wish to maintain the *status quo* and in that which desires the *status quo ante*, there is a myopic failure to consider the varied processes which control the Earth's history in broad scale and the history of local areas in detail as they develop in accordance with the on-going actions of Nature.

These politico-environmental positions and their popularity result from two quite comprehensible conditions which have controlled the general perception of the character of Earth's ecosystems. First, human perceptions of the nature of Earth's landscapes and the biological occupants thereof are those received, on the one hand, from personal observations and, on the other, from reported (by press, film, television, and other media) conditions as observed elsewhere around the Earth today and in recent years. We may call this an actualistic view of the Earth. Secondly, this actualistic view has support from the human recorded history of the last five millennia, more or less. During these five thousand years, the records have shown only minor variability, compared with the truly dramatic climatic changes involved in the many glacial and interglacial cycles (which have occurred on scales of some tens of thousands to somewhat more than one hundred thousand years) which have affected the Earth during the past two million-plus years. The period during which our civilization has evolved has been referred to as one of climatic metastability (Bowen, 1989) representing, as it has, a group of global conditions intermediate between those of continental scale glaciation (the most recent example being that of the Wisconsinan Ice Age, which reached its peak of intensity some 18-20 thousand years ago) and those prevailing on Earth during ice-free periods. Such an ice-free Earth was the case for more than two hundred million years between the middle of the Permian Period, some 265-270 million years ago, and the earlier portion of the Eocene Epoch, some 45 million years ago.

A majority of environmentalists, scientists, and politicians have failed to consider many of the questions associated with "Biodiversity, Temperate Ecosystems, and Global Change" within the historical context of a global ecosystem which is subject to tremendous variability on a time scale as short as thousands of years.

Indeed, news reports this year (1993) regarding analyses of the climatic record preserved in the cores of the Greenland Ice Cap suggest that great, global, climatic variations have naturally occurred on scales as short as centuries or even decades.

I aim here to show that wetlands, as one currently popular topic of environmental concern, should properly be considered within a longer time-frame than that of the lifetimes of one generation or even of ten generations.

WETLAND CATEGORIES

"Wetlands", as generally understood, fall into two broad categories, those whose origins are independent from climatic changes and those whose formation was consequent upon the many alternations between glacial and interglacial conditions that have occurred in the past two million-plus years.

1. *Non-Climatic "Wetlands".* For the purposes of comparison between climatically-influenced wetlands, whose histories, as discussed below, are by nature ephemeral (Bowen, 1991) and wetlands whose origins do not relate in any significant way to climatic changes, the nature and character of the latter require a brief summarization.

Non-climatic wetlands originate in a great diversity of manners. The first major group is wetlands that develop as drainage systems proceed through the cycle of erosion. By and large, these wetlands form in the downstream portion of the drainage system, where meandering streams produce cut-off ox-bows which gradually fill with the passage of time. In larger streams, the downstream flood plains may have a variety of wetlands (often called "backswamps") develop upon them as a consequence of the separation of the flood-plain regions from the main stream channel by the growth and formation of natural levees. Yet farther downstream, deltas, with their considerable variety of wetlands, develop for those streams which do not discharge in coastal zones that are strongly dominated by wave, storm, and tidal activity.

Tectonic processes are responsible for generating a considerable variety of wetlands. Readjustments of Earth's surface following earthquake activity may create wetlands by subsidence, as in the case of the Reelfoot Lake region of Tennessee, which was generated by the great New Madrid, Missouri, quakes of 1811-1812, or by the triggering of landslides which fall into valleys, blocking them and forming ephemeral lakes and wetlands, or by localized uplift, with upraised fault scarps providing a temporary damming effect. Some wetlands have formed along fault zones that have belts of crushed rock a kilometre or more in width as in the cases of the San Andreas Fault (California), Alpine Fault (New Zealand), and Great Glen Fault (Scotland). Occasionally, wetlands are consequent upon rifting processes, as in the African Rift System, Rhine Valley, and Rio Grande Rift (New Mexico) or from the uplifting of former lagoonal areas along coastlines (as in California or Alaska).

Volcanic activity may form wetlands by blockages of drainage routes through outpourings of lava (as has occurred in the past in the Grand Canyon and more recently in Japan, New Zealand, and Iceland). Other temporary wetlands have formed from blockage of established drainages by volcanic mud-flows ("lahars"). Small wetlands occur where springs and seeps of perched water tables in horizontally layered volcanic deposits have outlets. These most commonly occur, in the northern hemisphere, on the northern and eastern hillslopes.

Certainly, the most extensive group of non-climatic wetlands are those consequent upon solution of soluble rocks (carbonates, sulphates) beneath the surface, which results in the formation of karst topography with its bogs, sinks, disappearing streams, and related phenomena.

In semi-arid portions of the world, interior wetlands form where drainage from cooler, wetter uplands produces inland deltas, as in the cases of the Humboldt River (Nevada), Amu Darya (Russia), and Okovango (Botswana). Some such wetlands are called "oases".

Biologic activity results in formation of wetlands by constructions of the industrious beaver populations of North American and Eurasia, by the buildup of humus and peat on upland bogs and marshes in the British Isles, as well as portions of Canada and Scandinavia, and along coastal regions by the growth of mangroves in warm-temperate regions and salt-tolerant grasses in cooler coastal zones.

In this listing of the major groups of temperate wetlands whose origins do not relate directly to major climatic alternations, it should be noted once more that many of the cited examples were likely to persist only briefly before they pass through on-going natural processes into non-existence.

2. *Wetlands Formed Through Pleistocene Processes.* Wetlands whose origins are consequent upon the dramatic climatic alternations of the Pleistocene times of glaciation and interglaciation occur widely throughout the temperate portions of North America and Eurasia; they are much less extensive in comparable temperate regions of the southern hemisphere, for, proportionally, much less of the now-temperate portions of South America, Africa, Australia, and the larger islands was strongly affected by glacial activity during the Pleistocene.

Throughout most of Canada and the north-central and northeastern portions of the United States as well as in most of northern Europe and European Russia, the landscape is splattered with countless wetlands. Often these are peripheral to lakes and ponds; on other occasions, they occur scattered and sporadically through the upland and mountainous regions. All these wetlands derive from the irregularities of erosion and deposition associated with the growth of continental and mountain glaciers during the development of the major episodes of continental glaciation during the Pleistocene Epoch or with the melting and wastage of those same masses of ice. The landscape of deranged and unintegrated drainage left behind as a heritage of the effects of glacial-interglacial alternations is correctly regarded as "abnormal" as compared with the "normal" landscapes which evolve under global ice-free conditions such as have prevailed for more than two-thirds of Earth's history, during the past two thousand million years, for which we have a passable climatic record.

Pielou (1991) has pointed out that biotas are in disequilibrium, constantly responding to climatic vagaries. She (pp. 100-101) states that "*In the light of present knowledge...disequilibrium is much commoner than equilibrium*" and that the notion "*...that the living world is 'marvelously' and 'delicately' attuned to its environment is not so much a scientifically reasonable theory as a mystically satisfying dogma. Its abandonment might lead to a useful fresh start in environmental politics.*"

The glacially derived wetlands of northern North America and Eurasia are ephemeral. Through the normal processes of erosion, stream action, and sedimentation, the deranged and unintegrated drainages of the present will become integrated, thus removing a large portion of these wetlands from their role as stagnant, undrained water reservoirs. Another large group of these wetlands will become dry land through the normal progression of collecting the products of erosion in their drainage basins, filling in by the accumulation of sediment and organic matter, and thus building their surfaces above the local water table (Miller, 1990, p. 146).

Another small group of wetlands, formed as a consequence of wind action which resulted in interdunal wetlands in such regions as Nebraska and northern Indiana, will disappear in similar fashion through normal sedimentary and organic filling. The same processes apply to wetlands interior from the coastal dune complexes formed during Pleistocene times of higher sea-level. Such back-of-the-beach wetlands occur in the interior portion of Florida and other regions; they, too, are basically ephemeral.

Another very large group of temperate wetlands consists of those bogs, marshes, and fens that surround the borders of the digitated estuaries (or "drowned valleys") whose formation is entirely due to stream excavation during Pleistocene lower sea-level times (*i.e.*, the glaciations) followed by marine invasion with rising sea-levels of glacial melting as wastage. Often, on-going coastal erosion and longshore drift have produced barrier islands, spits, and baymouth bars which partially or wholly isolate each such estuary. So long as sea-level remains constant, the filling of these estuaries by sedimentation and organic activity and the consequent disappearance of the associated wetlands is inevitable, always providing that human activity does not attempt to stall or negate these natural processes. Also, many of the coastal wetlands of northern Europe and Canada will be drained and become dry land as those regions continue to rise as a response of isostatic uplift associated with the unloading from the mass of formerly-covering glacial ice.

TEMPERATE WETLANDS OF TODAY IN PERSPECTIVE

Wetlands of whatever size occurring at present in Earth's temperate regions can be considered from a great variety of viewpoints. Agriculturists and other developers, for the most part, wish to accelerate their natural disappearance. The recreationally-minded wish to preserve them as sites for hunting or bird-watching, or as locales of floral and faunal interest and uniqueness. Preservation of coastal wetlands will bring joy to those wishing to continue to have large catches of anadromous fishes, not to mention maintaining the life-styles of the Cajuns of Louisiana, the Geechees of South Carolina, and the Marsh Arabs of Iraq. Therefore, great numbers of people in the regions of the wetlands have vested interests either in their maintenance and persistence or, alternatively, in their accelerated elimination and disappearance. From the standpoint of Earth's history on the thousand to one hundred thousand year scales, we must recognize that a very large portion of all the temperate wetland Earth now has, Earth did not have 18,000 years ago, for much of the now-temperate region was then either covered by glacial ice or was in the largely arid to semi-arid periglacial regions.

From the foregoing, one deduces that, to a very large extent, the biotic inhabitants of today's temperate wetlands are opportunistic invaders which have occupied these regions that have been newly opened to colonization or, alternatively and more likely, in the majority of the taxa on the specific level, that they represent evolutionary radiation into the filling of the former biologic desert or semi-desert. And, as these communities have developed on a time scale involving only thousands of years, the normal progression of the agencies of erosion, sedimentation, drainage readjustments and integration, and humus pile-ups resulting in pond, marsh, and bog filling within a scale of hundreds to thousands of years may normally be expected to lead to a great reduction in the number of taxa evolved to occupy these highly specialized and ephemeral habitats.

The process of forming biological deserts or semi-deserts over portions of the Earth which we now regard as temperate, followed by the opportunistic invasion and localized evolution of taxa, must have taken place many times within the past two million years. Thus, the Earth's present 50 million or so species is a product of rapid post-glacial speciation from stocks that were sufficiently stable that they could survive the climatic alternations and environmental changes of the past two million years of the Ice Ages, while retaining sufficient genetic flexibility to radiate and diversify with the production of locally specialized taxa following deglaciation. Very likely, this process of widespread species extinctions during glacial periods and rapid evolution during climatic amelioration occurred more than a dozen times during the last two million years.

In sum, changing the scale at which Earth's systems are viewed from that of decades and centuries to a scale involving millennia and tens of thousands of years, leads to a recognition that Earth's landscapes are dynamic, and its biota is also dynamic, flexible, and opportunistic. Perhaps it is time for we ephemeral humans to consider accommodating ourselves to this dynamic global machine rather than spending our efforts to attempt to halt its historically established, continuingly changing, development.

REFERENCES

Bowen RL (1989) Climatic metastability and maximization of species diversity. Amer Assoc Advancement Sci Ann Mtg Abstracts Vol (Abs. 326)

Bowen RL (1991) Wetlands as geologic ephemeral: implications relating to biologic diversity. Geol Soc Amer Abstracts w Prog 23(2):7

Miller GT Jr (1990) Living in the environment. Wadsworth Publishing, Belmont California

Pielou EC (1991) After the ice age: the return of life to glaciated North America. University of Chicago Press, Chicago

Chapter 4: Predicting the Fate of Temperate Species in Response to Human Disturbance and Global Change

Ken Thompson
NERC Unit of Comparative Plant Ecology
Department of Animal and Plant Sciences
The University
Sheffield S10 2TN
UK

Mankind is currently conducting a number of large-scale, uncontrolled experiments on the biosphere. Some of these, including reduction of atmospheric ozone levels and climate warming induced by increased greenhouse gas emissions, have hardly begun and their outcomes are still uncertain. Others, including the increasing disruption, fragmentation and eutrophication of vegetation by agriculture and urbanisation have been underway for centuries, and in a less intense form for millennia. Currently, human land use has a far greater influence on vegetation than anthropogenic climate change. One can argue about the relative responsibility of climate and man for vegetational change in prehistory, but, in Europe at least, man has certainly been the greater influence for centuries. As Perring (1965) remarked, "*... the founding of the monasteries and their dissolution have had far greater effects on the pattern of vegetation than a slight shift in climate up or down 1°C in the annual mean.*"

Nevertheless climate warming will place additional strain on the already threadbare fabric of the world's natural and semi-natural ecosystems. How can we predict what additional effects will occur? Predictions of the consequences of climate change for plant communities range from the effects on global distribution of vegetation types (Cramer and Leemans 1993) to changed distributions of individual species in single countries (e.g. Holten 1993). Assuming that future climate scenarios

NATO ASI Series, Vol. I 20
Biodiversity, Temperate Ecosystems, and Global Change
Edited by T. J. B. Boyle and C. E. B. Boyle

are broadly correct, and that relationships of climate to vegetation types are reasonably well understood, predictions of changed vegetation distributions on a regional or global scale are likely to be reliable. Changed distributions of crop plants are also likely to be accurate; in a greenhouse world, man will continue to grow particular crops where the climate is suitable. Greatest difficulties arise in attempting to predict the future distributions of individual plant species on national, or smaller, scales (Hill 1991, Holten 1993). At this level the limiting factor is likely to be ability to migrate both far enough and fast enough to keep pace with rapid climate warming. Likely rates of climate warming require migration rates at least an order of magnitude greater than those achieved at the end of the last glaciation (Davis 1983, Huntley 1991). Furthermore, in the modern landscape many species are restricted to isolated fragments of semi-natural vegetation, which will severely curtail their ability to migrate. The inevitable consequence of climate warming for many species, then, seems to be at best a major reduction in range, and at worst extinction. However, not all species are equally threatened with this fate. Some species, e.g. *Pulsatilla vulgaris* (Wells and Barling 1971) and *Primula elatior* (Rackham 1975) are famously static, and rarely if ever spread to new sites. Recent history demonstrates that others possess remarkable powers of migration (Table 1).

Table 1. Rates of migration of selected species in the British Isles. Compare maximum rate achieved by trees at end of last glaciation of c. 500 m yr^{-1}

Species	Rate of migration (km/yr)	Period	Source
Epilobium ciliatum	40	1949 - 1959	Preston (1989)
Puccinellia distans	20	1980 - 1984	Scott (1985)
Reynoutria japonica	8	1920 - 1939	Conolly (1977)
Elodea nuttallii	20	1966 - 1981	Simpson (1984)
Veronica filiformis	6	1949 - 1961	Bangerter and Kent (1957, 1962)
Senecio squalidus	7	1879 - 1939	Kent (1960)

Quite apart from a capacity to exploit the new climates, species which will prosper in the greenhouse world will need to share the characteristics of those which have already demonstrated that they find the twentieth century a congenial environment. Equally, species resembling those which have failed to come to terms with modern land use will be unlikely to succeed, whatever their suitability or otherwise for the changed climate. The main task of this paper will be to discover the critical ecological attributes of these potential winners and losers.

DATA SOURCES AND ANALYSIS

Land-use is a major determinant of success and failure in modern floras (Hodgson 1986). One would therefore expect that countries with differing intensities of land-use would show increasing and decreasing species with differing patterns of ecological attributes. In an attempt to test this hypothesis, I analysed data from more than one European country. The recently completed *Botanical Society of the British Isles Monitoring Scheme* (Rich and Woodruff 1990) provides lists of species which have recently increased and decreased in each of the four separate countries of the United Kingdom (England, Wales, Scotland and Northern Ireland) and in the Republic of Ireland. The BSBI Monitoring Scheme was a 10 x 10 km square sample survey during 1987 and 1988 to assess the current status of the flora, and specifically to compare it with data collected for the *Atlas of the British Flora* (Perring and Walters 1962). Data in the *Atlas* cover the period 1930-1960, but most are post-1950. The deficiencies of the Monitoring Scheme, particularly as a means of discovering changes since 1960, have been discussed by Rich and Woodruff (1992), and the data analysed here derive from a revision in 1993 (Rich, pers. comm.). This revision attempts to eliminate apparent changes due to various forms of recording bias, and to list only those species showing genuine increases or declines since 1960. It is worth noting here that owing to the nature of the recording, the Monitoring Scheme can only detect changes in *range* of the order of at least 10 km, and is unable to detect changes in local abundance, however marked. Therefore species such as *Arrhenatherum elatius*, already present in all 10 km squares apart from high Scottish mountains in 1960, cannot be shown to have increased.

A limitation of the BSBI data is that one country (England) has a rather high population density, and therefore presumably intensity of land-use, while the other four have similar and much lower population densities. Additional data were therefore obtained from the Netherlands, a country with similar population density to England. Lists of species which have increased or decreased in the Netherlands between 1940 and 1990 were provided by the Dutch Centraal Bureau voor de Statistiek. The Dutch data are based on a 5 x 5 km grid and are therefore capable of detecting change on a finer scale than the British data. Thus several very common species are listed as increased in the Netherlands (including, for example, *Arrhenatherum elatius*). This difference between the Dutch and British data must be borne in mind during the following analysis, although in practice the story revealed is remarkably consistent. The autecological attributes of the species analysed were obtained largely from Grime, Hodgson and Hunt (1988) and are listed in Table 2.

Table 2. Autecological attributes used in the analysis

Attribute	Range of values
Life history	1 - 9
C radius	1 - 5
S radius	1 - 5
R radius	1 - 5
Canopy height	1 - 8
Lateral spread	1 - 5
Flowering period	number of months
Clonal growth	0 or 1
Seed bank	1 - 3
Wind dispersal	0 or 1
Dispersule weight	0 - 5

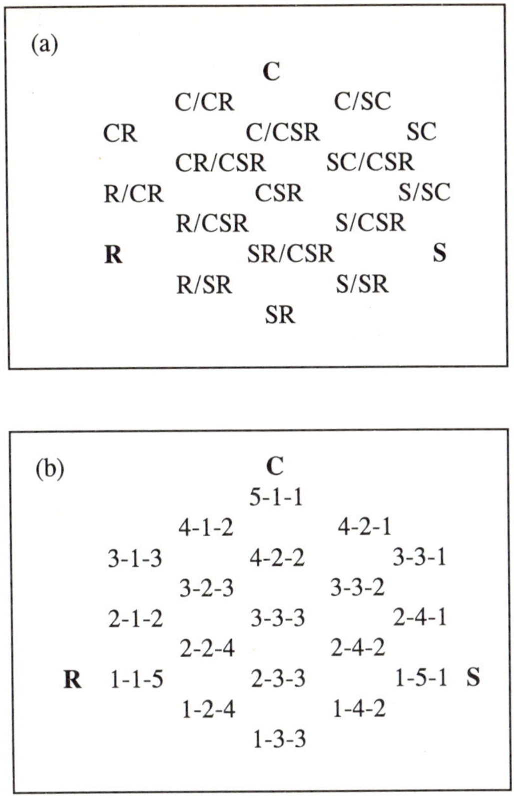

Figure 1. C-S-R plant strategy theory (Grime 1974) classifies plants into functional types according to their responses to gradients of stress and disturbance. The three main functional types are "competitors", "stress-tolerators" and "ruderals". Competitors are adapted to thrive in conditions of low stress and of low disturbance. Stress tolerators can survive nutrient and other stresses. Ruderals have a short life span and can survive by rapidly reproducing from seed on any fertile, open ground that may temporarily be produced by the action of disturbance. Ruderals, however, are sensitive to stress.

The location of species in the stress/disturbance matrix can be quantified by dividing the matrix into 19 zones. Each of these corresponds to one of the recognized C-S-R strategy types (a), the most extreme ones being the pure competitors (C), stress-tolerators (S), and ruderals (R). These three are located at the corners of a triangular framework within which the intermediate strategic types are distributed according to a nodal network. The combination of attributes displayed by a species occupying any given position can be represented as a C-S-R coordinate (b): a value 5 represents an extreme position most like the pure C, S or R strategy, and a value 1 represents the opposite position, most unlike the pure strategy.

The data, which are briefly described below, are essentially those in Grime, Hodgson and Hunt (1988), but some attributes differ in important respects. Life history is expressed on a scale from 1 (summer annual) to 9 (polycarpic perennial); higher numbers therefore represent increasing longevity. C, S and R radius are components of CSR strategy (Figure 1) and therefore represent, on a scale of 1 to 5, the extent to which the species is adapted to competition, stress or disturbance respectively (Grime 1974, 1979). C radius is thus primarily, though by no means exclusively, a function of plant height (Gaudet and Keddy 1988). S radius is largely an inverse function of maximum potential relative growth rate, while R radius has much in common with life history. Canopy height ranges from 1 (< 100 mm) to 8 (> 15 m), and lateral spread from 1 (annuals) to 5 (perennials > 1 m in diameter). Flowering period is simply the number of months the species is normally in flower. Dispersule weight varies from 0 (too small to be easily measured, e.g. orchid seeds, fern spores) to 5 (> 10 mg). The seed bank classes are 1 (seeds persist for < 1 yr.), 2 (1-4 yr.) and 3 (> 4 yr.). These data differ significantly from those in Grime, Hodgson and Hunt (1988) and are derived from a new seed bank database in preparation by K. Thompson and J. P. Bakker. Wind dispersal and clonal growth are both either present (1) or absent (0). Wind dispersal is here defined as the presence of obvious morphological adaptations for wind dispersal (pappus or wing), and does not include species such as orchids or ferns which are often assumed to be wind dispersed by virtue of their very tiny seeds or spores. Note that many attributes are expressed on markedly non-linear scales. Grime, Hodgson and Hunt (1988) contains other attributes which may have proved useful in the analysis but all except those listed above were excluded because they contained too many missing values.

Increasing and decreasing species were analysed separately for each country by discriminant analysis (SPSS). The analysis was carried out on only those species which also occur in the database in Grime, Hodgson and Hunt (1988). SPSS also excluded those species for which any of the independent variables was missing, and the effects of this on sample size, together with the results of the analysis, are shown in Table 3.

Table 3. Correlations between canonical discriminant functions and discriminating variables for each of six European countries. N is total number of species and N in analysis is those species with no missing values for any of the discriminating variables. Percentage of species correctly classified as "increasing" or "decreasing" by the discriminant function is also shown. A positive correlation indicates a larger value of the variable in decreasing species, and *vice versa*. For details of analysis, see text.

	Netherlands	England	Northern Ireland	Rep. of Ireland	Scotland	Wales
N	282	157	73	112	53	57
N included in analysis	178	105	43	70	32	27
S radius	0.873	0.868	0.239	0.127	0.306	0.002
R radius	-0.431	-0.293	0.181	-0.122	-0.080	-0.005
Canopy height	-0.399	-0.438	-0.252	0.094	0.012	0.221
C radius	-0.389	-0.436	-0.495	-0.331	-0.132	-0.139
Life history	0.256	0.194	-0.335	-0.150	0.097	0.037
Seed bank	-0.172	-0.099	-0.055	-0.075	0.317	0.098
Wind dispersal	-0.085	-0.127	-0.441	-0.089	-0.303	-0.112
Clonal growth	0.080	0.010	-0.227	0.031	-0.155	-0.121
Dispersule weight	-0.078	-0.090	-0.038	-0.246	0.015	0.101
Flowering period	-0.025	-0.088	0.017	-0.007	0.032	0.442
Lateral spread	0.013	-0.001	-0.214	-0.199	-0.270	0.017
% of species correctly classified	83.15	79.05	83.72	77.14	81.25	88.89

The discriminant function was about equally good at classifying increasing and decreasing species in each of the six countries. The importance of the individual discriminating variables varied enormously, however. The discriminating variables are ordered in Table 3 in their order of importance in the Netherlands, which serves to highlight three features. First, the order of the variables is almost the same in the Netherlands and in England, and in both is very different from the other four countries. S radius, R radius and canopy height occupy the first three places in both countries (canopy height and C radius are highly correlated). Second, those variables which might loosely be classed as "regenerative" in character (seed bank, wind dispersal, dispersule weight, flowering period and seed weight) make little or no contribution to the discriminant function. Third, only in the Netherlands and England does a single variable, S radius, stand out as much more significant than all the others. Indeed, in the Netherlands and in England 75% of species can be correctly classified on the basis of S radius alone. In contrast, no single discriminating variable stands out in either Northern Ireland, the Republic of Ireland, Scotland or Wales, and no consistent pattern emerges from these four countries.

DISCUSSION

The results of the analysis for England and the Netherlands confirm earlier work which suggested that fast-growing species (i.e. species with low S radius) are at an advantage in the modern English landscape (Hodgson 1986). The second place of R radius in the Netherlands suggests that the ruderal element in the flora is more successful and that therefore the landscape is perhaps more disturbed than in England. In contrast the higher ranking for canopy height in England is suggestive of a greater role for dereliction in the English flora. Not too much emphasis should be placed on either of these variables, however, since neither makes a large contribution to the power of the discriminant function.

The other four countries show no clear pattern. Partly this may reflect the low sample size, especially in Scotland and Wales. Mainly, however, it reflects the fact that in these less intensively disturbed countries the causes of success and failure of individual species vary locally both within and between countries, and no single attribute or suite of attributes confers success. The difference between the two groups of countries can be clearly seen in Figure 2, which relates the mean S radius of increasing and decreasing species in the six countries to their human population densities.

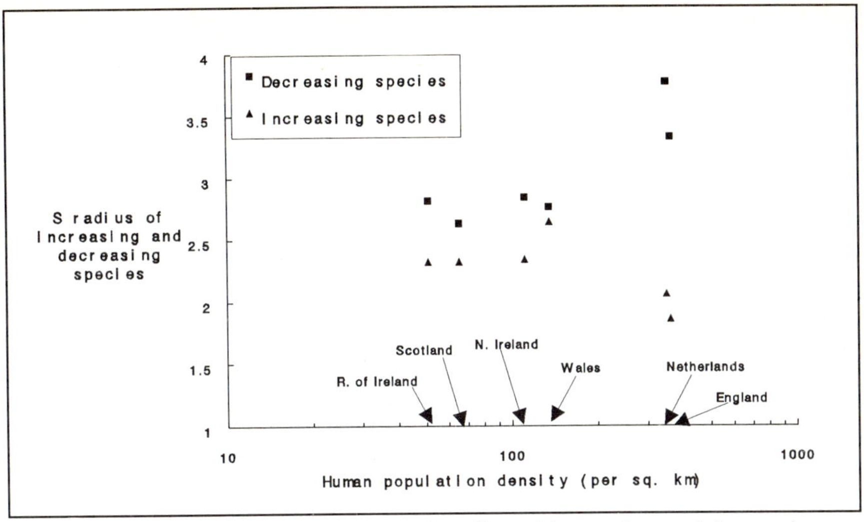

Figure 2. Relationship between mean S radius of increasing and decreasing species and human population density in six European countries. S radius of two groups not significantly different in Scotland, N Ireland or Wales. Two groups are significantly different in R of Ireland ($p = 0.049$), England and the Netherlands (both $p < 0.001$)

In the less populous countries the S radii hardly differ, while in the two densely populated countries the S radius of decreasing species is almost twice that of the increasing species. Human population density is a convenient metaphor for a wide range of human impacts on the landscape, both urban and agricultural, and it is probably pointless to speculate at any length on which aspects of human activity are most responsible for the pattern in Figure 2. Two major processes can be recognised, however. The first is the increasing restriction of slow-growing plants of infertile, relatively undisturbed habitats to fragmented islands of suitable habitat, many of them in nature reserves, surrounded by a sea of unsuitable landscape. Continued attrition of these fragments by road and house-building, pollution and agriculture, combined with inevitable chance extinctions from small, isolated fragments, have inescapably led to the contraction of slow-growing species. Second, the increasing abundance of disturbed, fertile, man-made habitats, which support communities exclusively comprising fast-growing ruderal and competitive species, has led to the continuing expansion of these species. The result, graphically demonstrated in Figure 2, is an increasing polarisation of the English and Dutch floras into an expanding, fast-growing component, and a declining, slow-growing component characteristic of an older, semi-natural landscape. In the less populous northern and western fringe of the British Isles this dichotomy can hardly be discerned.

A surprising feature of the results is the apparent lack of importance of regenerative attributes in discriminating between increasing and decreasing species in England and the Netherlands. This is despite earlier work which appeared to demonstrate a major role for effective dispersal in the exploitation of both man-made habitats (Grime 1986) and the English landscape in general (Hodgson and Grime 1990). Perhaps the movement of propagules (both seeds and vegetative fragments) by construction, mining, quarrying, agriculture and the sheer volume of human traffic is now so pervasive that dispersal has ceased to be a barrier to the spread of species in densely-populated landscapes. The example of a few effectively dispersed aliens which have achieved high rates of migration in the UK (e.g. *Senecio squalidus*, *Epilobium ciliatum*) should not obscure the fact that others have spread as quickly without producing seed at all (e.g. *Veronica filiformis*, *Reynoutria japonica*). One should also note that a relatively narrow definition of "wind dispersal" was employed in the analysis. If species with minute seeds or spores had been included, it seems very likely that the discriminant analysis would have classified wind dispersal as positively associated with *decreasing* species. Most orchids and pteridophytes are declining, despite a reputedly effective dispersal mechanism, and this serves to confirm that what happens after a propagule encounters a new site is more important than the ability to reach the site in the first place.

Implications for diversity

The inability of most slow-growing species to spread across densely-populated, lowland landscapes will inevitably lead to range contractions, and in many cases to extinction, in the face of climate warming. Species with small populations, often confined to reserves, will be particularly at risk. Climate warming will therefore impoverish the native floras of much of Europe and North America. On the other hand, new species will invade from other climates, and so many floras may not actually experience net declines in numbers of species. The effect on species richness at a regional or national scale may therefore be slight. At a local scale, however, what will be the effect of climate change on species diversity of plant communities? Will the new communities be less, or more, diverse than the ones they replace? Given that the responses of species to climate are individualistic (Huntley 1991), and that therefore many new communities will have no present-day analogues, this question is almost impossible to answer. An indication of the likely direction of change, however, can be derived from the species richness of the modern communities occupied by increasing and decreasing species. For northern England, extensive vegetation surveys (Grime, Hodgson and Hunt 1988) provide estimates of these. The data are analysed for the Netherlands and England only (Table 4).

For both the Netherlands and for England the results are strikingly similar. The communities presently occupied by decreasing species are significantly more species rich than those composed of increasing species. In both countries the species richness of the former communities is also more variable. This is because species richness of infertile semi-natural communities is bimodal, with species poor acid communities and species rich basic ones. Nevertheless the difference between the two groups persists despite this "tail" of species poor acid communities. It therefore seems likely that climate warming will cause a significant loss of local species richness, largely because infertile, basic communities will be replaced by less diverse communities of faster-growing species.

Relationship to previous work

Recently a consensus has begun to emerge that it is difficult, or perhaps even impossible, to predict the characteristics of potentially invasive or colonising species (Noble 1989; Roy 1990; Bengtsson and Baur 1993), although a few successful attempts exist (Ashton and Mitchell 1989; Rydin and Borgegård 1991). The work reported here provides some support for both points of view, and in particular highlights the importance of considering simultaneously the attributes of potential invaders *and* the invaded community. The view of many authors (Crawley 1987;

Table 4. Mean species richness (measured on a linear five point scale - see Grime, Hodgson and Hunt (1988)) of plant communities associated with increasing and decreasing species in the Netherlands and England. Differences significant for both countries at $p < 0.001$.

		Decreasing	Increasing
Netherlands	Mean	3.47	2.49
	Variance	1.70	1.07
	N	118	152
England	Mean	3.30	2.48
	Variance	1.49	0.91
	N	89	62

Mack 1989; Rejmanek 1989) that disturbance is crucial in facilitating invasion is consistent with the results reported here. In countries suffering high levels of human disturbance and eutrophication it clearly *is* possible to define the necessary attributes of potential invaders with considerable reliability. The high level of agreement between the results for England and the Netherlands gives one confidence that this approach could easily be extended to other densely-populated areas. On the other hand, the results for the more sparsely-populated countries suggest that in these circumstances it really is very difficult to discover consistent traits of invasive species, and it is therefore easy to see how the prevalent view has arisen.

It is probably also true that most previous studies have used too few species. The point has been well made by Roy (1990) that except for a very few characters such as seed weight or photosynthetic pathway, the comparative data on large numbers of species needed to reveal consistent patterns are simply not available. Even in England and the Netherlands, the "best" set of traits misclassifies a significant minority of species, and it is easy to see how meaningless or contradictory results could be obtained by choosing to work on pairs or small groups of species.

Practical implications

Climate and land-use are the major environmental "sieves" which determine which plant species are able to grow in any particular spot. I have implicitly assumed here that these sieves can be regarded as independent, i.e. that the likely future distribution of a species can be determined by first correlating its distribution with key climatic variables, then applying the criteria discussed above to determine whether or not the species is actually likely to move. Those species which are likely to migrate without assistance can be ignored. It is a subset of the remaining species, judged to be of sufficient rarity or conservation significance, which will require human intervention. Scarce resources may require that these species are further subdivided into those which will actually receive assistance and those which will not. The effort required, if a significant proportion of threatened species is to be aided, will be enormous.

It is possible, however, that responses to climate warming and land-use are not independent. Global climate models are not yet able to provide detailed predictions of the future climate of areas as small as, say, England. However, one aspect of future climates on which there is general agreement is that extreme climatic events will become more frequent (Ausubel 1991; Wigley 1985). Thus droughts, floods, fires and late frosts will all be more prevalent. All these will damage or destroy vegetation, freeing space and resources, and favouring the fast growing species which have already benefited from modern land-use. Quite apart from predicted responses to change in mean climatic variables, therefore, it seems possible that climate change itself will intensify exactly the same pressures which have already arisen from land-use.

Acknowledgements

I acknowledge the financial support provided by the NERC through its TIGER (Terrestrial Initiative in Global Environmental Research) programme, award number

GST/02/635. Thanks are also due to Tim Rich, for access to an unpublished revision of his report on the BSBI Monitoring Scheme, to C.L. Plate of the Centraal Bureau voor de Statistiek, the Netherlands, for unpublished data, and to John Hodgson for useful discussions.

REFERENCES

Ashton PJ, Mitchell DS (1989) Aquatic plants: Patterns and modes of invasion, attributes of invading species and assessment of control programmes. Pages 111-154 *in*: Drake JA et al. (eds) Biological Invasions, a Global Perspective. Wiley, New York

Ausubel, JH (1991) A second look at the impacts of climate change. American Scientist 79: 210-221

Bangerter EB, Kent DH (1957) *Veronica filiformis* Sm. in the British Isles. Proceedings of the Botanical Society of the British Isles 2: 197-217

Bangerter EB, Kent DH (1962) Further notes on *Veronica filiformis*. Proceedings of the Botanical Society of the British Isles 4: 384-397

Bengtsson J, Baur B (1993) Do pioneers have r-selected traits? Life history patterns among colonizing terrestrial gastropods. Oecologia 94: 17-22

Conolly AP (1977) The distribution and history in the British Isles of some alien species of *Polygonum* and *Reynoutria*. Watsonia 11: 291-311

Cramer WP, Leemans R (1993) Assessing impacts of climate change on vegetation using climate classification systems. Pages 190-217 *in*: Solomon AM, Shugart HH (eds) Vegetation Dynamics and Global Change. Chapman and Hall, New York

Crawley MJ (1987) What makes a community invasible? Pages 429-453 *in*: Gray AJ, Crawley MJ, Edwards, PJ (eds) Colonization, Succession and Stability. Blackwell, Oxford

Davis MB (1983) Holocene vegetational history of the eastern United States. Pages 166-181 *in*: Wright Jr HE. Late-Quaternary Environments of the United States vol 2. University of Minnesota Press, Minneapolis

Gaudet CL, Keddy PA (1988) A comparative approach to predicting competitive ability from plant traits. Nature 334: 242-243

Grime JP (1974) Vegetation classification by reference to strategies. Nature 250: 26-31

Grime JP (1979) Plant Strategies and Vegetation Processes. Wiley, Chichester

Grime JP (1986) The circumstances and characteristics of spoil colonization within a local flora. Philosophical Transactions of the Royal Society of London B314: 637-654

Grime JP, Hodgson JG, Hunt R. (1988) Comparative Plant Ecology: A Functional Approach to Common British Plants. Unwin Hyman, London

Hill MO (1991) Patterns of species distribution in Britain elucidated by canonical correspondence analysis. Journal of Biogeography 18: 247-255

Hodgson JG. 1986. Commonness and rarity in plants with special reference to the Sheffield flora. II The relative importance of climate, soils and land use. Biological Conservation 36: 253-274

Hodgson JG, Grime JP (1990) The role of dispersal mechanisms, regenerative strategies and seed banks in the vegetation dynamics of the British landscape. Pages 65-81 *in*: Bunce RGH, Howard DC (eds) Species Dispersal in Agricultural Habitats. Belhaven, London

Holten JI (1993) Potential effects of climatic change on distribuion of plant species, with emphasis on Norway. Pages 84-104 *in*: Holten JI, Paulsen G, Oechel WC (eds) Impacts of Climatic Change on Natural Ecosystems. Norwegian Institute for Nature Research and The Directorate for Nature Management, Trondheim

Huntley B (1991) How plants respond to climate change: migration rates, individualism and the consequences for plant communities. Annals of Botany 67 (Supplement 1): 15-22

Kent DH (1960) *Senecio squalidus* L. in the British Isles 2. The spread from Oxford. Proceedings of the Botanical Society of the British Isles 3: 375-379

Mack, RN (1989) Temperate grasslands vulnerable to plant invasions: characteristics and consequences. Pages 155-179 *in*: Drake JA, Mooney HA, di Castri F, Groves RH, Kruger FJ, Rejmanek M, Williamson MH. (eds) Biological Invasions: a Global Perspective. Wiley, Chichester

Noble IR (1989) Attributes of invaders and the invading process: terrestrial and vascular plants. Pages 301-313 *in*: Drake JA, Mooney HA, di Castri F, Groves RH, Kruger FJ, Rejmanek M, Williamson MH. (eds) Biological Invasions: a Global Perspective. Wiley, Chichester

Perring FH (1965) The advance and retreat of the British flora. Pages 51-59 *in*: Johnson CG, Smith LP (eds) The Biological Significance of Climatic Changes in Britain. Academic Press, London

Perring FH, Walters SM (1962) Atlas of the British Flora. Nelson London.

Preston CD (1989) The spread of *Epilobium ciliatum* Raf. in the British Isles. Watsonia 17: 279-288

Rackham O (1975) Hayley Wood: its History and Ecology. Cambs and Isle of Ely Naturalists' Trust, Cambridge

Rejmanek M (1989) Invasibility of plant communities. Pages 369-388 *in*: Drake JA, Mooney HA, di Castri F, Groves RH, Kruger FJ, Rejmanek M, Williamson MH. (eds) Biological Invasions: a Global Perspective. Wiley, Chichester

Rich TCG, Woodruff ER (1990) The BSBI monitoring scheme 1987-1988. Report to the Nature Conservancy Council, CSD report 1265. Nature Conservancy Council, Peterborough. Revised and updated by TCG Rich, 1993

Rich TCG, Woodruff ER (1992) Recording bias in botanical surveys. Watsonia 19: 73-95

Roy J (1990) In search of the characteristics of plant invaders. Pages 335-352 *in*: di Castri F, Hansen AJ, Debussche M (eds) Biological Invasions in Europe and the Mediterranean. Kluwer, Dordrecht

Rydin H, Borgegård S-O (1991) Plant characteristics over a century of primary succession on islands: Lake Hjälmaren. Ecology 72:1089-1101

Wells TCE, Barling DM (1971) Biological flora of the British Isles: *Pulsatilla vulgaris* Mill. Journal of Ecology 59: 275-292
Scott NE (1985) The updated distribution of maritime species on British roadsides. Watsonia 15: 381-386
Simpson DA (1984) A short history of the introduction and spread of *Elodea* Michx in the British Isles. Watsonia 15: 1-9
Wigley, TML (1985) Impacts of extreme events. Nature 316: 106-107

Chapter 5: Ground beetles in a changing environment : determinants of species diversity and community assembly

Michel Loreau

Department of Animal Biology, C.P. 160/13

Free University of Brussels

50, av. F.D. Roosevelt

B-1050 Brussels, Belgium

INTRODUCTION

The exponential increase of human activities has reached the point where man alters the natural environment on a global scale and threatens unintentionally to drive to extinction half of the species with which he shares his planet within a century; that is, at a rate about a million times greater than the rate at which new species are formed (May 1988). The most direct cause of species extinction is the destruction of their habitat, but even in protected habitats indirect factors, notably atmospheric or water pollution and climate change, might lead to profound changes in plant and animal communities.

Faced with this potential unprecedented ecological catastrophe, ecology is still ill-armed theoretically. Ecology is a young science confronted with levels of complexity that have no equivalent in other sciences. Community ecology in particular, whose object is to understand the processes governing the diversity and organization of species assemblages, is in the midst of what may be called a crisis. During previous years the prevailing view was the deterministic one developed by the Hutchinson-MacArthur school based on equilibrium competition theory (*e.g.* MacArthur 1972) - actually the first generally consistent theory of community organization. Despite partial successes, this theory failed to provide a satisfactory general explanation of the diversity and coexistence of species. The emphasis has now shifted to the significance of regional biogeographic processes (*e.g.* Ricklefs

NATO ASI Series, Vol. I 20
Biodiversity, Temperate Ecosystems, and Global Change
Edited by T. J. B. Boyle and C. E. B. Boyle

1987; Cornell & Lawton 1992) and non-equilibrium coexistence (*e.g.* Levins 1979; Chesson 1986; DeAngelis & Waterhouse 1987; Huston 1993). But this shift has not resulted in a new unified theory; the tendency has rather been to identify different kinds of communities (*e.g.* Schoener 1986a). If the competitive exclusion principle is analogous to Newton's first law of motion (Hardin 1960), then community ecology is still lacking the analogue of Einstein's theory of relativity.

What is a community, how does it assemble and how does it respond to environmental changes? These fundamental questions have not yet received final answers, and the present paper certainly does not pretend to provide them. But it attempts to shed light on these questions by examining factors that determine the diversity and structure of ground beetle communities (Coleoptera, Carabidae). Diversity is the hallmark of all life forms, but in no other group is it as spectacular as in insects, and more especially in beetles. Southwood (1978) identified three biological features responsible for this extreme diversity of insects: their small size, which allows them to occupy many special niches, their wings, which provide them with a great power of dispersal among temporary patches, and their metamorphosis, which allows them to partition time. The first two features, together with a moderate life span and intermediate positions in trophic chains, also contribute to the potential for non-equilibrium and non-competitive coexistence (*e.g.* Hutchinson 1961; Schoener 1986b). In this respect ground beetles are of special interest, because they are extremely diversified with about 40,000 described species; they are mostly predators, they span a wide range of sizes, and a number of large forest species have lost their ability to fly and can live for several years. The latter species can thus be viewed as making a transition with small vertebrates from an ecological viewpoint, and their study is likely to reveal more continuities between the ecological processes affecting small invertebrates and large vertebrates than is a comparison between extreme forms of these two groups.

REGIONAL VERSUS LOCAL PROCESSES

How local is a local community? Both the Hutchinson-MacArthur school emphasizing biotic interactions and the ecophysiological school emphasizing individual adaptations to abiotic factors have traditionally sought to explain the composition and diversity of communities in terms of local deterministic factors. For ground beetles, the most famous representative of the ecophysiological school was Thiele, who presented a remarkable synthesis on habitat selection by adaptations in physiology and behaviour in his 1977 book. A different perspective emerged from den Boer (*e.g.* 1977), who considered extinction of carabid populations a continually occurring process, compensated for by re-founding of populations through dispersal. On this view, the composition and diversity of local communities, especially in unstable biotopes, should be dependent on stochastic, regional dispersal processes. Empirical evidence in several taxonomic groups and theoretical arguments have now accumulated to show that regional and historical factors can profoundly influence local diversity (*e.g.* Ricklefs 1987; Cornell & Lawton 1992).

By way of example, Figure 1 summarizes the composition and species richness of ground beetle communities I studied intensively in three forest biotopes in Lembeek, Belgium: a young successional forest, which is intermediate between a meadow and a forest, a pine plantation, and a climax beechwood (Loreau 1984a). Though artificial, the pinewood harbours a carabid community which is in many respects intermediate between those of the other two biotopes; the three forests can thus be viewed as representing a succession to the climax stage. Ground beetles were sampled continuously using pitfall traps from 1978 to 1981. The total catch in pitfall traps was shown to be a good predictor of mean biomass, and hence of the ecological importance of species (Loreau 1992). In Figure 1 the species are ranked from the most important (rank 1) to the least important (highest rank) in terms of catch numbers, and their rough habitat type was determined from previous studies in Europe. In the climax beechwood all the main species were typical forest species, whose presence and abundance may be ascribed to local deterministic factors. But many of the rare species were typical for more open habitats and were probably occasional immigrants. On the other hand the successional forest contained a mixture of species of all habitat types. Since the biotope itself is heterogeneous, it is difficult in this case to separate regional from local effects, but it is likely that even the

HABITAT.XLS Chart 3

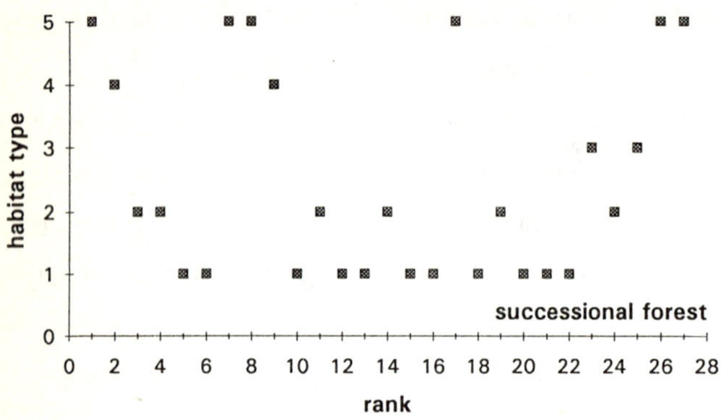

HABITAT.XLS Chart 2

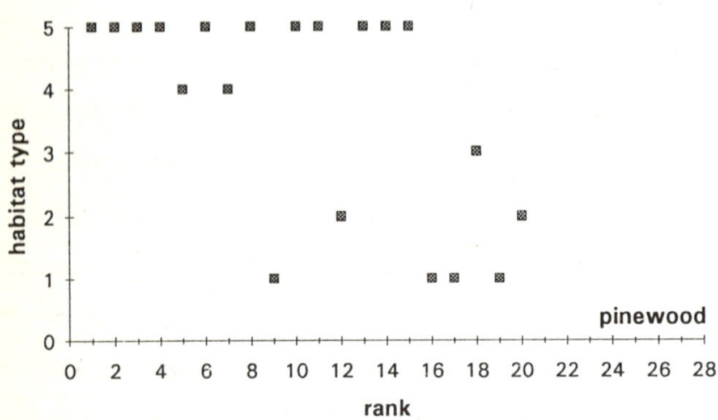

HABITAT.XLS Chart 1

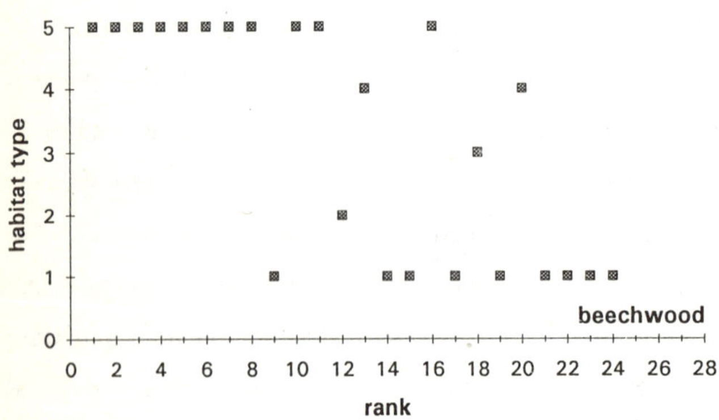

Figure 1. Habitat type of the species composing three forest carabid communities in Lembeek, Belgium, against their rank of ecological importance as obtained from pitfall trapping. Habitat types: open (1), intermediate (2), humid (3), non-specific (ubiquitous species) (4), and forest (5) habitats. Data from Loreau (1984a).

abundance of the dominant species was affected by immigration from the surrounding forests, where it was dominant too. Thus although regional processes affect local species diversity in all cases, the extent to which they affect the core species and community structure is likely to differ among sites, and to be greater in unstable biotopes (*e.g.* a successional stage) and lower in stable biotopes (*e.g.* a climax stage) (see also den Boer 1977). Another factor liable to affect the extent of regional influences is the area of the biotopes: the smaller their area, the greater the regional influences due to edge effects (*e.g.* Niemelä 1988). Other authors have recently presented evidence that biogeographic processes determine local ground beetle diversity. Thus Heijerman & Turin (1989) suggested that geographical position is more important than forest type in determining the composition of forest carabid communities in the Netherlands. Eyre & Luff (1993) were not able to ascribe differences in forest carabid communities from northeast England to environmental factors. However, caution must be exerted to avoid unwarranted generalizations. The relative importance of biogeographic factors in these studies may be explained by the fact that most of the forested areas considered were young, hence in a successional stage (Turin et al. 1991), or by the high similarity of the various forests in the region.

NICHE DIFFERENTIATION AND COMPETITION

The foregoing suggests that while local deterministic factors do not explain the whole diversity of communities, they do play a role in the presence and abundance of the main species. According to niche and competition theory, only those species will coexist which have sufficiently different niches to avoid competition for shared resources (*e.g.* Hutchinson 1957; MacArthur 1972). Community organization thus results from resource partitioning generated by interspecific competition. Can this theory explain the diversity of the main species in carabid communities?

I investigated this issue in great detail in the forest ground beetle communities from Lembeek. I first analyzed all the niche components likely to be involved in resource partitioning for the various species: annual and daily activity rhythms, type and size of prey eaten, and horizontal and vertical spatial distributions of activity. The two main components leading to niche differentiation (low niche overlaps) in the beechwood and the pinewood were found to be the annual activity pattern and the type of prey (Figure 2). There was a nice seasonal succession of periods of maximum activity of the various species (which also appears in Figure 5), as well as clear-cut differences in feeding preferences despite the fact that almost all species proved to be polyphagous predators on primary consumers. The combination of these two niche dimensions resulted in a remarkable overall niche differentiation, in contradiction with the view generally held before. Furthermore, temporal niche differentiation was much less in the successional forest, where the annual activity cycles of the dominant species appeared actually to converge rather than diverge; this finds expression in a consistently higher mean niche overlap whatever the number of species considered (Figure 2). These results are those predicted by competition theory: niche differentiation is achieved provided that the environment is stable enough for competitive relationships to stabilize in turn. They were further confirmed by tests against null models, but niche divergence was found to be statistically significant only among the four dominant species in the climax beechwood (Loreau 1989).

Horizontal spatial distribution of activity relates probably more to a species' habitat than to its niche (*sensu* Whittaker et al. 1973). For this component, niche overlap increased from the successional forest to the climax beechwood (Figure 2), which expresses the growing spatial homogeneity of the environment. Spatial distributions of ground beetles are usually contagious, but an interesting feature of large dominant species was that their distribution tended to be random at large spatial scales in those biotopes where they were most abundant, as if they saturated the environment (Loreau 1984b, Loreau & Nolf 1993). Detailed capture-recapture data further showed a considerable mixing of the population of the dominant species in the beechwood, *Abax ater* (Loreau & Nolf 1993).

A long-term (4 years) field experiment was then carried out to test the significance of intra- and interspecific competition in the beechwood (Loreau 1990). Intraspecific competition in the dominant species *A. ater* was tested by a convergence experiment in fenced plots with either normal or double population density; interspecific competition was tested by experimental introduction of a species

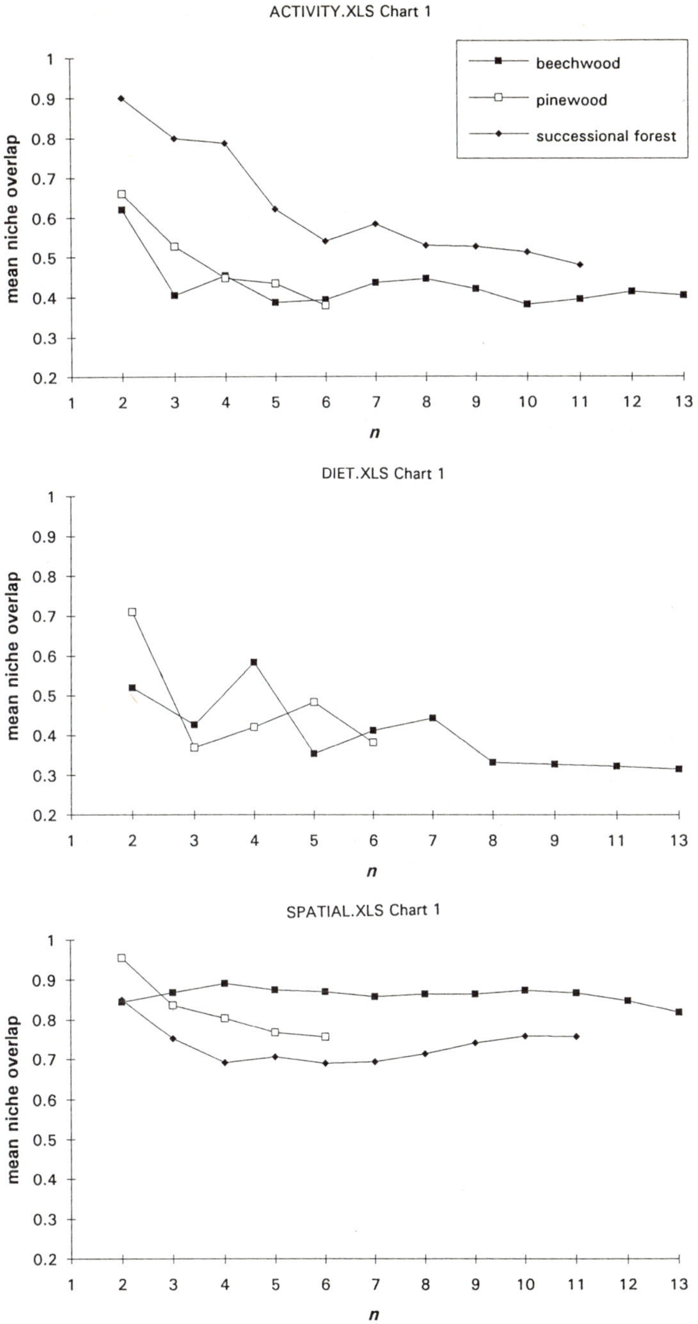

Figure 2. Mean niche overlap of the first *n* species in the three forest carabid communities of Lembeek (species ranked according to their ecological importance), for the three main niche components: annual activity cycle (top), diet (middle), and horizontal spatial distribution of activity (bottom). Data from Loreau (1984b, 1984c).

occupying a similar niche in the successional forest, *Pterostichus madidus*, either with or without *A. ater*, and by monitoring the effects of different densities of *A. ater* on the other local large-sized species. This experiment provided some evidence for competitive regulation in the dominant species *A. ater*, but no evidence for intra- or interspecific competition in the other species. The introduced populations of *P. madidus* did poorly, but persisted at low densities apparently without being affected by the presence of *A. ater*. The other resident species, among which was the second dominant species *P. oblongopunctatus*, also showed no sign of ecological release in the absence of *A. ater*. It must be noted, however, that the experimental plots were too small to harbour viable populations of the third and fourth dominant species, which are large-sized and highly mobile.

The overall conclusion suggested by these studies is that competition does play a role in ground beetle communities, but a much more restricted role than assumed in equilibrium competition theory. Competition seems to determine population density and dynamics only in a single or a few dominant species in the climax forest. This does not necessarily mean that competition is absent in the other species, but at least that, if present, it is too weak or variable to govern the presence, abundance and overall population dynamics of these species. In other, more unstable and heterogeneous biotopes, its role should be even more limited. This conclusion is not inconsistent with those of a recent extensive review of interspecific competition in ground beetles (Niemelä 1993).

The questions now are: how to explain that competition is significant in one species but not in others, and how to explain community-wide patterns such as niche differentiation in the climax forest? The answers to these questions, I suggest, lie in the coexistence of both equilibrium and non-equilibrium populations and the structuring influence of the lower trophic level, which will be examined in the next sections.

NON-EQUILIBRIUM VERSUS EQUILIBRIUM POPULATIONS

The equilibrium view of the world has a long tradition in science, and ecology is no exception. The idea that populations and communities reach some sort of

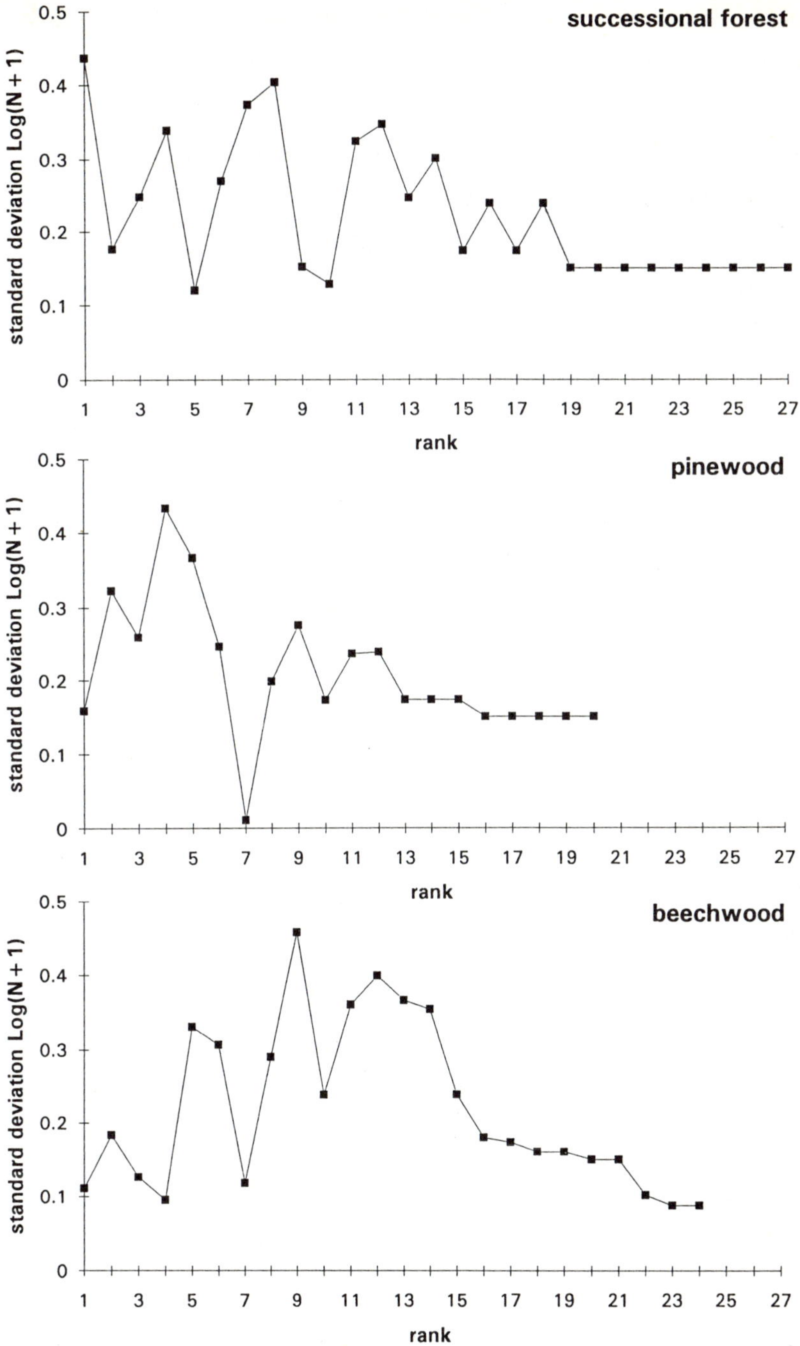

Figure 3. Population variability, as measured by the standard deviation of log-transformed year-catches (*N*), against the species' rank of ecological importance in the three forest carabid communities of Lembeek. Data from Loreau (1984a).

equilibrium or steady state has been one of the central tenets of the Hutchinson-MacArthur school. Others have strongly opposed this view, among them carabidologists like den Boer (*e.g.* 1968, 1981). Today non-equilibrium processes are accepted as significant components of species coexistence and diversity (*e.g.* Levins 1979; Chesson 1986; Huston 1993), and the opposition between equilibrium and non-equilibrium concepts has been largely resolved by noting that these generally apply to different scales (*e.g.* DeAngelis & Waterhouse 1987). What is perhaps less appreciated is the fact that equilibrium and non-equilibrium populations can coexist within a single community, in such a way that a community may consist of a core of equilibrium dominant species and a large set of non-equilibrium species (Caswell 1978; Hanski 1982; Loreau 1992).

This is strongly suggested by patterns of variability of population sizes in the carabid communities from Lembeek. If one plots population variability, as measured by the standard deviation of log-transformed year-catches, against the species' rank of ecological importance (Figure 3), the expected pattern for a non-equilibrium community is a widely and irregularly oscillating curve without detectable trend, because standard deviation after log transformation should be independent of the mean and variability should be distributed randomly. That is what was found in the successional forest (Figure 3). Note that the tail of the curve bends downward and becomes flatter due to low catch numbers (in particular the rarest species are those which are caught only once; thus they must have identical low variabilities because of identical series of zero catches). This reduced variability of small populations can be regarded as real when populations are censused exhaustively (*e.g.* Schoener & Spiller 1992), but has to be regarded as a statistical artefact when populations are only sampled as is the case here with pitfall trapping. On the other hand, the curve in the beechwood was very different from that "null" pattern: it tended to be bell-shaped with low variabilities in the dominant species. The situation was somewhat intermediate in the pinewood.

After elimination of the species with too low catch numbers (total 4-year catch 5 individuals) which contribute to the tail of the curves in Figure 3, the correlation

Table 1. Correlations between population variability [standard deviation Log(N+1) where N is the total year-catch], average ecological importance [average Log(N+1)], and body size in the three ground beetle communities from Lembeek. Highly significant correlations ($P < 0.01$) are indicated by **; other correlations are not significant ($P > 0.05$).

correlation	successional forest (10 df)	pinewood (4 df)	beechwood (12 df)
average - SD Log(N+1)	0.077	-0.519	-0.755**
body size - SD Log(N+1)	0.548	-0.388	-0.775**
average Log(N+1) - body size	0.496	0.502	0.519

between population variability and average ecological importance (average log-transformed year-catch) varied from zero in the successional forest to a highly significant negative value in the beechwood (Table 1). Body size also seemed to play a significant part in the pattern, since the correlation between population variability and body size varied from a positive value (just under the 0.05 significance level) in the successional forest to a significant negative value in the beechwood. The correlation between body size and ecological importance was positive though not significant in all cases. All this suggests that the succession to the climax stage is accompanied by a tendency for large dominant species to stabilize around an equilibrium state.

The results presented in Figure 3 actually underestimate the contrast between equilibrium and non-equilibrium populations because they are based on data from only four years of pitfall trapping. Over longer periods of time, the contrast is much more striking. Thus the average population density of the dominant species in the beechwood, *A. ater*, was remarkably constant over a 13-year period, while that of the second dominant species, *P. oblongopunctatus*, declined by three orders of magnitude over a 10-year period (Figure 4). Although population fluctuations alone are not sufficient to discriminate between equilibrium and non-equilibrium conditions, this kind of contrast is a strong argument for such a discrimination, all the more so since

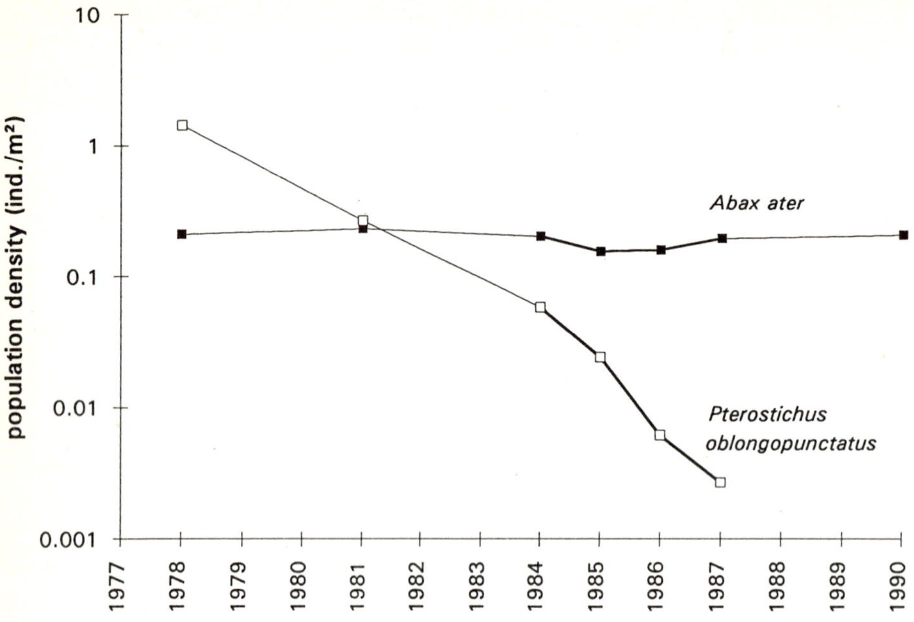

Figure 4. Long-term changes in the population density of the two dominant species in the beechwood of Lembeek, during their period of maximum density (late April to late June for *P. oblongopunctatus*, mid-May to late July for *A. ater*). Capture-recapture data from Loreau (1984d, 1990) and Loreau & Nolf (1993).

it is consistent with field experimental data on competition (see previous section) and does not seem to be consistent with alternative explanations. For instance, the population stability of *A. ater* cannot be ascribed to a spatial "spreading of risk" (den Boer 1968, 1981) because of the spatial homogeneity of the population and its environment (Loreau & Nolf 1993). Also widely fluctuating populations such as that of *P. oblongopunctatus* can hardly be assumed to be in a moving equilibrium with widely fluctuating resources because ground beetles are polyphagous and should not generally be strongly influenced by fluctuations of any particular prey species.

The tendency for large dominant species to reach equilibrium conditions in the climax forest while most other species and even dominant species in the successional forest do not, can be explained plausibly as follows. Most species are adapted to only part of the conditions experienced in their local environment and accordingly are sensitive to spatial and temporal variations in these conditions; the smaller their body size and the more unstable and heterogeneous the environment, the greater generally their sensitivity. Such species therefore reach neither an equilibrium nor generally a very high biomass. A few species can be especially well adapted to the whole range of conditions experienced in the local environment; this seems to be the case, for instance, with *A. ater* in the beechwood where its activity spans the whole biotope spatially and most of the year temporally. Favourable conditions for this to occur are a large body size and a stable and homogeneous environment. Such species accordingly reach a high biomass and stabilize at their carrying capacity through competition. These reflections are likely to apply to many other organisms; thus the tendency for large dominant species more often to approach equilibrium conditions and competitive regulation in stable environments while less abundant species are more often in non-equilibrium conditions and weakly affected by competition; this tendency is likely to be quite general.

On the other hand, the concrete realization of this tendency depends heavily on species- and site-specific characteristics. Therefore it comes as no surprise that no consistent effect of body size on population variability was found by Gaston & Lawton (1988) in a large-scale analysis of patterns in insect populations. What is more surprising is that they did find high positive correlations between population variability and abundance or ecological importance in all groups including carabids, in contradiction with the above arguments and results. These correlations are most probably due to the inclusion of species with low catch numbers in the correlations, and hence, as argued above, artefacts. I recalculated these correlations for carabids using the same data as in Gaston & Lawton (1988) (data from den Boer 1977) and did not find them to be significant in any of the four sites considered when species with too low catch numbers (total 8-year catch 8 individuals) are eliminated.

STRUCTURING INFLUENCE OF RESOURCES

The previous two sections make it clear that while competition can have a significant influence in one or a few dominant species in the climax stage, it cannot explain patterns at the level of whole ground beetle communities. Some of these patterns have been known for long and explained by abiotic factors. Thus most ground beetles from open biotopes are active in the spring while forest species tend to be active later in the year. This trend was explained by microclimatic factors: forests have buffered microclimates, they warm up and cool off less and later than do open biotopes; as a result they tend to harbour more autumn breeders than do open biotopes (Thiele 1977). Furthermore spring and autumn breeding are correlated with diurnalism and nocturnalism respectively, and this trend towards increasing night activity through the year was explained by the same microclimatic factors: the preference for a high humidity leads to nocturnalism - to ensure a high humidity - and to autumn activity, because nights are too cold in spring, while spring breeders must be diurnal and tolerate low humidity to enjoy high enough temperatures (Thiele 1977).

These explanations owe their success to their being straightforward and in keeping with the known microclimatic requirements of the various species. For this reason, however, they were accepted uncritically, without confrontation with alternative explanations. The main problem with such explanations is that they are unilateral; they merely express correlations between isolated aspects of a species' niche or habitat and isolated environmental factors. But there are manifold interrelations between the various components of a species' niche, which implies that the latter has to evolve and be considered as a totality (Loreau 1988). For instance, diurnalism was shown to be correlated also with body size, which prompts explanations in terms of biotic interactions (predation pressure, prey availability) (Southwood 1978; Loreau 1988).

Resource availability in fact seems to be a key determinant of the niche structure of carabid communities but has been rarely investigated. Ground beetles were regarded traditionally as broadly polyphagous, but, as mentioned above, their feeding preferences turned out to be much more differentiated than was assumed before, at least in forests. When I compared the niche structures of the carabid communities from the beechwood and the pinewood in Lembeek, I arrived at the conclusion that the only difference between them was the elimination of species with intermediate niches in the pinewood, and this appeared to correspond exactly to the impoverishment of the available prey spectrum and the simplification of the litter (Loreau 1984c). Thus the diversity of niches in a community seems to be directly dependent on the diversity of available resources.

A quantitative study of seasonal changes in prey availability further showed that available resources determined the temporal structure of the beechwood community. A distinct pattern emerged when body size was plotted against the period of maximum activity for the various species in the beechwood: large species tended to have a maximum body size towards late summer (Figure 5). This pattern was nicely explained by seasonal changes in the availability of the various prey types in which each carabid was most specialized: large insect larvae for *Carabus* species, annelids for *Abax* species (though *A. ater* was the most generalist species and also readily ate large insects and molluscs), molluscs for *Cychrus attenuatus*, and small and medium-sized insects for *P. oblongopunctatus* and *Nebria brevicollis*. The correspondence between the periods of maximum activity of large ground beetles, especially dominant species, and their preferred prey types was extremely good (Figure 5); even more so since within these periods the activity of dominant species was correlated with that of their preferred prey (Loreau 1988). Thus there was strong evidence that the seasonal niche structure is tightly adapted to available resources among large species in the climax forest. Note that this conclusion was not true for small species.

Much of the diversity and structure of carabid communities appears to be accounted for by the combined effects of abiotic factors and resource availability, and probably much more could further be accounted for were one to know better the ecology of larvae, which are likely to be the critical stages on account of their great ecological requirements and energetic needs. This suggests that a significant part of the organizing process of carabid communities occurs through the flow of energy and matter coming from the lower trophic level rather than through horizontal (competitive) interactions. The soil trophic web might therefore be organized in a bottom-up or donor-controlled fashion more than in a top-down or recipient-controlled fashion. This also conflicts with another central, if implicit, assumption of equilibrium competition theory, i.e., trophic interactions are recipient-controlled like in Lotka-Volterra systems (*e.g.* MacArthur 1972). It is now recognized that the control of trophic interactions is generally shared by the donor and recipient populations, with considerable variations from almost complete donor control to almost complete recipient control (*e.g.* Odum & Biever 1984; Hunter & Price 1992; Power 1992; Strong 1992). One important factor that bends trophic interactions towards donor control is spatial and temporal heterogeneity (Smith 1972; Arditi & Ginzburg 1989), and this is generally more significant in terrestrial than in aquatic systems. In particular, invertebrates that hunt at the soil surface such as adult ground beetles forage in an environment in which their prey are ephemeral; these prey either live permanently in the canopy (phytophagous insects) or find refuge in the soil, under bark, etc. (other insects, earthworms, molluscs). As a result, adult ground beetles generally are unlikely to control the abundance of particular prey species even though they may have a profound impact at the soil surface, and the potential for exploitation competition is low. The situation is different for the soil-dwelling larvae, and the most plausible mechanism for competition in ground beetles is food-dependent cannibalism at the larval stage (Heessen & Brunsting 1981).

CONCLUSIONS AND PERSPECTIVES ON BIOINDICATION

This (non-exhaustive) overview of determinants of species diversity in ground beetle communities shows that the assumptions on which equilibrium competition theory is based are not met on the scale of whole communities. The view presented here is that of communities as dynamic assemblages in which only some dominant species can reach equilibrium conditions and competitive regulation in sufficiently stable and homogeneous environments; diversity, especially of rare species, is influenced by regional processes, and the main local structuring factors are abiotic constraints and resource availability. I think that this view applies to many communities of other organisms and can accommodate a wide range of situations by letting the balance between local and regional processes, equilibrium and non-

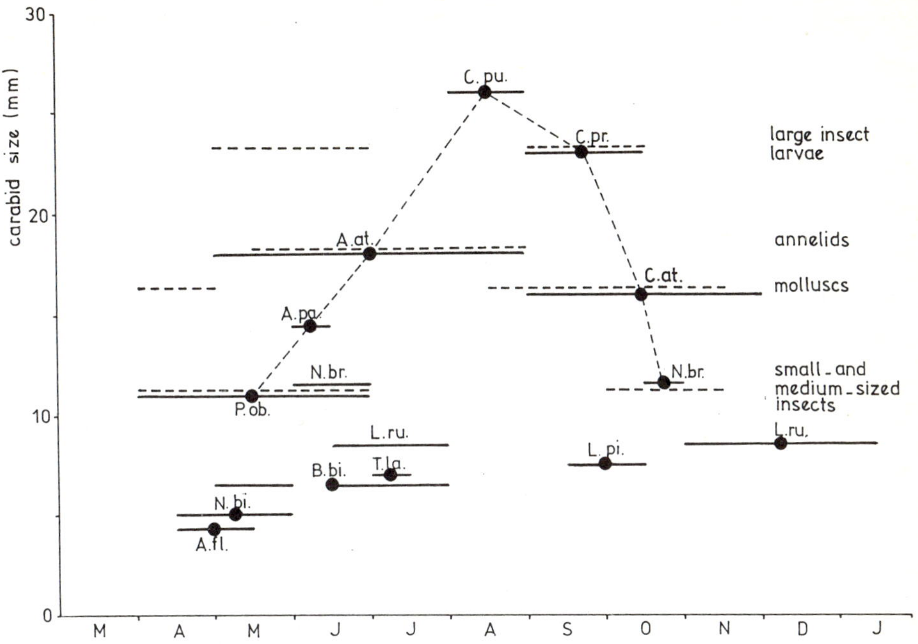

Figure 5. Body size against period of maximum activity (horizontal unbroken line) for the main species in the beechwood of Lembeek. The sloping dashed line joins the successive niche centres of the large species. The periods of maximum activity of the great prey types (horizontal dashed lines) are indicated at the same height as those of the four dominant species of which they are the preferred prey: *Abax ater* (A.at.), *Pterostichus oblongopunctatus* (P.ob.), *Cychrus attenuatus* (C.at.) and *Carabus problematicus* (C. pr.). Other species are *Abax parallelus* (A.pa.), *Carabus purpurascens* (C.pu.), *Nebria brevicollis* (N.br.), *Asaphidion flavipes* (A.fl.), *Notiophilus biguttatus* (N.bi.), *Badister bipustulatus* (B.bi.), *Trichotichnus laevicollis* (T.la.), *Lorocera pilicornis* (L.pi.) and *Leistus rufomarginatus* (L.ru.). Data from Loreau (1988).

equilibrium populations, donor and recipient controls of trophic interactions, vary from one extreme to the other.

This view is general, however, and as such imprecise. An ultimate goal of community ecology should be to formulate testable rules for the process of community assembly. Only then can community ecology relate itself directly to the practical problems raised by increasing environmental changes. Even strongly interacting organisms such as plants respond to long-term climate changes in a species-specific way, which leads to continuous changes in communities. Communities of weakly interacting organisms such as ground beetles are still more labile and change on the time scale of a generation, that is, roughly one year. Ground beetles in particular are highly diversified, they are sensitive to abiotic factors such as microclimate as well as resource availability, and their ecology and systematics is relatively well known. All this makes carabids and their communities potentially good biological indicators of environmental changes.

But much remains to be done, in my opinion, before we can use them efficiently as such. The most convincing application of ground beetles as bioindicators so far is in paleoecology and paleoarchaeology (*e.g.* Coope 1967; Osborne 1969). A lot of information has been accumulated on the habitat types of ground beetles, which allows reliable reconstructions of past environments. Since habitat type is largely determined by, and classified according to, plant communities, it is more disputable whether that kind of information can add much to the direct study of plant communities in present environments. The term bioindicator has been used rather loosely in recent literature, more to emphasize the fact that carabid communities are different in different environments than to show a direct practical use (*e.g.* various contributions in Stork 1990). Changes in carabid communities most probably do indicate changes in the environment, but which ones is not yet clear. In existing predictive models, carabid distributions are inferred mechanically from predicted changes in plant distributions and not the other way round (Rushton et al. 1993; Pizzolotto 1993). Large-scale databases of carabid distributions have been set up recently in several European countries, and have allowed investigation of long-term changes in distributions since the last century. Such changes were first attributed by Hengeveld (1985) to climate changes, but a much more convincing explanation is direct destruction of natural and semi-natural habitats by human activities (Desender & Turin 1989; Desender et al. 1993). For the time being, monitoring of ground beetle communities appears to be mainly useful for conservation purposes, but its use in bioindication of environmental changes is a reasonable hope for the future (*e.g.* Luff et al. 1992).

Progress in that direction demands more insight into the causal factors that govern the distribution and abundance of species and the assembly of communities. In particular, most studies so far have attempted to explain carabid distributions from correlations with some abiotic factors, but part of these correlations might in fact reflect only indirect effects, the direct causal factors being biotic, *e.g.* resource availability. This further suggests a new potential way of considering ground beetles as bioindicators, namely as indicators of the general state of the soil fauna, hence of ecosystem functioning. If this proved to be the case, ground beetle communities might contain less redundant information in relation to plant communities and would be all the more useful to consider. One of the major challenges for the years to come appears to be the development of an integrated research programme combining investigations on several ecological levels - from the individual organism to the ecosystem - and several spatial scales - from the local to the continental scale.

REFERENCES

Arditi R, Ginzburg LR (1989) Coupling in predator-prey dynamics: ratio-dependence. Journal of theoretical Biology 139: 311-326

Caswell H (1978) Predator-mediated coexistence: a non-equilibrium model. American Naturalist 112: 127-154

Chesson PL (1986) Environmental variation and the coexistence of species. Pages 240-256 *In* Diamond J, Case TJ. Community ecology. Harper & Row, New York

Coope GR (1967) The value of Quaternary insect faunas in the interpretation of ancient ecology and climate. Pages 359-380 *In* Cushing EJ, Wright HE. Quaternary paleoecology. Yale University Press, New Haven

Cornell HV, Lawton JH (1992) Species interactions, local and regional processes, and limits to the richness of ecological communities: a theoretical perspective. Journal of Animal Ecology 61: 1-12

DeAngelis DL, Waterhouse JC (1987) Equilibrium and nonequilibrium concepts in ecological models. Ecological Monographs 57: 1-21

den Boer PJ (1968) Spreading of risk and stabilization of animal numbers. Acta Biotheoretica (Leiden) 18: 165-194

den Boer PJ (1977) Dispersal power and survival: carabids in a cultivated countryside. Miscellaneous papers 14, Landbouwhogeschool Wageningen, the Netherlands

den Boer PJ (1981) On the survival of populations in a heterogeneous and variable environment. Oecologia (Berlin) 50: 39-53

Desender K, Dufrêne M, Maelfait J-P (1993) Long-term dynamics of carabid beetles in Belgium: a preliminary analysis of the influence of changing climate and land use by means of a database covering more than a century. *In* Desender K, Dufrêne M, Loreau M, Luff ML, Maelfait J-P. Carabid beetles: ecology and evolution. Kluwer Academic Publishers, Dordrecht (in press)

Desender K, Turin H (1989) Loss of habitats and changes in the composition of the ground and tiger beetle fauna in four West-European countries since 1950 (Coleoptera: Carabidae, Cicindelidae). Biological Conservation 48: 277-294

Eyre MD, Luff ML (1993) Carabid species assemblages of north-east England. *In* Desender K, Dufrêne M, Loreau M, Luff ML, Maelfait J-P. Carabid beetles : ecology and evolution. Kluwer Academic Publishers, Dordrecht (in press)

Gaston KJ, Lawton JH (1988) Patterns in the distribution and abundance of insect populations. Nature 331: 709-712

Hanski I (1982) Dynamics of regional distribution: the core and satellite species hypothesis. Oikos 38: 210-221

Hardin G (1960) The competitive exclusion principle. Science 131: 1292-1297

Heessen HJL, Brunsting AMH (1981) Mortality of larvae of *Pterostichus oblongopunctatus* (Fabricius) (Col., Carabidae) and *Philonthus decorus* (Gravenhorst) (Col., Staphylinidae). Netherlands Journal of Zoology 31: 729-745

Heijerman T, Turin H (1989) Carabid fauna of some types of forest in The Netherlands (Coleoptera: Carabidae). Tijdshrift voor Entomologie 132: 241-250

Hengeveld R (1985) Dynamics of Dutch beetle species during the twentieth century (Coleoptera, Carabidae). Journal of Biogeography 12: 389-411

Hunter MD, Price PW (1992) Playing chutes and ladders: heterogeneity and the relative roles of bottom-up and top-down forces in natural communities. Ecology 73: 724-732

Huston MA (1993) Biological diversity: the coexistence of species on changing landscapes. Cambridge University Press, Cambridge

Hutchinson GE (1957) Concluding remarks. Cold Spring Harbor Symposium of Quantitative Biology 22: 415-427

Hutchinson GE (1961) The paradox of the plankton. American Naturalist 95: 137-145.

Levins R (1979) Coexistence in a variable environment. American Naturalist 114: 765-783

Loreau M (1984a) Composition et structure de trois peuplements forestiers de Carabides. Académie royale de Belgique, Bulletin de la Classe des Sciences 70: 125-160

Loreau M (1984b) Les niches écologiques des Carabides en milieu forestier. I. Composantes spatio-temporelles. Académie royale de Belgique, Bulletin de la Classe des Sciences 70: 333-379

Loreau M (1984c) Les niches écologiques des Carabides en milieu forestier. II. Composante trophique et organisation générale des niches. Académie royale de Belgique, Bulletin de la Classe des Sciences 70: 480-525
Loreau M (1984d) Population density and biomass of Carabidae (Coleoptera) in a forest community. Pedobiologia 27: 269-278
Loreau M (1988) Determinants of the seasonal pattern in the niche structure of a forest carabid community. Pedobiologia 31: 75-87
Loreau M (1989) On testing temporal niche differentiation in carabid beetles. Oecologia (Berlin) 81: 89-96
Loreau M (1990) Competition in a carabid beetle community: a field experiment. Oikos 58: 25-38
Loreau M (1992) Species abundance patterns and the structure of ground-beetle communities. Annales Zoologici Fennici 28: 49-56
Loreau M, Nolf C-L (1993) Occupation of space by the carabid beetle *Abax ater*. Acta Oecologica 14 (in press)
Luff ML, Eyre MD, Rushton SP (1992) Classification and prediction of grassland habitats using ground beetles (Coleoptera, Carabidae). Journal of Environmental Management 35: 301-315
MacArthur RH (1972) Geographical ecology. Princeton University Press, Princeton, New Jersey
May RM (1988) How many species are there on Earth? Science 241: 1441-1449
Niemelä J (1988) Habitat occupancy of carabid beetles on small islands and the adjacent Åland mainland, SW Finland. Annales Zoologici Fennici 25: 121-131
Niemelä J (1993) Interspecific competition in ground-beetle assemblages (Carabidae): what have we learned? Oikos 66: 325-335
Odum EP, Biever LJ (1984) Resource quality, mutualism, and energy partitioning in food chains. American Naturalist 124: 360-376
Osborne PJ (1969) An insect fauna of late Bronze Age date from Wilsford, Wiltshire. Journal of Animal Ecology 38: 555-566
Pizzolotto R (1993) Ground beetles (Coleoptera, Carabidae) as a tool for environmental management: a geographical information system based on carabids and vegetation for the Karst near Trieste (Italy). *In* Desender K, Dufrêne M, Loreau M, Luff ML, Maelfait J-P. Carabid beetles: ecology and evolution. Kluwer Academic Publishers, Dordrecht (in press)
Power ME (1992) Top-down and bottom-up forces in food webs: do plants have primacy? Ecology 73: 733-746
Ricklefs RE (1987) Community diversity: relative roles of local and regional processes. Science 235: 167-171
Rushton SP, Wadsworth RA, Cherrill AJ, Eyre MD, Luff ML (1993) Modelling the consequences of land use change on the distribution of Carabidae. *In* Desender K, Dufrêne M, Loreau M, Luff ML, Maelfait J-P. Carabid beetles: ecology and evolution. Kluwer Academic Publishers, Dordrecht (in press)
Schoener TW (1986a) Overview: kinds of ecological communities - Ecology becomes pluralistic. Pages 467-479 *In* J Diamond, TJ Case. Community ecology. Harper & Row, New York

Schoener TW (1986b) Patterns in terrestrial vertebrate versus arthropod communities: do systematic differences in regularity exist? Pages 556-586 *In* J Diamond, TJ Case. Community ecology. Harper & Row, New York

Schoener TW, Spiller DA (1992) Is extinction rate related to temporal variability in population size? An empirical answer for orb spiders. American Naturalist 139: 1176-1207

Smith FE (1972) Spatial heterogeneity, stability, and diversity in ecosystems. Transactions of the Connecticut Academy of Arts and Sciences (New Haven, Connecticut) 44: 309-335

Southwood TRE (1978) The components of diversity. *In* LA Mound, N Waloff. Diversity of insect faunas. Symposium of the Royal Entomological Society no. 9, Blackwell Scientific Publications, Oxford

Stork NE (Ed.) (1990) The role of ground beetles in ecological and environmental studies. Intercept, Andover, Hampshire

Strong DR (1992) Are trophic cascades all wet? Differentiation and donor-control in specious ecosystems. Ecology 73: 747-754

Thiele HU (1977) Carabid beetles in their environments. Springer-Verlag, Berlin

Turin H, Alders K, den Boer PJ, van Essen S, Heijerman T, Laane W, Penterman E (1991) Ecological characterization of carabid species (Coleoptera, Carabidae) in The Netherlands from thirty years of pitfall sampling. Tijdschrift voor Entomologie 134: 279-304

Whittaker RH, Levin SA, Root RB (1973) Niche, habitat, and ecotope. American Naturalist 107: 321-338

Chapter 6: Fungal Species Diversity and Forest Ecosystem Functioning in The Netherlands

Thomas W. Kuyper
Biological Station
Centre for Soil Ecology
Kampsweg 27
9418 PD Wijster
The Netherlands

INTRODUCTION

Fungi constitute an important component in ecosystems, both in terms of species richness and in functional respects. However, the magnitude of fungal biodiversity is not well-known. A conservative estimate of the magnitude of fungal biodiversity by Hawksworth (1991) suggests that less than 5% of the species have been described. Even in comparatively well-investigated countries as the Netherlands the fungal biodiversity can only be roughly estimated. It is possible that this lack of fundamental knowledge of fungal species diversity has contributed to a general lack of awareness of the functional importance of fungal biodiversity in ecosystems. The growing interest in biodiversity, because of concern over extinction of species by habitat destruction and global change, has led to increased awareness of the functional importance of fungi. Ehrenfeld (1988) asked "*Before we fully appreciated the vital role that mycorrhizal symbiosis plays in the lives of many plants, what kind of value would we have assigned to the tiny, threadlike fungi in the soil that make those relationships possible?*" However, the below-ground connection, in which fungi play a very important role, is still not always sufficiently appreciated, as *e.g.* in the Sustainable Biosphere Initiative (Lubchenko *et al.*, 1991).

NATO ASI Series, Vol. I 20
Biodiversity, Temperate Ecosystems, and Global Change
Edited by T. J. B. Boyle and C. E. B. Boyle

In this contribution I will address the importance of fungal biodiversity in forest ecosystems in the Netherlands. More specifically I will deal with the following questions:

1. What kind of problems are encountered in the study of fungal species diversity? Even though biodiversity can be investigated at the ecosystem, species, and genetic level, knowledge of species diversity is fundamental for questions at the other levels.
2. What is the magnitude of macromycete diversity in forest ecosystems in the Netherlands and how is this diversity related to that of higher plants? What is the magnitude of macromycete species richness of different functional groups?
3. What kind of changes in fungal species richness have occurred in forest ecosystems in the Netherlands during the last decades and how can they be explained? These changes will be discussed per functional group.
4. How could these changes affect future functioning of forest ecosystems? In order to answer this question I will look at ecological equivalence and functional redundancy of species, and provide alternative explanations for high species diversity.

Considering the relatively large similarity in fungal species composition in different countries in northwestern Europe, I assume that many of the explanations given here pertain to a larger area, although there might be differences in pollution climate between the different countries. For that reason I will use several examples from forest ecosystems outside the Netherlands.

METHODOLOGICAL PROBLEMS IN THE STUDY OF FUNGAL DIVERSITY

Analysis of fungal biodiversity is problematic for a number of reasons (Arnolds, 1993). The organism proper is its vegetative structure, *i.e.* the mycelium, and these mycelia cannot easily be observed and identified. Identification usually relies on the presence of reproductive structures (sporocarps; conidiophores). Some fungal species form conspicuous macroscopic sporocarps (mushrooms; toadstools), whereas other species lack such structures and must be induced to sporulation in the laboratory. Fungi with macroscopic sporocarps are known as macromycetes, whereas the remaining fungi are considered micromycetes. This distinction between macromycetes and micromycetes is neither taxonomic nor ecological. Mutualistic symbionts (mycorrhizal fungi), saprotrophic, and parasitic fungi occur in both macromycetes and micromycetes. However, almost all of the ectomycorrhizal and the ligninolytic fungi belong to the macromycetes. The study of macromycetes and micromycetes is based on different methods and techniques. In this contribution I will only consider the macromycetes.

The first and foremost problem in the study of macromycetes is its dependence on reproductive structures. It is not clear how measurements on sporocarps relate to mycelial activities, but a positive correlation has repeatedly been found. There are reasons to suppose that formation of sporocarps is a more sensitive parameter than mycelia to changing ecological conditions (see below).

Other methodological problems in the study of macromycete diversity relate to the short life span of fruitbodies, the pronounced periodicity of most species, and yearly fluctuations due to weather conditions. These methodological problems make a special approach necessary, in which permament plots (usually 1,000 m^2) are investigated during a number of consecutive years (at least 3) and a number of times (3-5) within each year (Arnolds, 1992).

FUNCTIONAL GROUPS OF MACROMYCETES

Macromycetes have various important ecological functions in forest ecosystems. The three main groups that can be recognised are ectomycorrhizal symbionts, decomposers of litter and wood, and parasites on other organisms.

Mycorrhizal fungi are ubiquitous in forest ecosystems. Most forest trees form ectomycorrhiza, whereas the remaining tree species usually form vesicular-arbuscular mycorrhiza or ericoid mycorrhiza. Ectomycorrhizal fungi enhance water and nutrient uptake by increasing the soil volume that is exploited. Some ectomycorrhizal fungi are able to take up organic nitrogen and phosphorus. They can furthermore protect the tree against aluminium and heavy metals, and provide physical and/or chemical protection against root pathogens. Fungi are also very important as decomposers of organic material. Their hyphal growth form, enzymatic abilities, and tolerance towards acidic conditions give them a definite advantage over bacteria in many forest ecosystems. Only macromycetes (Basidiomycetes and a few groups of Ascomycetes) possess ligninolytic enzymes. The lignocellulosic matrix in litter and wood, constituting the bulk of organic material, can only be degraded by fungi. During the decomposition process, nutrients like nitrogen, phosphorus, and cations are mineralised, leading to efficient recycling of nutrients and carbon. Some macromycetes are known as tree parasites. They can enter trees via roots or above ground. Most of these are necrotrophic parasites, being able to persist by switching to a saprotrophic mode of life after the tree has been killed.

It is not always easy to classify a species as a wood necrotrophic parasite or a wood saprotroph. I will therefore combine both groups as the functional group of wood-inhabiting fungi. The terricolous litter and humus saprotrophic species will constitute a separate functional category.

MAGNITUDE OF MACROMYCETE SPECIES RICHNESS ON A NATIONAL SCALE

The number of macromycetes from the Netherlands amount to at least 3,400. This number is still increasing, as new species are added yearly to the Dutch list. Summary statistics about the numbers of species of various functional groups and in various habitats were provided by Arnolds and De Vries (1989). In their analyses 3,391 species of macromycetes are dealt with. Of these, 2,004 species (59%) show a clear preference for forests or comparable ecosystems in which (ectomycorrhiza-forming) trees and shrubs dominate, like roadside verges planted with trees, parks, young plantations along recreation areas, juniper shrub, willow scrub along rivers and in the dunes, etc. The high percentage of macromycetes that are characteristic for forest ecosystems is in striking contrast with the phanerogams, of which only 25% is characteristic for forested ecosystems.

However, it has been insufficiently appreciated that even in grasslands the absolute number of characteristic fungi (365) are somewhat higher than that of higher plants (345).

Species diversity of vascular plants and that of macromycetes is often not correlated. Habitats that are very rich in (ectomycorrhizal) fungi and poor in higher plants include pine forests (Jansen *et al.*, unpublished observation) and deciduous forests (*Quercus robur*: Jansen, 1984; *Fagus sylvatica*: Arnolds *et al.*, 1993) on poor sandy soils with a thin litter layer, and roadside verges with old individuals of *Quercus robur*, *Q. rubra* or *Fagus sylvatica* on poor soils (Keizer, 1993a). A positive correlation between diversity of vascular plants and macromycetes was noted for *Salix repens* shrub in dune ecosystems (Baartmans *et al.*, 1991). Diversity of phanerogams and macromycetes is also uncorrelated in grasslands (Arnolds, 1981) and alpine communities with ectomycorrhizal dwarf shrubs (Senn-Irlet, 1987).

The three guilds of macromycetes comprise more or less similar numbers of species. Arnolds and De Vries (1989) listed 652 species as ectomycorrhizal fungi, 636 as terrestrial saprotrophs on litter and humus, and 716 lignicolous saprotrophs and necrotrophic parasites.

The different functional groups (guilds) are dependent on different and independent carbon sources. They might partly compete for the same mineral nutrients. However, as the three guilds are considered to be energy-limited, competition between guilds is minimised. Data on competition between guilds are very scarce. The hypothesis of a negative effect of ectomycorrhizal fungi on litter decomposition by saprotrophic Basidiomycetes (Gadgil and Gadgil, 1975) has repeatedly been refuted in the field and in laboratory experiments (Dighton *et al.*, 1987; Kuyper, 1990). Correlations between species diversity of different guilds cannot be used to prove or disprove the hypothesis. Data on species richness of different guilds have repeatedly shown that diversity of different guilds is significantly positively correlated. Veerkamp and Kuyper (1993) noted that species numbers of all three functional groups were significantly positively correlated in forest reserves both in the Netherlands and in the Federal Republic of Germany.

Species richness of ectomycorrhizal fungi and terricolous saprotrophic fungi was positively correlated in roadside verges planted with *Quercus robur* or *Fagus sylvatica* (Keizer, 1993a) and in plantations of conifers (Jansen *et al.*, unpublished observation) and *Fagus sylvatica* (Arnolds *et al.*, 1993). However, Tyler (1985) noted that diversity of mycorrhizal and saprotrophic fungi was inversely correlated along a fertility gradient in beech (*Fagus sylvatica*) forest in Sweden, with ectomycorrhizal fungi being more common in sites with a mor humus profile and saprotrophic species more common in sites with a mull humus profile. The former high diversity of ectomycorrhizal fungi in the Netherlands in conifer and deciduous forests on very poor, sandy, acid soils with hardly any accumulation of litter, was accompanied by a low diversity of terricolous saprotrophic and lignicolous fungi. From this survey we might therefore tentatively conclude that, even though positive correlations between diversity of different guilds have repeatedly been found, designation of protected areas and nature reserves because of high diversity of one guild will not automatically protect the other guilds, and that at least some management practices will not automatically favour all functional groups. Cessation of management in forests will undoubtedly favour lignicolous fungi, but might have adverse effects on ectomycorrhizal fungi, at least in the nitrogen-saturated forests in the Netherlands (Veerkamp and Kuyper, 1993), whereas removal of organic material by restoration of former management practices will stimulate ectomycorrhizal fungi but have adverse effects on saprotrophic fungi (Baar and Kuyper, 1993; Keizer, 1993b).

Diversity of ectomycorrhizal fungi

During the last decades a number of reports have documented a decrease in species diversity and abundance of sporocarps of ectomycorrhizal fungi in northwestern and central Europe (Arnolds, 1991). From recent studies on the changing macromycete flora in the Netherlands (Arnolds and Jansen, 1992) the following conclusions regarding the ectomycorrhizal flora can be drawn (Table 1).

Table 1. Changes in the ectomycorrhizal flora, based on a comparison of relative frequency of 58 well recognisable species before and after 1970 (after Arnolds and Jansen, 1992).

	Increase	No change	Decrease
With conifers	0	1	15
With deciduous trees on poor soils	0	3	20
With deciduous trees on rich soils	0	1	8
With conifers and deciduous trees	0	7	3

Not all ectomycorrhizal fungi are equally sensitive. The decrease of diversity is larger for species exclusively associated with conifers than for species associated with deciduous trees, whereas species associated with both are least affected. The decrease in forests on poor, acid, sandy soils with a low base content took place at an earlier time than in those on richer, clayey soils. Those forests on poor soils, which were initially the richest in ectomycorrhizal fungi, are the most sensitive to this decline. However, the decline of ectomycorrhizal fungi associated with old trees in forests on clayey soils has been drastic during the recent years as well.

Due to differential sensitivities of species, a complete disappearance of specialised species has locally been noted, whereas only some generalist species remained. In some conifer stands, species that were originally considered specific for deciduous trees have established. Hydnaceous and cantharelloid fungi, and species of the genera *Cortinarius* and *Tricholoma* have been most sensitive, which suggests that decline of ectomycorrhizal fungi is not only a consequence of decreasing tree vitality. Several species have now completely disappeared from forests and have only refuges in roadside verges planted with old trees (Keizer, 1993a).

The most important cause for this dramatic decline has been atmospheric nitrogen deposition, with an additional role for acid deposition. Jansen (1991) noted a significantly negative correlation between the number of ectomycorrhizal species associated with *Pseudotsuga menziesii* and deposition of total acid and sulphur oxide; no correlation with deposition of ammonia or nitrogen oxides was observed. Termorshuizen and Schaffers (1991) found negative correlations between the number of ectomycorrhizal species associated with *Pinus sylvestris* and deposition of sulphur dioxide, nitrogen oxides, and ammonia in some years but not in other years. Large differences in the mycorrhizal flora between vital and non-vital stands were observed in the Federal Republic of Germany (Schlechte, 1986), where 21 and respectively 3 species were found on plots of similar size, and Czechoslovakia (Fellner, 1989). The latter author provided a scheme in which three phases of pauperisation of ectomycorrhizal fungi were distinguished: (1) inhibition of fruitbody production; (2) reduction of species diversity; (3) partial and finally total destruction of mycorrhizal species diversity. This scheme was further refined by Arnolds (1991). Besides a direct change of the soil nutrient status and an unbalanced nutrient status for the tree, the nitrogen input has also contributed to increased litter accumulation, for which most ectomycorrhizal fungi are sensitive (Baar and Kuyper, 1993). If correlated with below-ground parameters, the decrease of diversity and abundance of ectomycorrhizal fungi will have a negative impact on forest tree vitality.

Several studies have confirmed a correlation between above-ground reproductive parameters and below-ground vegetative parameters. Jansen and De Nie (1988) noted in stands of *Pseudotsuga menziesii* that species diversity was significantly positively correlated with diversity of mycorrhizal types, and that abundance of mycorrhizal root tips was positively correlated with abundance of sporocarps of ectomycorrhizal fungi. Termorshuizen and Schaffers (1989) noted that in stands of *Pinus sylvestris* the abundance of mycorrhizas was positively correlated with the abundance of sporocarps of ectomycorrhizal fungi. I therefore assume that changes in the ectomycorrhizal flora, based on sporocarps of ectomycorrhizal fungi, are correlated with below-ground changes with possible implications for mycorrhizal functioning. However, data from forest fertilisation (Kuyper, 1989) and experiments with removal of litter and humus layers (Baar *et al.*, 1993) have shown that fruitbody formation is a much more sensitive parameter to changing environmental conditions than the mycorrhiza. The scheme provided by Fellner (1989) is also based on the observation that decrease of below-ground ectomycorrhizal diversity lags behind the decrease of sporocarps of ectomycorrhizal fungi. However, no simple physiological explanation for the different sensitivities of sporocarp formation, extramatrical mycelium, and ectomycorrhizal root tips for excess nitrogen is available (Wallander and Nylund, 1992).

Conventional wisdom on evolution of mutualistic symbioses suggests that most ectomycorrhizal fungi are associated with a diverse suite of hosts and express only little specificity, indicating a kind of functional redundancy. Species with broad host ranges might also provide a strong potential for linking trees in a mycorrhizal network (Newman, 1988). Such networks can affect vegetation succession by providing positive feed-backs between compatible tree species (bootstrapping, Perry *et al.*, 1989). However, as reviewed by Molina *et al.* (1992), ectomycorrhizal fungi might vary from narrow to broad host ranges. Species with only a restricted host range and/or ecological specificity might have very different ecologies in space and time and have unique contributions to ecosystem resilience.

To date there is not much evidence how resilient the ectomycorrhizal flora is with regard to disturbance, and how much functional redundancy there is within the ectomycorrhizal flora. The standard model of mycorrhizal functioning assumes that both vesicular-arbuscular and ectomycorrhizal fungi function along largely similar lines (Tinker *et al.*, 1992), where mycorrhizas enhance phosphorus uptake and other beneficial effects of mycorrhizas are consequent on this better phosphorus status of the plant. The fact that trees have multiple fungal partners and that a number of species coexist on the same root system would then have no functional consequences.

Evidence for different functional roles of ectomycorrhizal fungi is scarce and partly circumstantial. Circumstantial evidence comes from large variation in mycorrhizal morphology. Agerer *et al.* (1986) suggested that in some cases form and function could be related: smooth mantles and uptake of organic nutrients; rhizomorphs and water-uptake under dry conditions, etc. Field evidence for such differences was obtained by Dighton *et al.* (1990) for mycorrhizal fungi of birch (*Betula pendula* and *B. pubescens*). They showed that three taxa of ectomycorrhizal fungi differed in phosphorus uptake rate, with a species of *Hebeloma* being much more efficient than species of *Laccaria* or *Lactarius*. However, it is not clear how many species have unique or (broadly) overlapping physiological roles. Species diversity of ectomycorrhizal fungi usually shows a log normal or log distribution, with a few (very) abundant and many rare species. The occurrence of ectomycorrhizal mats in conifer forests in North America also indicates that some forests are dominated by one or a very few species (Griffiths and Caldwell, 1992). Intuitively, it is difficult to imagine that numerically rare species play irreplaceable roles in forest ecosystems. Furthermore, the differences shown on morphological and physiological grounds are often based on functional differences between genera. It is quite striking that a large amount of species diversity relates to different species within a genus. Within a genus, and especially between closely related species within a genus, demonstration of functional differences might be very difficult. Functional equivalence would probably imply that competitive interactions between species result in deadlock and that both species occupy separate spatial domains. Augustin (1974) noted that, on very small spatial scales in a mixed beech-oak forest, *Russula nigricans* and *R. foetens* tended to completely exclude each other; the same was noted in a mixed forest in the Federal Republic of Germany for *R. fellea*, *R. ochroleuca* and *R. vinosa* (Agerer and Kottke, 1981), in a spruce plantation in Switzerland for *R. nigricans* and *R. ochroleuca* (Bieri *et al.*, 1992), and in an evergreen broad-leaved forest in Japan for *R. densifolia* and *R. laurocerasi* (Murakami, 1987). Even though lack of overlap in mycelial domains in closely related congeneric taxa is probably not uncommon, complete separation of all mycelial domains is unlikely, considering the large differences in mycelial growth forms (Ogawa, 1981). From data in Dahlberg (1991) it seems likely that there is quite extensive overlap between mycelial domains of different species. Such overlap also follows from the fact that even in small root samples several mycorrhizal types (5-10) occur together.

I therefore conclude that functional differences between mycorrhizal fungi do indeed exist, especially on genus level. The existence of functional differences on lower levels has not convincingly been demonstrated but is probably unnecessary to explain high species richness within sites. However, considering that almost all ectomycorrhizal fungi in the Netherlands are seriously decreasing, this decline might, even if there is large functional redundancy between species and genera, have grave consequences for the functioning of forest ecosystems.

Diversity of terricolous saprotrophic fungi

Changes in the terricolous saprotrophic mycoflora in the Netherlands have been relatively minor in comparison with changes in the ectomycorrhizal and lignicolous mycoflora (Arnolds and Jansen, 1992). The number of species have remained more or less equal during the last decades and the main change that has been noted is that oligotraphent species are being replaced by eutraphent species and that species that are characteristic for a mor profile are replaced by species that are characteristic for a moder profile. Increased input of atmospheric nitrogen is the most likely factor that causes this shift.

The guild of terricolous saprotrophic species is fairly heterogeneous and comprises primary and secondary decomposers of fresh leaves and needles, herbaceous stems, roots, mosses, etc. However, a more accurate assignment to functional subgroups and from that an assessment of functional redundancy is not easy. Several authors (Tyler, 1985; Veerkamp and Kuyper, 1993) noted that species richness is positively correlated with a fertility gradient, implying that in forests with a mull humus profile the number of species is higher than on a mor humus profile. This seems intuitively strange as bacteria are considered to play a large role in decomposition in mull profiles, whereas on mor and moder sites fungi play an almost exclusive role. However, species diversity might not be the best measure to assess the importance of terricolous saprotrophic fungi for ecosystem functioning. The number and the dry matter production of sporocarps have also to be taken into account (Hering, 1982).

Swift (1976) noted that some saprotrophic species are resource specific, whereas other species are resource non-specific, being found on a large number of substrates. Resource specific saprotrophic basidiomycetes grow mainly on rather undecomposed leaves and are usually component restricted (Cooke and Rayner, 1984) as well. The chemical composition of these leaves could exert selective influence, which explains host specificity of resource specific fungi, as found in some leaf-inhabiting species of the genera *Marasmius* and *Mycena* (Table 2). Such species can have small mycelial domains. For *Marasmius androsaceus* Holmer and Stenlid (1991) noted that up to 18 different individuals could be found within an area of 2.8 m^2.

Resource non-specific, often also not component-restricted fungi, grow on already comminuted litter and partly humified material. Species of the genera *Clitocybe*, *Collybia*, and *Mycena* (pro parte) belong to this group. It is this group in which the mycelia can form large rings and persist for a long time. Such mycelia can have large domains. For *Megacollybia platyphylla* Thompson and Rayner (1982) noted the occurrence of 22 different individuals within an area of 20,800 m^2.

Ecological specialisation among these species seems rather restricted. The coexistence of large numbers of resource non-specific and non-restricted saprotrophs of genera of *Clitocybe* and *Collybia* is not well explained to date. Swift (1976) suggested that in the later stages of the decay process, when energy instead of nutrients becomes the limiting factor, the ecological similarity between species increases. He introduced the concept of the equivalent species, which are functionally equivalent and competitively similar. If such species interact, deadlock will occur. Anderson (1987) also noted that terricolous saprotrophic communities comprise few species with very specific roles, but mainly contain a large number of species with broadly overlapping niches. This would imply large functional redundancy on the stand level, even though some of these equivalent

Table 2a. Component restricted, resource specific species of *Marasmius*.

Name	Substrate (leaves)
M. buxi	*Buxus*
M. hudsonii	*Ilex*
M. capillipes	*Populus, Salix*
M. epiphylloides	*Hedera*
M. tremulae	*Populus*
M. recubans	*Fagus*
M. corbariensis	*Olea*

Table 2b. Component restricted, resource specific fungi of *Mycena*.

Name	Substrate (leaves)
M. capillaris	*Fagus*
M. juncicola	*Juncus, Carex*
M. riparia	*Scirpus, Carex*
M. polyadelpha	*Quercus*
M. smithiana	*Quercus*
M. lohwagii	*Pteridophytes*
M. terena	*Salix*

species might have major local effects. Statements of equivalence and functional redundancy are therefore scale-dependent.

However, not all cases of deadlock can be interpreted as indications of ecological equivalence. Kuyper and Bokeloh (1993) compared the oligotraphent *Clitocybe vibecina* with the eutraphent *C. metachroa*. Although interactions between these species resulted in deadlock, large differences in decomposition (of cellulose and lignin) and mineralization of nitrogen were noted. *Clitocybe metachroa* was a more efficient decomposer of lignin and had higher mineralization rates of nitrogen, a large part of which was nitrified.

Not all coexisting species might be ecologically and functionally equivalent. Whereas many interactions between species of *Clitocybe* resulted in deadlock, most interactions between species of *Clitocybe* and *Collybia* were quite specific (Kuyper and Verschoor, unpublished observation). *Collybia dryophila* and *C. distorta* grew over the mycelium of several *Clitocybe* species, whereas *C. maculata* and *C. peronata* grew under the mycelium of neighbouring *Clitocybe* species. In the latter case competitive replacement of *Clitocybe* by *Collybia* was noted. Such patterns of mycelial growth could also produce niche differentiation in vertical layers, where different species exploit different parts of the humus profile. Such a niche differentiation was noted between *Marasmius androsaceus*, which mainly forages in the L-layer, and *Mycena galopus*, which mainly forages in the upper F-layer, under the influence of grazing by Collembola (Newell, 1984).

The relationship between species diversity and process rates is not clear. For dung (which was rich in lignin) the decomposition rate by mixtures of species was not significantly different from the decomposition rate by the species with the lowest rate (Table 3a). However, ammonification and nitrification in mixed cultures of species of *Clitocybe* were sometimes enhanced in comparison with monocultures (Table 3b).

Hence changes in species richness or species composition of the community of terricolous saprotrophic fungi cannot easily be translated to ecosystem processes. There are indications (Nohrstedt *et al.*, 1989) that increased nitrogen input can have a negative effect on litter decomposition, especially on originally acidic sites with a mor or moder humus profile, where the larger changes in species composition took place. The mechanism of retarded litter decomposition is not well known but a correlation between mineral nitrogen and repression of ligninolysis has been noted for some lignicolous fungi, and a similar reaction for some terricolous saprotrophic fungi seems likely.

Table 3a. Decomposition of rabbit faeces by single species and combinations of species (after Wicklow and Yocom, 1981).

Species	% weight loss
Sordaria fimicola	24.0
Sordaria humana	20.7
Podospora curvicolla	16.6
Podospora setosa	14.0
Coprinus sp.	24.4
Ascodesmis sphaerospora	16.0
S. fimicola + *S. humana*	20.8
P. curvicolla + *P. setosa*	13.9
Sordaria + *Podospora* (4 species)	15.5
All 6 species	15.3

Table 3b. Production of ammonium and nitrate ($mmol.l^{-1}$) by single species of *Clitocybe* and combinations of species (after Kuyper and Verschoor, unpublished observation).

Species	ammonium	nitrate
C. vibecina	5.81	0.16
C. metachroa	0.55	0.45
C. inversa	1.89	0.88
C. odora	0.58	0.11
C. vibecina + *C. metachroa*	3.34	1.07
C. vibecina + *C. inversa*	7.80	0.09
C. vibecina + *C. odora*	1.44	0.75
C. metachroa + *C. inversa*	1.39	0.27
C. metachroa + *C. odora*	1.54	0.55
C. inversa + *C. odora*	0.58	0.11

Diversity of lignicolous saprotrophic and necrotrophic fungi

Large changes in the guild of lignicolous saprotrophic and necrotrophic macromycetes have occurred in the Netherlands during the last decades (Table 4). Especially the facultative and necrotrophic parasites have shown an increase in diversity and abundance in forest ecosystems. Saprotrophic wood-inhabiting fungi on deciduous wood (or without preference for a certain type of wood) are generally also increasing, but those on conifer wood have fared less well generally and several of them have distinctly declined during the last decades. These changes can be explained by an increase in afforested areas, increase in forest and tree age, less accurate management practices, a decrease in tree vitality that resulted in an increased input in coarse woody debris and a change in chemical quality of the wood due to nitrogen and acid deposition. Simultaneously, species that were considered typical for deciduous wood have established on conifer wood, possibly as a consequence of nitrogen-enriched air pollution.

Table 4. Changes in the lignicolous saprotrophic and necrotrophic mycoflora, based on a comparison of relative frequency of 32 well recognisable species before and after 1970 (after Arnolds and Jansen, 1992).

	Increase	No change	Decrease
Parasites	6	2	0
Saprotrophs on deciduous wood	8	6	4
Saprotrophs on conifer wood	1	1	4

Several lignicolous species have established and conspicuously expanded in the Netherlands during the last decades, the most spectacular example being *Schizopora carneolutea* (Keizer, 1990). This species, which was first recorded in the Netherlands in 1975 is now already common in large parts of our country. The reasons for such a successful establishment are unclear; it is difficult to imagine that this species, which mainly occurs on fallen branches of deciduous trees, found an empty niche, as such wood can be colonised by many different species, amongst which is the closely related and very common *S. paradoxa*. Some field observations suggest competitive replacement of the latter species by the former, although this has not sufficiently been substantiated. Another possible explanation could be that this species is able to form spores during a larger part of the year. This latter factor has been suggested as an explanation for the recent expansion of *Calocera pallidospathulata* in England (Ing, 1991).

The group of wood-decaying macromycetes is subdivided into two main physiological groups, *viz.* white-rot and brown-rot fungi. Whereas brown-rot fungi cannot degrade lignin (they can only chemically modify it), white-rot fungi have ligninolytic capabilities. Between both groups there are also differences in the way in which cellulose is degraded, *viz.* enzymatically by exo- and endoglucanases in white-rot fungi, and with an iron-hydroxyperoxyde system in brown-rot fungi (Montgomery, 1984).

Within the groups of brown-rot and white-rot fungi no large differences in decay rates have been found. Wood decay rates were depressed, or at least not enhanced by the presence of a larger number of species (Käärik, 1975; De Vries and Kuyper, 1988). This negative effect of increased species diversity on process rates is probably related to mycelial domains. Coates and Rayner (1985) showed experimentally that high inoculum densities, leading to very small mycelial domains, effectively retarded decomposition of beech logs. Enzymatic synergisms by cellulolytic and ligninolytic fungi are mainly found in cell-free cultures.

As the existence of high species diversity cannot likely be explained by functional differences, several other hypotheses have been forwarded. A high species diversity on wood, belonging to different tree species, can be explained by chemical and/or physical differences of the wood (Hintikka, 1982) that could select for a small number of adapted species. The importance of this factor is likely to diminish during later stages of decay. A high species diversity on wood of one tree species can also be maintained by microhabitat segregation. Such a microhabitat segregation was shown for wood of *Juniperus communis* (Barkman *et al.*, 1983) and *Taxus baccata* (De Vries and Kuyper, 1990). On old branches of *Taxus* that obtain low light levels parasitic *Nectria* species appear which cause die off of these branches. On these branches *Hyphodontia nespori*, *Trechispora cohaerens*, *T. microspora*, and *Dacrymyces stillatus* can then establish. *Cerocorticium confluens* can only establish after the bark has disappeared. As dead branches remain on the plant for a long time, these microhabitats might be too dry for some species. *Amylostereum laevigatum* is found on old trees in moist microhabitats, and *Ascocorticium anomalum* occurs only on decorticated branches lying on the soil. Only in the later stages of decomposition can saprotrophic Agaricales be found. Specific associations between lignicolous fungi might also increase species diversity on wood. Several lignicolous macromycetes combine a saprotrophic mode of life on wood with a parasitic mode of life on mycelium of other macromycetes, *e.g. Lenzites betulinus* is associated with *Trametes versicolor*, and *Trametes gibbosa* with *Bjerkandera adusta* (Rayner and Boddy, 1988). Whereas these species combinations are characteristic for intensively managed forests, such specific interactions also occur in pristine conifer forests in Fenno-scandia. Many rare saprotrophic species occur only on wood that has been colonised by specific other, less rare saprotrophs, *e.g. Junghuhnia collabens* on wood decayed by *Phellinus ferrugineofuscus*, *Piloporia sajanensis* follows *Trichaptum laricinum*, and *Antrodiella citrinella* follows *Fomitopsis pinicola* (Renvall and Niemelä, poster presented at the Congress of European Mycologists, 1992).

A high species diversity can be maintained by species that are functionally and ecologically equivalent. Swift (1976, 1984) introduced the concept of unit-communities, *i.e.* spatially discrete communities of fungi on a restricted substrate. There might be intense competition within such communities but not between communities, the species composition of which is partly determined by chance. The species that compose such unit communities are competitively equivalent and upon interaction they often form demarcation zones, similar to those that are formed when two conspecific individuals interact. Even in cases where competitive replacement occurs, species diversity might remain high, as competitive relationships between saprotrophic species can be intransitive. Chapela *et al.* (1988) noted that *Phallus impudicus* was able to replace *Psathyrella hydrophila*, *Megacollybia platyphylla* could replace *P. impudicus*, whereas *P. hydrophila* could replace *M. platyphylla*.

Considering this competitive equivalence there could be no simple coupling of species diversity and decay processes on a small scale. On a larger scale, however, the existence of several ecologically and functionally equivalent species might well enhance decomposition.

CONCLUSION

The macromycete flora in the Netherlands has shown dramatic changes during the last decades. Ectomycorrhizal fungi have shown a very strong decline, with only a few generalist species surviving. Terricolous saprotrophic species diversity has not much changed in forests, although eutraphent species are replacing oligotraphent species. Lignicolous saprotrophic and necrotrophic macromycetes have tremendously increased. The main explanation for these changes is the atmospheric nitrogen input. Acidification, lowering of ground-water table, and changes in forest management have played a less important role.

This change in the species composition and general decline of the macromycete flora is accompanied by changes in forest vitality. Trees show signs of nutrient imbalances, are more sensitive to drought, and are more susceptible to parasites. Litter accumulates at a higher rate in such forests. A causal relationship between these changes in the macromycete flora and low forest tree vitality seems likely.

There has sometimes been a view of high species diversity being a reflection of ecosystem well-being. With regard to the macromycetes such a view must be regarded as simplistic and naive. There is no logical coupling between species diversity and process rates. Furthermore, pool sizes and process rates might well be less important than synchronisation and synlocalisation of fungal and forest tree activities.

The role played by individual species in the functioning of forest ecosystems has hardly been elucidated. It remains an important task to assess this role for the resilience of forests with regard to environmental stress. For the time being it is not well possible to determine if, and how much, functional redundancy in the guilds of macromycetes occurs.

Acknowledgements

This is contribution 509 of the Biological Station Wijster. I thank my colleague E. Arnolds for constructive criticisms of an earlier draft of this manuscript.

REFERENCES

Agerer R, Brand F, Gronbach E (1986) Die exakte Kenntnis der Ectomykorrhizen als Voraussetzung bei Feinwurzeluntersuchungen im Zusammenhang mit dem Waldsterben. Allg Forst Z 41:497-503, 505

Agerer R, Kottke I (1981) Sozio-ökologische Studien an Pilzen von Fichten- und Eichen-Buchen-Hainbuchen-Wäldern im Naturpark Schönbuch. Z Mykol 47: 103-122

Anderson JM (1987) Interactions between invertebrates and microorganisms: noise or necessity for soil processes? Pages 125-145 *in* Fletcher M, Gray TRG, Jones JG (Eds), Ecology of microbial communities. Cambridge University Press, Cambridge

Arnolds E (1981) Ecology and coenology of macrofungi in grasslands and moist heathlands in Drenthe, the Netherlands. Vol. 1. Bibl mycol 83:1-407

Arnolds E (1991) Decline of ectomycorrhizal fungi in Europe. Agric Ecosyst Environ 35:209-244

Arnolds E (1992) The analysis and classification of fungal communities with special reference to macrofungi. Pages 7-47 *in* Winterhoff W (ed), Fungi in vegetation science. Kluwer Academic Publishers, Dordrecht

Arnolds, E (1993) Problems in measurement of species diversity of macrofungi. Submitted

Arnolds E, Jansen E (1992) New evidence for changes in the macromycete flora of the Netherlands. Nova Hedwigia 55:325-351

Arnolds E, Opdam A, Van Steenis W, De Vries B (1993) Mycocoenology of stands of Fagus sylvatica L. in the northeastern Netherlands. Phytocoenologia (in press)

Arnolds E, De Vries B (1989) Oecologische statistiek van de Nederlandse macrofungi. Coolia 32:76-86

Augustin A (1974) Über die Verteilung einiger Täublingsarten auf kleinstem Raum. Z Pilzk 40:215-220

Baar J, Kuyper TW (1993) Litter removal and effects on ectomycorrhizal fungi. Pages 275-286 *in* Pegler DN, Boddy L, Ing B, Kirk PM (eds), Fungi of Europe: Investigation, Recording and Conservation. Royal Botanic Gardens Kew

Baar J, Ozinga W, Kuyper TW (1993) Fruitbody abundance and population structure of the ectomycorrhizal fungus Laccaria bicolor in stands of Pinus sylvstris after removal of litter and humus layers. Myc Res (in press)

Baartmans M, Biemans A, Verbruggen R (1991) Paddestoelen op Terschelling. Een mycocoenologisch onderzoek naar de mycoflora in *Salix repens* struwelen op Terschelling. M. Sc. report Agricultural University Wageningen

Barkman JJ, Jansen AE, De Vries BWL (1983) De betekenis van dood hout voor de schimmelflora. Ned Bosbouwtijds 55:57-64

Bieri C, Lussi S, Senn-Irlet B, Hegg O (1992) Zur Synökologie der Makromyzeten in wichtigen Waldgesellschaften des Berner Mittellandes, Schweiz. Mycol helv 5:99-127

Chapela I, Boddy L, Rayner ADM (1988) Structure and development of fungal communities in beech logs four and a half years after felling. FEMS Microbiol Ecol 53:59-70

Coates D, Rayner ADM (1985) Fungal population and community development in cut beech logs. III. Spatial dynamics, interactions and strategies. New Phytol 101:183-198

Cooke RC, Rayner ADM (1984) Ecology of saprotrophic fungi. Longman London

Dahlberg A (1991) Ectomycorrhiza in coniferous forest: structure and dynamics of populations and communities. Ph. D. thesis Swedish University of Agricultural Science Uppsala

De Vries BWL, Kuyper TW (1988) Effect of vegetation type on decomposition rates of wood in Drenthe, the Netherlands. Acta bot neerl 37:307-312

De Vries BWL, Kuyper TW (1990) Holzbewohnende Pilze auf Eibe (*Taxus baccata*). Z Mykol 56:87-94

Dighton J, Mason PA, Poskitt JM (1990) Field use of ^{32}P to measure phosphate uptake by birch mycorrhias. New Phytol 116:655-661

Dighton J, Thomas ED, Latter PM (1987) Interactions between tree roots, mycorrhizas, a saprotrophic fungus and the decomposition of organic substrates in a microcosm. Biol Fert Soils 4:145-150

Ehrenfeld D (1988) Why put a value on biodiversity. Pages 212-216 *in* Wilson EO (ed), Biodiversity. National Academy Press, Washington

Fellner R (1989) Mycorrhiza-forming fungi as bioindicators of air pollution. Agric Ecosyst Environ 28:115-120

Gadgil RL, Gadgil PD (1975) Suppression of litter decomposition by mycorrhizal roots of *Pinus radiata*. NZ J. For. Sci. 5:35-41

Griffiths RP, Caldwell BA (1992) Mycorrhizal mat communities in forest soils. Pages 98-105 *in* Read DJ, Lewis DH, Fitter AH, Alexander IJ (eds), Mycorrhizas in ecosystems. CAB, Wallingford

Hawksworth DL (1991) The fungal dimension of biodiversity: magnitude, significance, and conservation. Mycol Res 95:641-655

Hering T (1982) Decomposing activity of basidiomycetes in forest litter. Pages 213-225 *in* Frankland JC, Hedger JN, Swift MJ (eds), Decomposer basidiomycetes: their biology and ecology. Cambridge University Press, Cambridge

Hintikka V (1982) The colonisation of litter and wood by basidiomycetes in Finnish forests. Pages 227-239 *in* Frankland JC, Hedger JN, Swift MJ (eds), Decomposer basidiomycetes: their biology and ecology. Cambridge University Press, Cambridge

Holmer L, Stenlid J (1991) Population structure and mating system in Marasmius androsaceus. New Phytol 119:307-314

Ing B (1991) BMS Mapping Scheme. *Calocera pallidospathulata*. The Mycologist 5:48

Jansen AE (1984) Vegetation and macrofungi of acid oakwoods in the northeast of the Netherlands. PUDOC Agricultural Research Reports 923:1-162

Jansen AE (1991) The mycorrhizal status of Douglas fir in The Netherlands: its relation with stand age, regional factors, atmospheric pollutants and tree vitality. Agric Ecosyst Environ 35:191-208

Jansen AE, De Nie HW (1988) Relations between mycorrhizas and fruitbodies of mycorrhizal fungi in Douglas fir plantations in The Netherlands. Acta bot neerl 37:243-250

Käärik A (1975) Decomposition of wood. Pages 129-174 *in* Dickinson CH, Pugh GJF (eds), Biology of plant litter decomposition. Academic Press, London

Keizer PJ (1990) The expansion of *Schizopora carneolutea* (Basidiomycetes) in Europe, in particular in the Netherlands. Persoonia 14:167-171

Keizer PJ (1993a) The ecology of macromycetes in roadside verges planted with trees. Ph. D. thesis Agricultural University Wageningen

Keizer PJ (1993b) The influence of nature management on the macromycete flora. Pages 251-269 *in* Pegler DN, Boddy L, Ing B, Kirk PM (eds), Fungi of Europe: Investigation, Recording and Conservation. Royal Botanic Gardens Kew

Kuyper TW (1989) Auswirkungen der Walddüngung auf die Mykoflora. Beitr Kenntn Pilze Mitteleur 5:5-20

Kuyper TW (1990) De rol van ectomycorrhiza-schimmels in de nutriëntenkringloop. Coolia 33:32-37

Kuyper TW, Bokeloh DJ (1993) Ligninolysis and nitrification in vitro by two species of saprotrophic Basidiomycetes. Submitted

Lubchenko J, Olson AM, Brubaker LB, Carpenter SR, Holland MM, Hubbell SP, Levin SA, MacMahon JA, Matson PA, Melillo JM, Mooney HA, Peterson CH, Pulliam HR, Real LA, Regal PJ, Risser PG (1991) The sustainable biosphere initiative: an ecological research agenda. Ecology 72:371-412

Molina R, Massicotte HB, Trappe JM (1992) Specificity phenomena in mycorrhizal symbioses: community-ecological consequences and practical implications. *in* MF Allen (ed), Mycorrhizal functioning: an integrative plant-fungal process, Pages 357-423. Chapman and Hall London

Montgomery RAP (1984) The role of polysaccharidase enzymes in the decay of wood by Basidiomycetes. Pages 51-65 *in* Frankland JC, Hedger JN, Swift MJ (eds), Decomposer Basidiomycetes: their biology and ecology. Cambridge University Press, Cambridge

Murakami Y (1987) Spatial distribution of Russula species in *Castanopsis cuspidata* forest. Trans Br mycol Soc 89:187-193

Newell K (1984) Interaction between two decomposer basidiomycetes and a collembolan under Sitka spruce: distribution, abundance and selective grazing. Soil Biol Biochem 16:227-233

Newman EI (1988) Mycorrhizal links between plants: functioning and ecological significance. Adv ecol Res 18:243-270

Nohrstedt H-Ö, Arnebrant K, Bååth E, Söderström B (1989) Changes in carbon, respiration, ATP and microbial biomass in nitrogen fertilized pine forest soils in Sweden. Can J For Res 19:323-328

Ogawa M. (1981) Ecological characters of ectomycorrhizal fungi and their mycorrhizae. Proc XVII IUFRO World Congress, Pages 89-95

Perry DA, Amaranthus MP, Borchers JG, Borchers SL, Brainerd RE (1989) Bootstrapping in ecosystems. BioScience 39:230-237

Rayner ADM, Boddy L (1988) Fungal decomposition of wood. John Wiley Chichester

Schlechte G (1986) Zur Mykorrhizapilzflora in geschädigten Forstbeständen. Z Mykol 52:225-232

Senn-Irlet B (1987) Ökologie, Soziologie und Taxonomie alpiner Makromyzeten (Agaricales, Basidiomycetes) der Schweizer Zentralalpen. Ph. D. thesis University of Berne

Swift MJ (1976) Species diversity and the structure of microbial communities in terrestrial habitats. Pages 185-222 *in* Anderson JM, Macfadyen A (Eds), The role of terrestrial and aquatic organisms in decomposition processes. Blackwell, Oxford

Swift MJ (1987) Organization of assemblages of decomposer fungi in space and time. Pages 229-253 *in* Gee JHR, Giller PS (Eds), Organization of communities, past and present. Blackwell, Oxford

Termorshuizen AJ, Schaffers AP (1989) The relation in the field between fruitbodies of mycorrhizal fungi and their mycorrhizas. Agric Ecosys Environ 28:509-512

Termorshuizen AJ, Schaffers AP (1991) The decline of ectomycorrhizal fungi in stands of *Pinus sylvestris* L. in the Netherlands: possible causes. Nova Hedwigia 53:267-289

Thompson W, Rayner ADM (1982) Spatial structure of a population of *Tricholomopsis platyphylla* in a woodland site. New Phytol 92:103-114

Tinker PB, Jones MD, Durall DM (1992) A functional comparison of ecto- and endomycorrhizas. Pages 303-310 *in* Read DJ, Lewis DH, Fitter AH, Alexander IJ (eds), Mycorrhizas in ecosystems. CAB, Wallingford

Tyler G (1985) Macrofungal flora of Swedish beech forest related to soil organic matter and acidity characteristics. For Ecol Manag 10:13-29

Veerkamp MT, Kuyper TW (1993) Mycological investigations in forest reserves in the Netherlands. *in* MEA Broekmeyer, W Vos, H Koop (eds), European Forest Reserves. PUDOC, Wageningen (in press)

Wallander H, J-E Nylund (1992) Effects of excess nitrogen and phosphorus starvation on the extramatrical mycelium of ectomycorrhizas of *Pinus sylvestris* L. New Phytol 120:495-503

Wicklow DT, Yocom DH (1981) Fungal species numbers and decomposition of rabbit faeces. Trans Brit mycol Soc 76:29-32

Chapter 7: Effects of former land use on plant species diversity and pattern in European deciduous woodlands

Martin Hermy
Institute of Nature Conservation
Kiewitdreef 5
B-3500 Hasselt [1]
Belgium

INTRODUCTION

The forest environment offers to living creatures a habitat that buffers from extremes of temperature, exposure to wind, and desiccation (Crawford 1989). Although plant communities of woodlands provide examples of varying ecological strategies (cf. Grime 1979), perhaps the most typical is the large number of stress-tolerant species, species adapted to the low light intensity and the scarcity of mineral nutrients. They are mostly perennial species with a Spartan utilization of captured resources and an ability of survival for long periods with little growth or reproduction. In reproduction, emphasis is on few and relatively large seeds (cf. Salisbury 1942, Harper 1977), which makes recruitment under forest conditions more likely.

Woodlands are exposed to a variety of disturbances that range from the fall of a single tree to the clearing and reclamation of complete woodlands for other land-uses (*e.g.*, agriculture, urban infrastructure). The former is not considered as a vital change, yet the slow growth and the longevity of most tree species make even this recovery process a matter of decades. The replacement of woodlands by other forms of land-use is a catastrophic event that makes a recovery to the original conditions improbable, even over centuries. Woodlands in Europe are, par excellence, the outcome of centuries of interaction between nature and men.

University of Gent,
K.L.Ledeganckstraat 35, 9000 Gent [1]

NATO ASI Series, Vol. I 20
Biodiversity, Temperate Ecosystems, and Global Change
Edited by T. J. B. Boyle and C. E. B. Boyle

Much of the land, particularly in western Europe, has been inhabited and extensively cultivated since pre-Roman times (Pounds 1990). Many European woodlands are characterized by centuries of severe human influence in terms of exploitation and fragmentation (Ellenberg 1963, Falinski 1985, Vos & Stortelder 1992, Pott & Hüppe 1991, Tack *et al.* 1993). Both have major effects on woodland composition, diversity and structure (*e.g.* Rackham 1980, Peterken 1981, Peterken and Game 1984, Hermy 1985, Tack *et al.* 1993, Foster 1992). In many European countries land-use has changed repeatedly over the last millenium. In terms of former land-use two historical types of woodland have been recognized (Peterken 1974, 1981, Rackham 1980, Hermy & Stieperaere 1981) in western Europe. Primary woodland refers to land that has remained wooded throughout the historical period and is more or less a direct descendant of original, natural woodland. Secondary woodland then refers to land that has been cleared and used as arable land, pasture or for other purposes, before reverting to woodland, usually in historical times (Peterken & Game 1984). For practical reasons woodlands cannot be proved to be primary, unless palynological evidence is available from the site. We therefore prefer the term "ancient" woodland, suggesting the constant presence of woodland since at least a threshold date (for Flanders, for practical reasons, the late eighteenth century; in England about 1600). Woodlands originating after this date are then called recent woodlands; they often originated from planting. Ancient woodland thus consists of both primary and ancient secondary woodland, but in usually unknown proportions. In many parts of Europe, woodlands are strongly isolated from one another by a sea of often hostile other land-uses. Recent woodlands are often not only isolated from ancient woodlands, but they are also isolated in time. They have no direct links to the original woodlands in earlier times.

The aims of this paper are (a) to examine the qualitative and quantitative floristical differences between ancient and recent woodlands in Flanders and Europe; (b) to compare the number of species in plots from recent and ancient forests; (c) to assess other than historical factors affecting the distribution and number of woodland species.

MATERIAL AND METHODS

The material used in this paper comes from a variety of sources. Various authors have pointed to the existence of ancient woodland species (*e.g.* Waterbolk 1951, Roisin & Thill 1952, Froment & Tanghe 1967, Pigott 1969, Pollard 1973, Peterken 1974, Pigott 1977, Van den Wijngaard 1977, Rackham 1980, Hermy & Stieperaere 1981, Trepl 1982, Dzwonko & Loster 1989, Whitney & Foster 1988). I used these sources here to make an overview of ancient woodland plant species. Most studies have revealed qualitative floristical differences between ancient and recent woodlands. Although quantitative floristical differences may be expected as well, they have not been examined much, since these studies all start from qualitative species lists. The bulk of the quantitative data comes from a historical-ecological study of woodlands in Flanders (the western part of Belgium and some adjacent parts of Brabant, Henegouwen and N.France) (Hermy & Stieperaere 1981, Hermy 1985, 1992, Tack *et al.* 1993). Here, I combined floristical and phytosociological approaches.

From 1977 to 1983 I described 640 plots, selected in uniform vegetation according to the representative principle of Braun-Blanquet (Westhoff & Van der Maarel 1978). They were distributed over 183 deciduous woodlands. I excluded plantations on grasslands (Hermy 1985). Cover of all plant species in each plot was visually assessed using the decimal dominance-abundance scale of Londo (1975). From 1986 onwards qualitative floristical data on 231 woodlands have been collected (Tack *et al.* 1993). Within the latter, the Forest of Ename (Bos t'Ename) has been intensively studied using data from a variety of sources (archives, old maps, soils, palynology, floristics and vegetation, mapping of individual species).

I identified ancient woodlands using several historical sources, but mainly old maps (Ordnance Survey maps from the nineteenth century onwards). For land-use before the nineteenth century, land-use information came from estate maps from 1610 onwards (England), the map of Ferraris (1770-1779), and historical references to individual woods in medieval and later documents.

THE FLORA OF ANCIENT AND RECENT WOODLANDS: A EUROPEAN OVERVIEW

Plant species with a bias to ancient woodlands were described at least from England, Belgium, the Netherlands, Poland and Czechoslovakia (Hermy 1992). A review of that literature provides us with a list of about 90 woodland species having a strong affinity for ancient woodland (Table 1). This list is incomplete since no data seem available from a large part of Europe. Regional variation in the association of plant species and ancient woodland occurs (Peterken 1974, Rackham 1980). From Table 1 it is clear that ancient woodland species occur most on mesophile conditions of Querco-Fagetea-forests.

In the relative stabile conditions of the forest we may expect that short-distance dispersal is the rule. This should particularly hold for the plant species of the undergrowth. Indeed, some 40 % of these European ancient woodland species disperse by ants. Although the amount of anemochores is still considerable (about 20 %) (Figure 1), the success of wind dispersal for field layer species within woodland is limited.

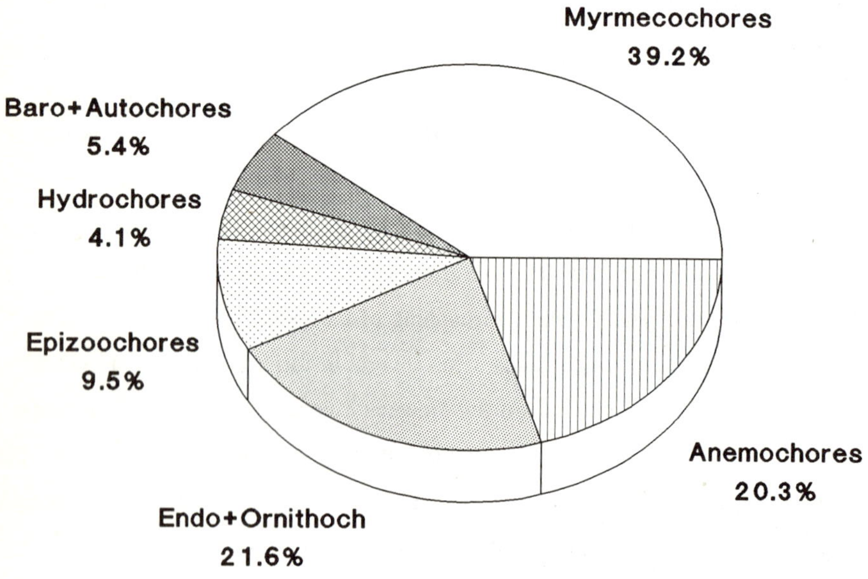

Figure 1. Distribution of dispersal modes in European ancient woodland species (see Table 1).

Table 1. List of ancient woodland species as mentioned from Europe and their indicator values

Species names	Lit	L	F	R	Life form	Leaf	Species names	Lit	L	F	R	Life form	Leaf
Acer campestre	3	5	5	7	P	S	Hordelymus europaeus	1	4	5	7	H	W
Actaea spicata	2	3	5	6	H,G	S	Lamiastrum galeobdolon	6	3	5	7	G	I
Adoxa moschatellina	2	5	6	7	G	V	Lathraea squamaria	2	3	6	7	Gvp	S
Agropyron caninum	1	6	6	7	H	S	Lathyrus vernus	1	4	5	8	G,H	S
Allium ursinum	2	2	6	7	G	V	Luzula luzuloides	1	4	5	3	H	W
Anemone nemorosa	10	x	5	x	G	V	Luzula pilosa	5	2	5	5	H	W
Anemone ranunculoides	1	3	6	8	G	V	Luzula sylvatica	2	4	5	4	H	W
Asarum europaeum	2	3	5	7	H,G	I	Lysimachia nemorum	3	2	7	7	C	W
Asperula odorata	4	2	5	6	H	W	Maianthemu bifolium	6	3	5	3	G	S
Athyrium filix-femina	2	3	7	x	H	S	Melampyrum nemorosum	2	5	4~	6	Thp	S
Bromus ramosus benekenii	1	5	5	7	H	S	Melica nutans picta	3	5	4	6	H,G	S
Campanula trachelium	5	4	6	8	H	S	Melica uniflora	4	3	5	6	G,H	S
Carex digitata	3	3	5	x	H	W	Melittis melissophyllu	1	5	4	6	H	S
Carex laevigata	2	4	9	5	H	W	Mercurialis perennis ovat	7	5	4	7	H,C	S
Carex pendula	2	5	8	6	H	S	Mespilus germanica	1	(6)	4	6	P	S
Carex pilosa	1	4	5	5	H,G	W	Milium effusum	5	4	5	5	H	W
Carex remota	4	3	8	x	H	S	Narcissus pseudo-narcis	3	8	6	4	G	V
Carex strigosa	3	3	7	7	H	W	Neottia nidus-avis	4	2	5	7	Glvp	S
Carex sylvatica	4	2	5	6	H	W	Orchis mascula	3	6	4	8	G	S
Chrysospleni alternifolium	3	4	8	7	H	S	Oxalis acetosella	6	1	5	4	G,H	W
Chrysospleni oppositifolium	2	6	9	5	H	W	Paris quadrifolia	7	3	6	7	G	S
Conopodiu majus	2	8	5	4	G	S	Phyteuma spicatum	2	x	5	6	H	S
Convallaria majalis	7	5	4	x	G	S	Polygonatum multiflorum	6	2	5	6	G	S
Cornus mas	1	6	4	8	N,P	S	Polystichum aculeatum	1	3	6	6	H	W
Cornus sanguinea	4	7	5	7	N	S	Primula elatior	7	6	6	7	H	S
Corydalis solida	1	3	5	7	G	V	Primula vulgaris	4	6	5	7	H	V
Corylus avellana	5	6	x	x	N	S	Pteridium aquilinum	5	6	5~	3	G	S
Crataegus laevigata agg.	4	6	5	7	N,P	S	Pulmonaria officinalis obs	1	4	6	8	H	W
Dactylis polygama	1	5	5	6	H	W	Pulmonaria officinalis	1	5	5	8	H	W
Dactylorhiza maculata	1	7	8	x	G	S	Pyrus communis ag	1	(6)	5	8	P	S
Daphne mezereum	2	4	5	7	N,Z	S	Ranunculus auricomus ag	5	5	x	7	H	W
Dryopteris affinis (pseud)	1	3	6	5	H	S	Ranunculus lanuginosus	2	3	6	7	H	S
Dryopteris carthusiana	1	5	x	4	H	S	Sanicula europaea	5	4	5	8	H	W
Epipactis purpurata	1	2	6	8	G	S	Scilla non-scripta	7	5	5	7	G	V
Equisetum sylvaticum	1	3	7	5	G	S	Sorbus torminalis	4	(4)	4	7	P,N	S
Euonymus europaea	4	6	5	8	N	S	Symphytum tuberosum	2	4	6	7	G	S
Euphorbia amygdaloides	3	4	5	8	C	1	Tilia cordata	4	(5)	5	x	P	5
Euphorbia dulcis	1	4	5	8	G,H	S	Tilia platyphyllos	1	(4)	6	x	P	S
Festuca heterophylla	1	5	4	5	H	S	Vaccinium myrtillus	2	5	x	2	Z	S
Gagea lutea	1	4	6	7	G	V	Veronica montana	3	4	7	5	C	W
Geum rivale	2	6	8	x	H	W	Vinca minor	2	4	5	7	C	I
Helleborus viridis	1	3	5	8	H	S	Viola mirabilis	1	4	5	8	H	S
Hepatica nobilis	2	4	4	7	H	1	Viola reichenbachia	4	4	5	7	H	W

Explanation of abbreviations overleaf

Table 1 (continued)

Abbreviations:

Lit.: number of citations as ancient woodland species in literature (see Hermy 1992)

indicators: x: indifferent or varied
L: light indicator value, according to Ellenberg et al. (1992):
1: in deep shade; 3: shade plant (mostly < 5% relative light intensity);
8: light loving plants (rarely under 40 % relative light intensity);(): tree seedling in field layer

F: moisture indicator value, according to Ellenberg et al. (1992):
< 3: dry-site indicaors; 5: mainly on fresh soils, absent from both wet and dry soils;
> 9: wet-site indicators; =: indicator of inundated soils; ~ : indicating fluctuating water tables

R: soil reaction ligure, according to Ellenberg et al. (1992):
< 3: indicator of acid soils; 5: moderate acid soils; 7: weakly acid to weakly basic conditions
9: basic soils and calcareous soils

Life form, according to Ellenberg et al. (1992):
C: chamaephyte; G: geophyte; H: hemicryptophyte; N: nanophanerophyte; P: phanerophyte;
T: therophyte; Z: woody chamaephyte; hp: hemi-parasite; vp: holo-parasite

Leaf persistence, according to Ellenberg et al. (1992):
I: evergreen; W: leafs remaining green during winter; S: summergreen; V: green only from early spring until early summer

Former land-use and population sizes: some examples from Flanders

However, dispersal is only one requisite for a successful establishment of a plant species. The surrounding vegetation strongly affects germination and growth. Particularly in recent woodlands, often with a dominance of high competitive species (*e.g. Urtica dioica, Molinia caerulea*), this is a difficult matter. Fragmentation of habitats (Saunders *et al.* 1991, Opdam *et al.* 1993) and consequently the growing isolation of woodlands will further decrease the establishment success of many species in (recent) woodlands. Due to the predominance of short distance dispersal, typical woodland species may be more susceptible to these processes. From these considerations we may expect that:

1. Recent woodlands will tend to have smaller populations of ancient woodland species, if they have them at all; indeed, the development of a population takes

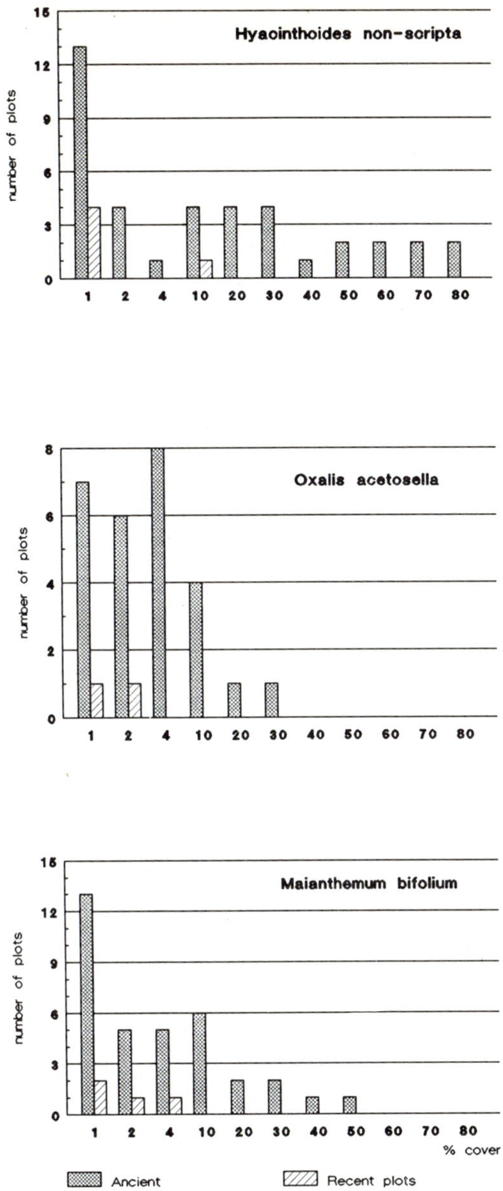

Figure 2. Cover distribution of three common woodland species in plots from ancient and recent woodlands (n=640) in western Belgium.

much time. This is suggested by the distribution of the performance of species *like Hyacinthoides non-scripta, Maianthemum bifolium*, and *Oxalis acetosella* in recent and ancient woodland plots (Figure 2). Above a certain amount of cover I observed ancient woodland species only in plots from ancient woodland: *Oxalis acetosella* (> 2%), *Maianthemum bifolium* (> 4%), *Hyacinthoidesnon-scripta* (> 15%) and *Pteridium aquilinum* (> 40%). High cover of *Urtica dioica* (> 60%) was particularly indicative of recent woodland.

2. A negative effect on ancient woodland species (and their diversity) of strongly competitive species may be found; the more ancient woodlands are invaded by these species, the more the typical woodland plants will decrease (Figure 3). A highly fragmented forest cover and intensive agriculture will accentuate this effect. Pigott (1982) found in field experiments that nitrogen and phosphate had a negative effect on cover of *Anemone nemorosa*; in contrast, the establishment of *Urtica dioica* and other "nitrophilous" species was stimulated by phosphate rather than by nitrogen (Pigott & Taylor 1964, Pigott 1971).

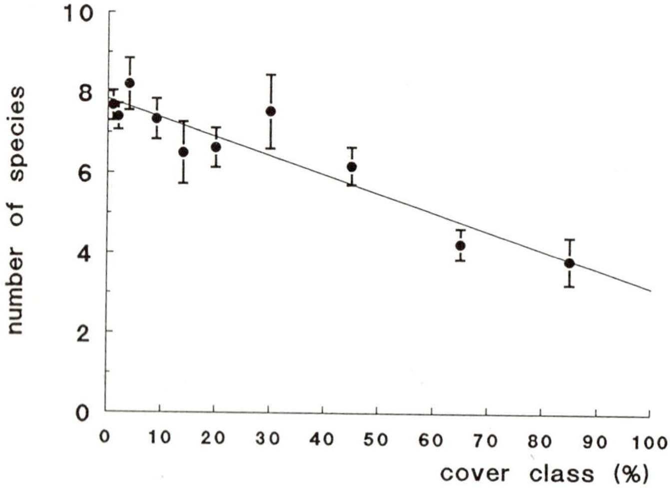

Figure 3. The relation between the mean number of true woodland species per plot (n=308) and the percentage cover of Urtica dioica. Errors bars are shown and a line is fitted to the data. Data come from 640 plots from about 180 woodlands in western Belgium (Hermy 1985).

Effects of other ecological variables

Of course the occurrence of woodland species is not only determined by former land-use. Soil characteristics (*e.g.*, moisture, soil reaction and richness in nutrients, soil texture) all have an impact on the performance of species (see Table 1). Area, as a function of these variables, seems to effect the occurrence of woodland as well. Opdam (1991) concluded from several bird studies, that the local extinction rate is negatively related to the area of the habitat fragment (which is proportional to the size of the local population). Plants, being less mobile and often of considerable longevity, need much more time before clear responses will be shown (mostly regeneration is first affected). Yet positive correlations between individual species (particularly the rarer species) and forest area have been found in northwest Germany (Zacharias & Brandes 1990). In Flanders the probability of occurrence of some woodland species *e.g. Maianthemum bifolium* & *Luzula pilosa* increases both with the size of the woodland and at the same time with the proportion of the wood that is ancient. Other species, like *Hyacinthoides non-scripta* only showed an effect related to the proportion of the wood that is ancient (Figure 4).

Isolation and colonization capacity

Ancient woodland species all have a slow colonization capacity in common, but none responds in the same way. For *Hyacinthoides non-scripta* and *Anemone nemorosa*, Pigott (1982) recorded rates of 6 to 10 m and 1 to 2 m in a century. In combination with the absence of persistent seed banks (Brown & Oosterhuis 1981) this means that elimination by a conversion to non-woodland conditions (reclamation, even if temporal), to conifers or by competitive exclusion will cause severe problems for reestablishment and development to their former population size. However the (re)colonization success may be larger as the forests are less isolated. Peterken & Game (1981) found that colonization by *Mercurialis perennis* depends mainly on the proximity to refuges in ancient woods and wood-relic hedges. In the eastern part of the Netherlands, Van Ruremonde & Kalkhoven (1991) showed that the probability of occurrence of *Lonicera periclymenum* and the density of *Ilex aquifolium* decreased with an increasing isolation of the forest patches. In the Bos t'Ename in S. Flanders we found that the distribution of *Corylus avellana*, *Mercurialis perennis* and other typical woodland species was clearly linked to the presence of former non-cleared fringes of woodland vegetation in the nineteenth century, when the wood was almost entirely reclaimed for agriculture (Figure 5). The subsequent recolonization of the forest area, after abandonment of the arable land, showed also a clear relation to the slope of the woodlots (downwards quicker than upwards). Probably at least occasionally their presence is merely a consequence of survival under non-woodland conditions (even in arable fields for some geophytes, *e.g.*, *Hyacinthoides non-scripta*). It is well known that certain woodland species may survive for decades outside woodlands (Dierschke 1968).

"Biodiversity" of ancient and recent woodlands

"*We cannot even estimate the number of species of organisms on Earth to an order of magnitude, an appalling situation in terms of knowledge*" (Raven 1992). Even in the western world, biodiversity, expressed as species diversity, is badly known. Few case studies seem to be available and most of them are very incomplete (Table 2). The collection of complete species lists of sites is a difficult task (the available knowledge is scattered over many people and sources and the flora and information is changing with time). The knowledge of the diversity of ancient and recent woodlands is usually restricted to "woodland" vascular plant species, and even to collect this information is a time-consuming task (Peterken & Game 1984, Tack *et al*. 1993).

Peterken & Game (1984) found that ancient woodlands had a higher woodland vascular plant species richness than recent woodlands. Based on plots of about 150 m2 we found that plant species richness was not significantly different between plots from ancient and recent woodlands. However, the number of true (see appendix) woodland species differed significantly (Table 3). The probability of finding a certain number of true woodland species in recent and ancient woodlands was highly different (Figure 6). So the number of true woodland species may be used as an indicator for the presence of ancient woodland. Of course, other woodland features (*e.g.*,

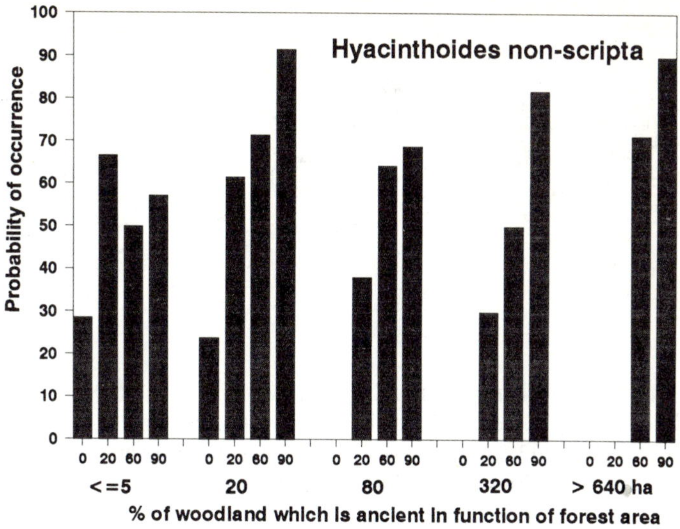

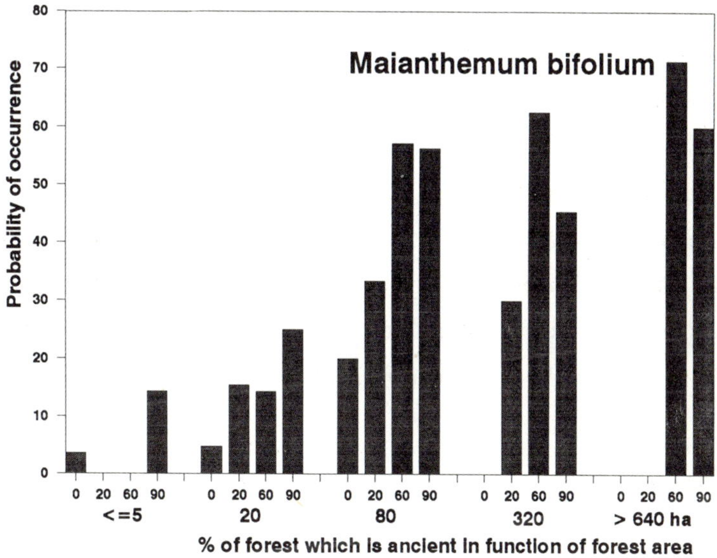

Figure 4. Incidence functions for Hyacinthoides non-scripta and Maianthemum bifolium on 231 woodland patches in Flanders. Hyacinthoides responds only to the proportion of the woodland that is ancient (site history); the probability of occurrence per area class remains more or less the same. Maianthemum responds both to area and site history. Data from a historical-ecological study of woodlands in Flanders (Tack *et al.* 1993).

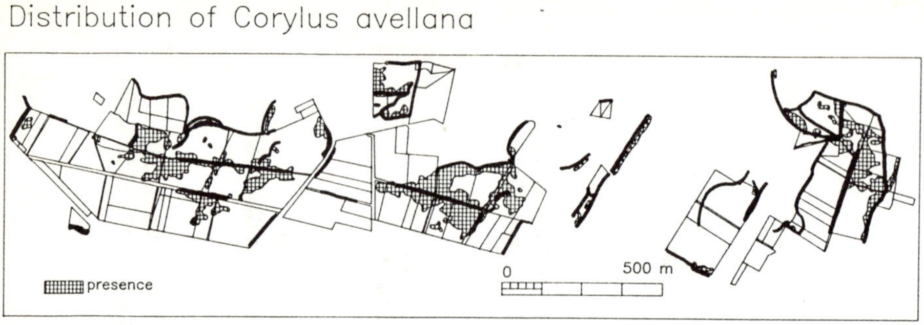

Figure 5. The modern distribution of Corylus avellana in the forest of Ename (near Oudenaarde) is clearly linked to the distribution of unreclaimed linear elements. This woodland was reclaimed between 1851 and 1896 except for some small lots and many linear elements, mostly along rides (thick black lines).

woodland banks, other archaeological and palynological evidence may provide further evidence for ancient character of the woodland) (see Rackham 1980). So, former land-use seems to determine which species are present (quality element) and not the total number of species.

Regional species richness differences in woodlands

Forests often differ regionally, usually because the regions vary in topography, soil texture, (meso)climate and many other features (*e.g.*, former land-use). In Flanders, with a large fluctuation in forest area in the last centuries, recent research suggests (Tack *et al.* 1993) a high correlation between the present, total number of

Table 2. Observed biodiversity in some European woodlands

Site Country	Hayley Wood GB	Bos t'Ename Belgium	Monks Wood GB	New Forest GB	Bialowieza Poland
Area (ha)	49	61	157	37900	125000
Plants	284	381	372	?	990
Bryophytes	81	?	97	?	254
Fungi	375	?	337	?	>2000
Lichens	53	?	34	278	334
Mammals	17	24	24	?	62
Birds	59	116	115	?	228
Beetles	?	?	1017	1539	ca.2000
Butterflies	?	?	459	1234	ca.1000
Amphibians & reptiles	?	10	10	12	19
Other	175	?	1398	279	3825
Total	1044	531	3863	3342	>10712

From various sources (see Tack *et al.* 1993)

Table 3. Plant species richness in plots from ancient and recent woodlands

		N	Mean	SD	Sign.
Total number of species	Ancient woodland	320	25.3	10.52	ns
	Recent woodland	246	24.4	10.44	
Number of species in field layer	Ancient woodland	320	14.7	8.08	ns
	Recent woodland	246	14.4	8.68	
Number of species in woody layer	Ancient woodland	320	6.9	2.73	**
	Recent woodland	246	5.9	2.54	
Number of true woodland species++	Ancient woodland	320	7	3.26	***
	Recent woodland	246	4.9	3.14	

++: see appendix
N: number of plots
**: $P < 0.001$; ***: $P < 0.0001$; ns: not significant

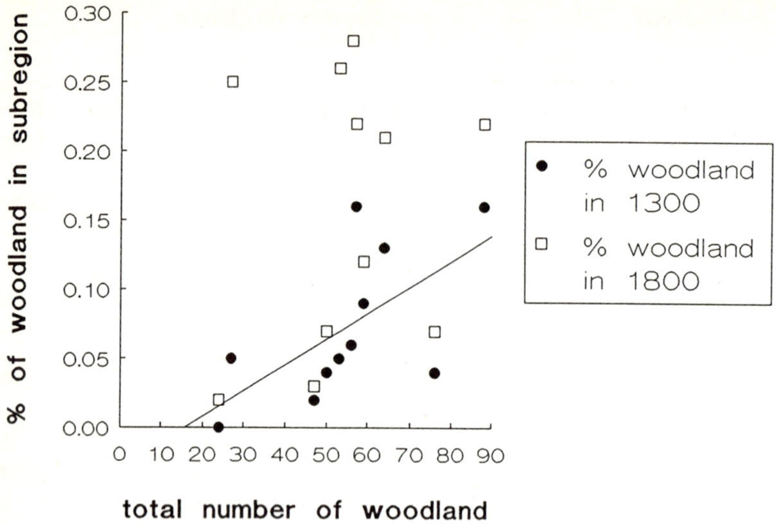

Figure 6. The probability of occurrence of the richness in true woodland species in plots from ancient and recent woodlands in the western part of Belgium. Data are based on 640 plots from 180 woodlands. Recent woodland has been further subdivided depending on the time (in years) of the land-use.

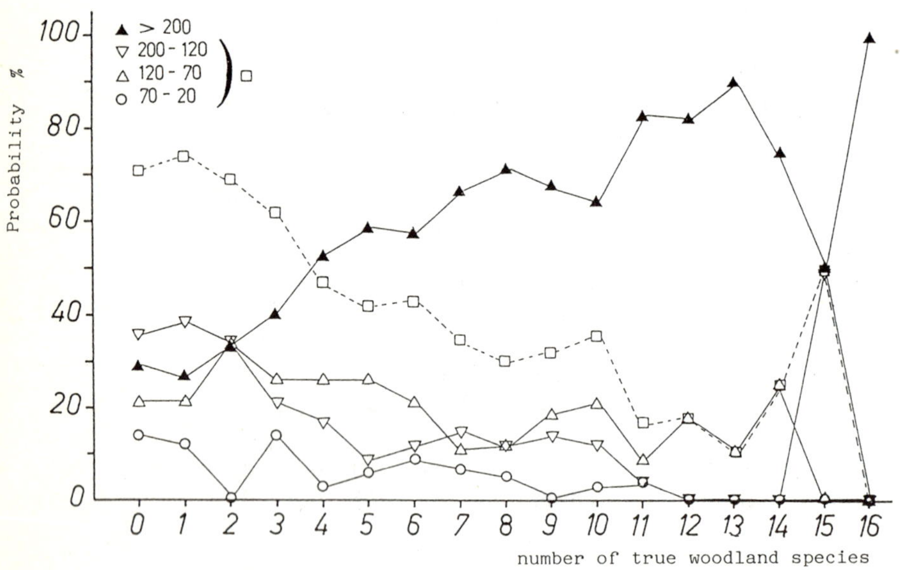

Figure 7. The relationship between the total number of woodland species (max. = 203) in subregions of Flanders and the afforestation index (in %) in 1300 and in 1800 (from Tack *et al.* 1993). The present woodland species diversity is strongly affected with the woodland cover in the Middle Ages.

field layer woodland species and the ratio between wooded area in about 1300 and the area of the subregion (Figure 7). Correlation with nineteenth century afforestation rate is not significant. Although this is no evidence for causal relations, it is an additional argument for the importance of the past to the present vegetation cover and biodiversity. It does emphasize the actual reality of the past, as it is archived in our woodlands.

DISCUSSION

An integrated historical-ecological research on forests and on the landscape of which they are part, may yield an insight into the spatial and temporal distribution patterns of species, woodland communities and woodlands. Modern woodlands can be understood only in the light of their history (Rackham 1980, Tallis 1991, Hermy 1992). Historical ecology provides (1) a background for understanding the development of the modern landscape (cf. Foster 1992), (2) information on the response of species and communities to changes in land-use and other disturbances and (3) insights that may be useful for the important issues of conservation, management and restoration of woodland habitats.

Historical ecology offers a good example of interdisciplinary research. The study presented here greatly benefitted from the knowledge of the history of the area and the sites. It is part of an interdisciplinary research on woodlands in Flanders (Tack *et al.* 1993), where data from archaeology, pedology, palynology, history, vegetation science, and landscape ecology have been combined to understand the present and past structure and composition of woodlands. Other recent examples are offered by Berglund (1988), Pott & Hüppe (1991) and Vos & Stortelder (1992). In my opinion the challenges of future research lie not only in further specialization but also in building bridges between various disciplines.

Historical ecology will help to understand from where we came and partially where we go. It is essentially a comparative research. To increase its accuracy and to test generated hypotheses, it should be coupled with field experiments looking at the colonization and recruitment possibilities of woodland species.

Former land-use determines to a large extent which woodland species are present, but not necessarily how many species (Peterken & Game 1984). The number of true woodland plant species (including ancient woodland species) is significantly higher in ancient than in recent forests. However (re)colonization of recent forests does occur at a slow rate. This depends on the presence of relic populations in the surroundings. Population size of true woodland species in recent forests strongly differs from ancient forests. Ancient woodland species may be regarded as extinction-prone (Peterken 1977), because of their limited colonization capacities and their absence in persistent seed banks. A review of the European ancient woodland species suggests that short distance dispersal is the rule in ancient woodland species. Short-distance dispersal and, more generally, any dispersal of plant diaspores is a requisite to a successful establishment, although it is not the only bottleneck. Particularly in small forests in a hostile environment (*e.g.* agricultural land), typical woodland species may decrease or be excluded by vigorous competitive species (which are better adapted to the new environmental disturbances).

Increasing the knowledge of dispersal in time and space seems another challenge for future research. Theoretical dispersal modes are well known, but in the field difficulties remain.

The existence of ancient woodland species with their poor colonization capacities severely restricts the feasibility of woodland reconstruction, as is suggested by Figure 7. Complete recreation in a western European context, without planting of ancient woodland species, will take many centuries; Peterken (1977) suggested more than 800 years. If we consider as a starting point the present agricultural land, which often has become very nutrient rich through fertilizing, the feasibility reduces even further. Particularly on these soils, we observe after abandonment an enormous expansion of competitive species (*e.g. Urtica dioica, Phalaris arundinacea*). The island character of many woodlands in western Europe additionally will have a negative impact on dispersal success. Apart from that, projects with long realization periods are not acceptable in a modern society. A noteworthy exception to this rule are nuclear power projects yielding radio-active wastes at the risk of many future generations of humankind. If we consider woodland restoration projects, it is important to decide on the goals of recovery. Four management options are available (Cairns 1986):

a) restoration to original conditions;

b) rehabilitation of some of the original conditions;

c) development of an alternative ecosystem (*e.g.* ponds); and

d) neglect or natural reclamation.

In view of the complexity of ancient woodlands only the last 3 options are open. Another possibility might be to use recent woodlands as receptor sites for introducing extinction-prone woodland species. At least this would have the advantage of starting with an already closed canopy of more or less full grown trees. For woodlands, it seems clear that woodland functions to society (*e.g.* recreation, economic purposes) are more easily restored than ecosystem form and quality (*e.g.* diversity of ancient woodland species). The introduction of rare, endangered or threatened species needs more consideration, but is perhaps more critical for grasslands and marshes than for woodlands. Table 2 leaves us with a double challenge: filling in the gaps and making it a challenge to funding organisations to sponsor time and money consuming biodiversity research. The Convention on Biological Diversity offers an unique opportunity for biodiversity research. If conceived well, biodiversity prospecting may be an important tool for biodiversity research and conservation particularly in tropical forests (WRI 1993). Tropical forests currently benefit from a world wide attention. Finding funding for biodiversity research in temperate ecosystems will be much more difficult.

Acknowledgement

This paper greatly benefitted from the insights and field experiences from Guido Tack and Paul van den Bremt. Without them this would have been impossible.

REFERENCES

Berglund BE (1988) The cultural landscape during 6000 years in southern Sweden - the Ystad project. Ecological Bulletins 41

Brown AHF, Oosterhuis L (1981) The role of buried seeds in coppice woods. Biol. Conserv. 21: 19-38

Cairns J Jr (1986) Restoration, reclamation, and regeneration of degraded or destroyed ecosystems. Pages 465-484 *in* Soulé ME (Ed.), Conservation biology. Sinauer Ass. Inc. Publ., Sunderland

Crawford RMM (1989) Studies in plant survival. Ecological case histories of plant adaptation to adversity. Blackwell London

Dierschke H (1968) Zur synsystematischen und syndynamischen Stellung einiger Calthion-Wiesen mit *Ranunculus auricomus* L. und *Primula elatior* (L.) Hill im Wümme-Gebiet. Mitt Flor soz Arbeitsgem N.F. 13,59-70

Dzwonko Z, Loster S (1989) Distribution of vascular plant species in small woodlands on the Western-Carpathian foothills. Oikos 56:77-86

Ellenberg H (1963) Vegetation Mitteleuropas mit den Alpen in ökologischer Sicht. Ulmer Stuttgart

Ellenberg H, Weber HE, Düll R, Wirth V, Werner W, Paulissen D (1992) Zeigerwerte von Pflanzen in Mitteleuropa. Scripta Geobotanica 18, Verlag Goltze, Göttingen

Falinski JB (1986) Vegetation dynamics in temperate lowland primeval forests. Junk Dordrecht

Foster DR (1992) Land-use history (1730-1990) and vegetation dynamics in central New England, USA. J. Ecol. 80: 753-772

Froment A, Tanghe M (1967) Recherches sur l'écosystème forêt. Répercussion des formes anciennes d'agriculture sur les sols et la composition floristique. Bull Soc Roy Bot Belg 100:335-351

Grime JP (1979) Plant strategies and vegetation processes. Wiley & Sons New York

Harper JL (1977) Population biology of plants. Academic Press London

Hermy M (1985) Ecologie en fytosociologie van oude en jonge bossen in Binnen-Vlaanderen. Doctoraatsproefschrift University of Gent Belgium

Hermy M (1992) Compositional development of deciduous forests from non-forest precursors in northern Belgium: evidence from historical ecology. Pages 437-444 *in* Teller, A Mathy P, Jeffers JNR (eds). Responses of forest ecosystems to environmental changes. Elsevier Applied Sciences, Essex

Hermy M, Stieperaere H (1981) An indirect gradient analysis of the ecological relationships between ancient and recent riverine woodlands to the south of Bruges (Flanders, Belgium). Vegetatio 44: 43-49

Londo G (1976) De decimale schaal voor vegetatiekundige opnamen van permanente kwadraten. Gorteria 7:101-106

Opdam PFH (1991) Metapopulation theory and habitat fragmentation: a review of holarctic breeding bird studies. Landscape Ecology 5: 93-106

Opdam P, van Apeldoorn R, Schotman A, Kalkhoven J (1993) Population responses to landscape fragmentation. Pages 147-171 *in* Vos CC, Opdam P (eds). Landscape ecology of a stressed environment. Chapman & Hall, London

Peterken GF (1974) A method for assessing woodland flora for conservation using indicator species. Biol. Conserv 6: 239-245

Peterken GF (1977) Habitat conservation priorities in British and European woodlands. Biol. Conserv. 11: 223-236

Peterken GF (1981) Woodland conservation and management. Chapman & Hall London

Peterken GF, Game M (1981) Historical factors affecting the distribution of *Mercurialis perennis* in central Lincolnshire J Ecol 69:781-796

Peterken GF, Game M (1984) Historical factors affecting the number and distribution of vascular plant species in the woodlands of central Lincolnshire. J Ecol 72: 155-182

Pigott CD (1969) The status of *Tilia cordata* and *T. platyphyllos* on the Derbyshire limestone. J Ecol 57:491-504

Pigott CD (1971) Analysis of the response of *Urtica dioica* to phosphate. New Phytol 70: 953-966

Pigott CD (1982) The experimental study of vegetation. New Phytol 90: 389-404

Pigott CD (1984) The flora and vegetation of Britain: ecology and conservation. New Phytol 98: 119-128

Pigott CD, Taylor K (1964) The distribution of some woodland herbs in relation to the supply of nitrogen and phosphorus in the soil. J Ecol 52: 175-185

Pollard E (1973) Hedges 7. Woodland relic hedges in Huntingdon and Peterborough. J Ecol 61:343-352

Pott R, Hüppe J (1991) Die Hudelandschaften Nordwestdeutschlands. Westfälisches Museum für Naturkunde, Münster.

Pounds NJG (1990) An historical geography of Europe. Cambridge University Press Cambridge

Rackham O (1980) Ancient woodlands. Arnold London

Raven P (1992) The nature and value of biodiversity. Pages 1-5 *in* WRI, IUCN, UNEP. Global Biodiversity Strategy. IUCN, Gland, Switzerland

Roisin P, Thill A (1952) Aperçu de la végétation forestière de quelques bois de la région Sablo-Limoneuse (District Picardo-Brabançon). Bull Soc Roy Forest Belg 59-60:5-58

Salisbury EJ (1942) The reproductive capacity of plants. Bell London

Saunders DA, Hobbs RJ, Margules CR (1991) Biological consequences of ecosystem fragmentation: a review. Conservation Biology 5: 18-32

Tack G, van den Bremt P, M Hermy (1993) Bossen van Vlaanderen, historische ecologie. Davidsfonds, Leuven

Tallis JH (1991) Plant community history. Long term changes in plant distribution and diversity. Chapman & Hall London

Trepl L (1982) Zur anthropogenen Beinträchtiggung stadtnaher Wälder. Das Beispiel der Eilenriede bei Hannover. Tuexenia 2:195-208

van den Wijngaard JKR (1977) Een bostypering van de Veluwse bossen. Ned Bos Tijdschr 49:66-78

van Ruremonde RHAC, Kalkhoven JTR (1991) Effects of woodlot isolation on the dispersion of plants with fleshy fruits. J Vegetation Science 2: 377-384

Vos W, Stortelder A (1992) Vanishing Tuscan landscapes. Landscape ecology of a submediterranean-montane area (Solano Basin, Tuscany, Italy). Pudoc Wageningen

Waterbolk HTj (1951) De tegenwoordige plantengroei van Drente. *in* Poortman J (red), Drente, Meppel

Westhoff V, Van der Maarel E (1978) The Braun-Blanquet approach. *in* Whittaker R (ed) Classification of communities. Junk Den Haag

Whitney GG, Foster D (1988) Overstorey composition and age as determinants of the understorey flora of woods of central New England. J Ecol 76:867-876

World Resources Institute (1993) Biodiversity prospecting: using genetic resources for sustainable development. WRI, INBio, Rainforest Alliancen ACTS

Zacharias D, Brandes D (1990) Species area-relationships and frequency - Floristical data analysis of 44 isolated woods in northwestern Germany. Vegetatio 88: 21-29

APPENDIX:

True woodland species usually occur in woodlands and their survival depends on the presence of woodland conditions. Tree and scrub species have not been included, since it often was impossible to distinguish between what is spontaneous and what is planted. The species in the list have been taken from a species data base containing 433 species and made for a general phytosociological study (Hermy 1985) of woodlands in western Europe. Species between brackets do not occur in the sampled plots in Flanders.

(Aconitum vulparia)
Adoxa moschatellina
Allium ursinum
Anemone nemorosa
(Anemone ranunculoides)
Arum maculatum
Athyrium filix-femina
Blechnum spicant
(Calamagrostis arundinacea)
Campanula trachelium
Cardamine amara
(Carex brizoides)
Carex elongata
(Carex laevigata)
(Carex ornithopoda)
Carex pendula
Carex remota
Carex strigosa
Carex sylvatica
(Carex tomentosa)
Chrysosplenium alternifolium
Chrysosplenium oppositifolium
Circaea lutetiana
(Conopodium majus)
Convallaria majalis
Corydalis claviculata
Deschampsia flexuosa
Dryopteris carthusiana
Dryopteris dilatata
Dryopteris filix-mas
Elymus caninus
Epilobium montanum
(Equisetum sylvaticum)
(Euphorbia amygdaloides)
(Euphorbia stricta)
Fragaria vesca
Gagea lutea (1)
(Gagea spathacea)
Holcus mollis
Hyacinthoides non-scripta
Impatiens noli-tangere
Lamium galeobdolon
Lathraea clandestina (2)
(Lathraea squamaria)
Listera ovata
Luzula pilosa
(Luzula luzuloides)
(Luzula sylvatica)
Lysimachia nemorum
Maianthemum bifolium
Melampyrum pratense
(Melica nutans)
Melica uniflora
Mercurialis perennis
Milium effusum
Moehringia trinervia
Narcissus pseudonarcissus
(Orhis fuchsii)
(Orchis mascula)
(Orchis purpurea)
Ornithogalum umbellatum (1)
Oxalis acetosella
Paris quadrifolia
(Peucedanicum gallicum)
(Poa chaixii)
Polygonatum multiflorum
(Polygonatum verticillatum)
Polypodium vulgare (s.l.)
Potentilla sterilis
Primula elatior
Primula vulgaris
Pteridium aquilinum
Ranunculus auricomus
Ranunculus ficaria
Ribes nigrum
Ribes rubrum
Ribes uva-crispa
Rubus caesius
Rubus fruticosus coll. (3)
Sanicula europaea
Sedum telephium
(Stellaria nemorum)
(Trientalis europaea)
Vinca minor
(Viola hirta)
Viola odorata
Viola reichenbachiana
Viola riviniana

(1) *Ornithogalum umbellatum* & *Gagea lutea* optimally occur on well aerated sandy river banks occasionally outside woodlands.
(2) *Lathraea clandestina* was strongly associated with poplars also if it occurred outside woodlands.
(3) *Rubus fruticosus* coll. is considered as a group of species of which some frequently have been observed in non-woodland habitats. However, as a whole this taxon mostly was only abundant on woodland sites and edges.

Chapter 8: Impact of Global Change on Genetic Diversity of Temperate Ecosystems

Gene Namkoong
USDA - Forest Service &
Forest Sciences Department
University of British Columbia
Vancouver, B.C.
Canada

The genetic diversity that permits populations to evolve and respond to future stresses and selective pressures is the result of its past evolutionary history. Forces that affected its population size and changes in gene and genotypic frequencies have molded the levels and structure of its genetic variation and hence the future evolutionary capacities of its present populations. Therefore, the genetic diversity is both a cause and an effect of global change and is a measure of the dynamic changes underlying genetic variation that both permits and constrains the future evolution of temperate ecosystems. Neither the global climate nor the genetic variation in ecosystems are independent of the effects of the other, and the history of temperate ecosystems involves the joint evolution of genes and environment.

In some cases, the predicted global change may be reasonably considered in the foreseeable future to be merely a change in environment that is not qualitatively different from any other environmental shift that the earth's biota has met before. If that were the case, the task of predicting the impacts of global change would involve only having to continue our usual population genetic studies to understand and predict genetic changes and to recommend what might be done to improve adaptability or to assuage deleterious consequences. Unfortunately, this is not a simple problem since even under mild selection pressures, the influence of population size, population differentiation and inter-population gametic migration, and the joint dynamical effects

NATO ASI Series, Vol. I 20
Biodiversity, Temperate Ecosystems, and Global Change
Edited by T. J. B. Boyle and C. E. B. Boyle

of multi-species evolution, all compound the problem of determining the fate of populations and their genes in gradual evolution. The direct and indirect effects of genes that affect populations are not simple one-to-one relationships. Since most traits of significance are under multiple locus control (i.e., epistasis is important) and individual genes usually affect several traits (i.e., pleiotropy is important) the relationships between genes and phenotypes is complex. It is not even a matter of studying one gene-to-many effects or many genes-to-one effect, it is many-to-many, and indirect gene effects can significantly affect evolution. In addition, our knowledge of heritabilities and of the correlations between traits is usually insufficient to predict how rapidly and what directions the joint evolution of multiple traits will take when moderate rates of change can be expected.

The problem we are to address in this conference further complicates the already difficult analysis of moderate change rates with possibly very large and rapid changes forced by new and largely uncontrollable events. It has been made abundantly clear that the physical environment may be drastically different in several factors simultaneously, such that large scale fragmentations and extinctions may occur. For species that may be site specialists or are dependent on particular mutualists, and those that are little managed and exist in small or fragmented populations, rapid changes can force at least local extinctions. The threat of global change not only implies a magnitude and rate change but also that multiple factors will change. In itself, the impacts of change in multiple factors can be analyzed by their net effect on fitness components. This in turn can then be linearized for local changes and predictions made for survival and trait changes. However if changes are not in single directions such as warmer and drier, but are both warmer and drier and cooler and moister, then disruptive instead of directional or stabilizing selection forces will occur. In fact, the patterns of change in several variables cannot be expected to be tightly coordinated. Thus a new level of concern is that we may find that several factors may change simultaneously or sequentially, requiring populations to adapt to multiple challenges where single populations might have been formerly forced to meet only a single set of changes. Areas may become warmer in the summer or in the winter and each may exist with heavier or lighter rainfall, or all four combinations may occur.

Four more environmental factor combinations may occur if cooler conditions are feasible, and in fact, the full panoply of factorial combinations may fall within the purview of feasible futures for a single population. Either because we cannot predict the probabilities of which future conditions may be realized, or because multiple future states will exist, the problem presented by the physical environment is multidimensional and would require anticipation of wider variations than presently exist in all of those dimensions.

To address the problem of the impact of global change on genetic diversity, it is first necessary to consider what evolutionary dynamics may affect the amounts and structure of genetic variation in traits that affect adaptation. If genetic variation exists in how trees respond to environmental challenges, then the limits of adaptability may be either limited or very wide. Knowledge of how that genetic variation may be compromised would be useful for the purposes of management. In this paper, the evolution of genetic diversity is first briefly considered then the kinds of challenges that global change may present is examined for how it might impact evolution.

GENETIC DIVERSITY

The persistent existence of genetic variation in traits that affect viability or fecundity and fertility is a long standing problem in population genetics. Crudely stated, the problem is that after many generations of selection, why have not all genetic sources of variation been fixed at some optimum either genetically or phenotypically. Heterosis at many individual loci does not seem to be supported by data (Bush *et al.*, 1987) nor is it tenable theoretically (Lande, 1976). Thus the maintenance of genetic variance requires that effects at the individual level of organization, or at the population level, exist that can either stabilize the variance or regenerate it at a rate that counters the effects of genetic erosion such that a steady state variance exists. One of the major forms of argument is that there is a strong antagonistic pleiotropy between fitness and other effects of a single gene. Thus, genes favouring early growth may be balanced against their effects in later life stages such that stabilizing selection persistently maintains an optimum phenotype. If many loci similarly affect the traits, then paired loci with alleles of opposite effect would tend to be fixed. Though a few loci may maintain overdominance, additive variance would be lost. Since additive genetic variance is present at substantial levels this seems unlikely.

Another way that a type of antagonistic pleiotropy could exist is if there is genotype by environment interaction whereby different alleles are favoured in different environments. Genetic variation can be maintained (Barker and Thomas, 1987; Via and Lande, 1985) and depending on how seed and pollen may migrate, different patterns of allele distributions can be maintained (Gregorius and Namkoong, 1983; Namkoong and Gregorius, 1985). In all of these cases, largely additive genetic variance can be maintained by a mutation-selection balance especially where there is partial dominance in the recurring mutations (Turelli, 1985). Thus, even persistent directional selection will not eliminate additive genetic variance (Hill and Rasbash, 1986) in closed breeding populations, and diversifying selection can increase the amounts and types of genetic variance contained in single populations. Thus, even with completely random intercrossing among diversely selected populations, high levels of genetic diversity can persist. If populations are limited in their gene exchanges, much more rapid diversity can be accumulated and, with recurring mutations, the diversity can be rapidly increased between populations. The state of genetic diversity then is a resultant not only of mutation-selection balance, but also of their balance with interpopulational gene migration.

For most temperate forest tree species, very high levels of genetic variance in isozyme allele frequencies and in phenotypic growth traits commonly exists but the patterns of variation are not usually congruent (Namkoong and Kang, 1990). Not all phenotypic variation could be of adaptive significance nor can any sense of perfect adaptability ever be assumed. With generally believed wide pollen and seed migration, the genetic variance in large populations seems to be primarily distributed within rather than between populations. However, there are clear differences in the degree and patterns of provenance variation between species such as Douglas fir and White Pine in the interior of western North America, implying different forms of vulnerability to population loss. Nevertheless, there would therefore presumably be substantial levels of genetic variation within many temperate tree populations to allow them to respond rapidly to mild selection pressures. While it often also occurs that there is more populational differentiation in growth traits than in isozyme allele frequencies (Lagerkrantz and Ryman, 1990), the data analyses do not permit direct comparison (Lewontin, 1984) and it remains reasonable to assume that adaptationally significant differences among populations exists. Populations of temperate zone trees which can survive can nevertheless be expected to respond with much the same set of initial alleles regardless of the sub-sampling that may occur for regenerating future forests. For those species that often exist in small populations or in fragmented isolates containing "private" alleles, sub-sampling for future regeneration can have a large initial biassing effect. An obvious solution for conservation is to sample amply and widely to ensure the capture of extant diversity and, in the future, to enforce diversity by directionally diverse selection. This option will be limited to those few species for which we can afford intensive interventions for their genetic management. Even then, it will require anticipatory breeding and population development to avoid substantial economic productivity losses. The question for the impacts of global change for less intensively managed species are whether the changes are within what can be considered normal limits of environmental change and if there are critical changes to population genetic dynamics that would alter our expectations of normal evolutionary response.

IMPACTS OF GLOBAL CHANGE

Two kinds of impact seem most critical to consider. The impact of global change on population survival and the impact on those populations that remain in situ. Impacts that threaten stand survival affect not only the existence of the species or of any special local adaptabilities, they also affect their participation in the migratory network that holds species together. As noted above, the migratory network influences the genetic dynamics within populations and determines the stability of genetic variances within any component population. The reduction of gene migration not only reduces the effective population size, it reduces the level of diversifying selection and hence, reduces the equilibrium genetic variance.

The second kind is conditional on population survival but simple reductions in census size would be expected to reduce the effective population size. More subtle effects on sexuality, and the disproportionate representation of particular genotypes in reproduction and seedling success would further reduce the effective population size. Within smaller populations, selection for specific responses may then reduce the total genetic variation within the remnant populations. There is also a kind of interaction between the two effects in that with reduced gametic migration local selection effects will not be moderated by selection for other environmental conditions and hence, there could be lower selection intensities for response plasticities. Thus, individual tree homeostasis may be reduced at the same time that stand level homeostasis is reduced by the loss of genetic variance.

If whole populations are eliminated by direct habitat alteration such as deforestation, the remaining stands may have substantially reduced effective population sizes as well as reduced opportunities for gametic or individual migration. More or less continuously distributed populations may be fragmented into partially isolated islands and the structure of diversity and selection within the context of a selection-mutation-migration balance would be altered. Thus, it is not only the direct loss of populations, local adaptations, and "private" alleles that is of significance, but it is also the effect on population dynamics within the remainder that is affected. Species that are now robust with respect to global change may soon be included in the list of those that are sparsely distributed and endangered. Within the remnant but confined stand areas, long lived trees would occupy a finite portion of the available space for reproductive individuals and would be expected to repeatedly participate in reproduction. Hence, the sample of parents and the possibilities of parent-offspring mating, restrict the gametic sampling to comparatively fewer genotypes than if, over the same generation interval, independent sampling were to exist (Hill, 1972). For tree species, the imbalances that occur in fecundity, sexuality, and mating success (Muller-Stark, 1985; Chaisurisri and El-Kassaby, 1993; Moran and Griffin, 1985; Schoen and Cheliak, 1987) would themselves reduce the ratio of effective population size to census number. Hence to conserve populations, the number of adult trees must be substantially larger than any minimum population size considered necessary for avoiding populational stochastic events. In natural regeneration, the possibilities of mate selection also exist (El-Kassaby and Ritland, 1992) that could alter the dynamics of gene frequency change and can generate population dynamical stabilities and instabilities (Selgrade and Ziehe, 1987).

For any well managed breeding population, these problems can be readily managed and population sizes of a few tens of individual parents can be sufficient for maintaining high levels of genetic variance in quantitative traits more or less indefinitely (Hill and Rasbash, 1986). However, for populations that are not so intensively managed, adult population sizes five to ten times greater than needed for minimum breeding population sizes could be required for even the simpler dynamics to be well approximated. If it can be expected that normal stochastic variations in population sizes require initial population sizes to be greater than a few tens of adults, the sensitivity of genetic variance to bottlenecks requires that the minimum effective population size anticipated over many generations be at least that large. However, direct experience with maintaining small populations of trees is absent and since seed production is often very large, the possibilities of genetic rescue can assuage these risks. Even in the face of inbreeding depression, a purging of deleterious alleles over a few generations may allow even very small populations to survive and to regenerate genetic variance. The problem then becomes one of determining the extent to which these populations may participate in any general species evolution and if they would be fulfilling the various ecological and economic functions that may be demanded of them.

Since global change will present different challenges to populations both in the magnitude as well as in the particular selection pressures, the effects of simultaneous changes in multiple environmental variables have to be considered for their effects on multiple response traits. With either hard selection or soft but intensive selection, as would have to be anticipated for drastic environmental change, effective population size reductions would have to be anticipated. Under these conditions the theory of multivariate selection response for infinite or very large populations may be inadequate. Nevertheless, the equations of Lande (1976) are useful in indicating that the response to selection in any one trait depends on the effects of selection on correlated traits as well as on direct selection effects. Any one environmental change has different intensities of selection effect on several traits, and even those traits that are not directly affected may change because of their genetic correlation with a directly selected trait. Thus, environmental stresses that affect only seedling growth, can affect a selection for root/shoot ratios that affect mature tree growth even though older trees may be immune to factors that affect juvenile trees. Similarly, the genetic correlations between growth and crown form as occurs in Douglas fir (King, 1986) can induce correlated responses in either trait. Early growth may also affect the age of reproduction and if reproduction is genetically correlated with diameter growth (El-Kassaby and Barclay, 1992), there may also be a response in adult diameter growth. The joint effects of direct and indirect responses to selection can therefore be different than if traits were independent, but can nevertheless be predicted.

Since the response of populations in the multivariate traits depends on their genetic correlations as well as on the selection pressures, the nature and stability of those correlations can determine the consistency of response. However, it is not obvious that different populations of the same species share the same genetic correlation matrices, and in fact, the genetic correlations at the provenance level of organization can be quite different from the genetic correlations of individuals within populations. In Norway spruce, the genetic correlations among growth rhythm traits when estimated with family performances can sometimes be opposite in sign from the correlations at the provenance level of observation (Ekberg, *et al.*, 1993). Thus, in addition to the problems of estimating the genetic covariances, there may be real differences in genetic associations between traits within and among populations. This implies that the genes affecting the joint trait behaviour are different and hence that selection can have drastically different effects depending on how populations are inter-mated.

With intensive selection as is implied by large environmental shifts, simple responses may be quite deviant from that which is realized in both the short and the long terms. The nature of indirect responses is further complicated by the effects of environmental shifts on other organisms with which trees interact in competitive, mutualistic, and in host-pest forms of interaction. It is not necessary that all of these changes will be either positive or negative with respect to tree species vigour and reproduction, but it would be naive to believe that none would be seriously deleterious to fitness.

CONCLUSIONS

The threat of global change lies not only in the direct loss of populations via local extinctions and the loss of special adaptabilities and productivities, it lies also in the altered dynamics of the remnant populations. Especially with fragmentation of previously contiguous populations the inter-populational and intra-populational dynamics are altered. In the absence of other populations to serve as sources of mates and the absence of wider allelic sampling, the long term effects of smaller population sizes would include a reduction in the additive genetic variance maintained by selection-mutation balance. The smaller population sizes would maintain a lower level of variance if they continued to exist, and random allele loss would be higher. In addition, without population interbreeding among different sites, the multivariate responses described above, would not be subject to multiple diversifying selection pressures. In the approximations used by Via (1987) for evolution in two environments, selection for plasticity and the maintenance of genetic variance in plasticity would be lost, and the opportunities for breeding or for evolution to generate more widely adaptable genotypes would be lost. Thus, not only would specific environmental adaptations be lost, but the future capabilities for evolving wide adaptability at both the individual or populational levels could be diminished as well.

Within the context of isolated populations, the threats to future evolution warrant some concern. As individual populations, in-situ conservation of stands does not provide long term security for the survival, well being, or use of the species involved. With multiple environmental factors changing and affecting multiple species, and individuals and populations responding with constrained capabilities in multivariate trait space, it must be anticipated that more species will be affected adversely than would find that their evolutionary potential has improved. It is possible that, by diversifying sampling and inducing more diverse responses among populations, greater diversity among populations can be created than now exists.

Thus, in-situ conservation can influence future evolution even if no single population's probability of survival and growth is enhanced. The structure and connectivity of component populations of a meta-population strategy is necessary to construct, if multiple futures and uncertainty are to be met constructively.

If alternate populations can be considered as parts of the genetic system that would be impacted by global change, then one of the sets of alternatives would logically be those that might exist in surrogate environments. One of the simpler possibilities is that only a single future environmental shift will be important such as a general warming and moistening from coastal areas that may shift inland. In that case, if populations existed in the surrogate environments then surrogate populations would only have to migrate or to have their genes migrate at a rate and distance to match the environmental shift. As indicated by many others (e.g., Davis and Zabinski, 1991), this may not be a reasonable possibility. Thus even if such suitable alternatives may exist, the physical problem of matching population to its environment may prove insurmountable. An alternative is to mount a set of anticipatory plantings of potential surrogate populations as a sort of pre-adaptational provenance series. A different form of alternative would be to select for divergent behaviour in multiple and segregated populations to expand the inter-population variance. It would then also be possible to create different correlation sets by hybridization and thereby to increase the generalized variance of the total genetic variances and covariances. While such programme may not require plantation management and those attendant costs, they would require more intensive genetic management than hitherto devoted to tree species.

REFERENCES

Barker JSF, Thomas RH (1987) A quantitative genetic perspective on adaptive evolution. Pages 3-23 in Loeschcke V (ed). Genetic Constraints on Adaptive Evolution. Springer-Verlag, Berlin

Bush RM, Smouse PE, Ledig FT (1987) The fitness consequences of multiple locus heterozygosity: the relationship between heterozygosity and growth rate in pitch pine (*Pinus rigida*). Evol. 41: 787-798

Chaisurisri K, El-Kassaby YA (1993) Estimation of clonal contribution to cone and seed crops in a Sitka spruce seed orchard. A. Sci. For. (in press)

Davis MB, Cabinski C (1991) Changes in geographical range resulting from greenhouse warming effects on biodiversity of forests. *in* Peter RL, Lovejoy TJ (eds). Global Warming and biological diversity. Yale University Press, New Haven

Ekberg I, Erikson G, Namkoong G, Nilsson C, Norell L (in press) Genetic correlations for growth rhythm and growth capacity at ages 3-8 years in provenance hybrids of *Picea abies*. Scad. J For. Res.

El-Kassaby YA, Barclay HJ (1992) Cost of reproduction in Douglas-fir. Can. J. Bot. 70: 1429-1432

El-Kassaby YA, Ritland K (1992) Frequency-dependent female reproductive success in a polycross of Douglas fir. Theor. and Appl. Gen. 83: 752-758

Gregorius HR, Namkoong G (1993) Conditions for protective polymorphism in subdivided plant populations. Theor. Pop. Biol. 24: 252-267

Hill WG (1972) Effective size of populations with overlapping generations. Theor. Pop. Biol. 3: 278-292

Hill WG, Rasbach J (1986) Models of long term artificial selection in finite populations. Genet. Res. 48: 41-50

King JN (1986) Selection of Traits for Growth, Form, and Wood Quality in a Population of Coastal Douglas Fir From British Columbia. Ph.D. Thesis. Univ. Alberta

Lagerkrantz U, Ryman B (1990) Genetic structure of Norway Spruce (*Picea abies*). Concordance of morphological and allozymic variation. Evol. 44: 38-43

Lande R (1976) The maintenance of genetic variability by mutation in a polygenic character with linked loci. Genet. Res. 26: 221-235

Lewontin R (1984) Detecting population differences in quantitative characters as opposed to gene frequencies. Amer. Nat. 123: 115-124

Moran GF, Griffin AR (1985) Non-random contribution of pollen in polycross of *Pinus radiata* (D.Don). Silv. Gen. 34: 117-121

Muller-Stark G (1985) Reproductive success of genotypes of *Pinus sylvestris L.* in different environments. Pages 118-137 *in* Gregorius H-R (ed.) Population Genetics in Forestry. Lect. Notes in Biomath. Vol. 60. Springer-Verlag, Berlin

Namkoong G, Gregorius H-R (1985) Conditions for protected polymorphism in subdivided plant populations. 2. Seed versus pollen migration. Amer. Nat. 125: 521-534

Namkoong G, Kang HC (1990) Quantitative genetics of forest trees. Pages 139-188 *in* Janick J (ed). Plang Breeding Reviews, Vol. 8. Timber Press

Schoen DJ, Cheliak WM (1987) Genetics of the polycross II. Male fertility variation in Norway spruce. Theor. and Appl. Gen. 74: 554-559

Selgrade JF, Ziehe M (1978) Convergence to equilibria in a genetic model with differential viability between the sexes. J. Math. Bio. 25: 477-490

Turelli M (1985) Effects of pleiotropy on predictions concerning mutation-selection balance for polygenic traits. Genetics 111: 165-195

Via S (1987) Genetic constraints on the evolution of phenotypic plasticity. Pages 47-71 *in* Loeschcke V (ed). Genetic Constraints on Adaptive Evolution. Springer-Verlag, Berlin

Via S, Lande R (1985) Genotype-environment interaction and the evolution of phenotypic plasticity. Evol. 39: 505-522

Chapter 9: Strategies for conserving genetic diversity in the face of global change

Hans-Rolf Gregorius

Abteilung für Forstgenetik und Forstpflanzenzüchtung

Universität Göttingen

Büsgenweg 2

37077 Göttingen

Fed. Rep. of Germany

INTRODUCTION

The deliberations in the present paper will be concerned with providing a basis for finding answers to the following three questions:

- What, if anything, can and should be conserved in the face of global changes, and how should this be done?
- In a strict sense, conservation addresses an identity that is to be protected from loss and thus maintained unchanged. What then is the significance for conservation programmes of the identity of a resource that is to be preserved?
- Is the preservation of identity compatible with general biological, ecological, and evolutionary principles?

Even though the focus will be on strategies of conserving genetic variation *within* biological species (*i.e.* intraspecific genetic variation), the broader field of conservation biology shall also be taken into consideration. The latter includes problems of conservation of genetic variation *between* biological species (*i.e.* interspecific genetic variation) and is generally treated under the heading of preservation of species diversity. The totality of intra- and interspecific genetic variation defines biodiversity. The necessity to distinguish between the two kinds of genetic variation rests on the fundamental fact that it is not possible inheritably to combine information on adaptive characteristics between biological species. Adaptation of biocoenoses to their abiotic environment evidently does not include the

NATO ASI Series, Vol. I 20
Biodiversity, Temperate Ecosystems, and Global Change
Edited by T. J. B. Boyle and C. E. B. Boyle

option of merging advantageous traits fixed in different species into a single species. Therefore, adaptation to variable environments, which plays a central role in the stability of biological resources, follows principles which differ between the species level and higher levels of biological organization, such as biocoenoses and ecosystems or landscapes.

With the exception of the purely static forms (vegetative propagation, cryopreservation, etc.), conservation includes phases of alteration or modification of original states. These dynamic phases seem to contradict the very meaning of the term "conservation" (agricultural seed genebanks provide a well known example for static conservation with intermittent dynamical phases; see *e.g.* Wilkes 1992, p. 18). On the other hand, if the identity of an object is required to reflect only a certain set of its characteristics (including such basic aspects as a collection's origin) rather than its totality, conservation methods could indeed be dynamic as long as they do not affect this set. Thus, basically, *conservation measures might be meaningless if they do not conserve a specified identity of the object* (population, species, ecosystem, landscape, etc.).

Turning to genetic resources, specification of the (genetic) identity to be preserved might be a complicated task, particularly in the face of a globally changing environment with unpredictable effects on the local conditions (*cf. e.g.* Miller 1993, Scholz 1993). Including genetic information on traits potentially involved in adaptation into the resource's identity might hinder adaptive processes by fixing that which requires dynamics. This could endanger the physical persistence of a resource whether conserved *in situ* (and thus dynamically) or statically. As was repeatedly emphasized, predominantly statically conserved resources run the risk of building up a considerable adaptive lag with respect to a changing environment, and this lag does not show up until regeneration of the resource (see *e.g.* Ziehe *et al.* 1989, p. 181). In particular, natural regeneration involving sexual processes is the only mode of reproduction realizing, via recombination, the genetic adaptive potential inherent in a population's gene pool. Static conservation should therefore be considered as an emergency measure which should be complemented or replaced by dynamical methods of conservation as soon as possible.

The accumulation of maladaptedness can be prevented by specifying the resource identity to include genetic information which codes for adaptively neutral or adapted traits that are not affected by the environmental change. In essence, this requires distinction between constant and (non-cyclically) varying environmental factors, and the adaptations to the former should be included into the resource identity in order to support adaptability to the latter factors. In this context, genetic variation involved in adaptational processes is a means rather than an object of conservation.

For example, if the declaration of a genetic resource was motivated by the presence of valuable genetic variants which, at the same time, represent important adaptations to special local conditions, the declaration can be maintained if there is no reason to assume that the global changes affect these local conditions. Yet, if the local conditions are affected, and if the variants turn out to be maladapted, resource management should not aim at maintaining these variants artificially in order not to prevent necessary adaptive processes: these variants need to be removed from the resource's identity. If, however, these genetic variants provided the only motivation for the declaration, the resource would of course have lost all of its identity, which could in turn justify the risk of preserving them artificially.

In general, however, the decision on which traits are more likely than others to show adaptive neutrality with respect to the envisaged global changes is quite difficult and is further complicated by the unpredictability of local environmental dynamics. Preservation of the adaptability of the resource will in this situation gain priority over preservation of the genetic *status quo*. As mentioned above, genetic variation is then assigned the role of a means of conservation, so that primarily those characteristics of the genetic structure relating to adaptability become the object of conservation programmes.

Since (evolutionary) adaptation implies genetic change, specification of the identity of a resource to be conserved dynamically may raise both conceptional and practical problems. The concept of "identity" will therefore be a prominent topic in the following. It will be demonstrated that this concept allows for a system theoretical representation of the genetic system, which helps to make the position of adaptation in conservation biology more transparent on the intraspecific level, as well as on higher levels such as the ecosystem.

ADAPTATION AND SYSTEM IDENTITY

With reference to changing conditions, it is implicit (at least tacitly) in the usage of the term "adaptation" that some invariant entity (called "identity" in the following) exists that would suffer in some sense from the change, unless it can adjust. The existence of such an invariant and persisting entity is indeed indispensable, since one would otherwise not be able to recognize who adapted to what. This basic reasoning gives rise to the following heuristic definition:

> ***Adaptation*** *refers to all processes within a system that serve the preservation of its identity under changing external influences. Here* ***identity*** *primarily refers to the system's invariants. Accordingly, an* ***adaptive system*** *is characterized by the mechanisms that enable adaptation with respect to the system identity.*

In a strictly evolutionary sense, the essence of a system's identity reduces to genealogical continuity (retrospectively defined by common ancestry, including "origin", and prospectively defined by the capacity to have common descendants) among the members of a collection. This assumes that these members have the capacity to replicate. Since persistence is an indispensable condition for the preservation of a system's identity, extinction marks the ultimate loss of its identity (and, in particular, its genealogical continuity). Formation of stable hybrid populations (in which at least one gene locus is fixed for alleles characteristic of one population, and a second locus is fixed for the alleles characteristic of the other population) provides an example, where one would not speak of extinction, but where nevertheless the identities of the component populations terminate at least as a consequence of the genealogical discontinuity (for more details on the problem of hybridization in conservation biology see *e.g.* Templeton 1991).

On the other hand, the fact that two populations may fuse so as to result in a stable hybrid population assumes that these populations belong to the same biological species. *A biological species, in turn, can be conceived of as the largest potential reproductive community fulfilling the condition of genealogical continuity.* For this reason, the minimum identity to be respected in any conservation strategy must comprise those species characteristics resulting from the species' genealogical continuity. Most demographic characteristics (see *e.g.* Lande 1988 or Gregorius 1991), for example, belong to this category. Hybridization between two populations may indeed provide the adaptive potential required to preserve the species characteristics (and thus the identity shared by both populations) in a changing environment.

Specification of identity

For special objectives of conservation on the population level, genealogical continuity is not a sufficient (though necessary) characterization of identity, since the intention is to maintain a particular genetic variant or a combination of these as a declared part of the system's identity. If it turns out that under changed environmental conditions the collection persists as a genealogical unit but loses these variants, this indicates that, according to the above definition, the adaptive capacity cannot be assumed to extend to these conditions without a change in identity. It follows from this reasoning that *the identities of non-specific populations differ from one another chiefly by the adaptations to their habitats*; changes in habitat may trigger the evolution of new adaptations and thus of new identities. Therefore, the assignment of the rank of identity should be restricted to those adaptations that have a direct bearing on persistence (and thus on genealogical continuity) with respect to a stable habitat characteristic.

Invariance, as one of the fundamental conditions for an identity, suggests that traits fixed by random drift and thus by non-adaptive processes in a population may, at least in principle, also be assigned the rank of identity. However, in general, invariance, in spite of the possibility of gene flow and hybridization among conspecific populations, requires that the entire species be fixed for the respective trait, which classifies this trait as a species rather than a population characteristic.

It is important to note that large genetic diversity can by itself be an adaptation to a special type of habitat (see *e.g.* Levins 1968, p. 10, or Gregorius 1991) and may therefore be part of the identity of a population. The condition of invariance is fulfilled if the amount of genetic diversity remains unchanged, as can be achieved without retention of the qualitative genetic variation by substitution of new for extant genetic types. Habitat protection with emphasis on preservation of habitat characteristics, such as spatial or temporal heterogeneity, may in this case be the appropriate action in order to preserve a population's identity (Gregorius 1991). This could also be an appropriate conservation method if adaptive potential for unpredictably changing environments is to be provided.

Processes leading to the loss of a collection's identity may thus be viewed as an indication of adaptation or of maladaptedness, depending on whether one focuses on the effect of securing genealogical continuity by removing old or evolving new adaptations, or whether one simply notes the apparent lack of adaptive potential for maintaining the totality of the prior identity. The difference between the two points of view lies in the fact that, in the first case, even the characteristic adaptations are considered to serve solely the preservation of the ultimate identity, which is seen in genealogical continuity.

This relates to an important topic in population and ecological genetics, namely the difference between selection and adaptation. In his influential discussion of this topic, Dobzhansky (1968) came to the conclusion that adaptation is necessarily related to population persistence, while this need not be true for selection (selection refers to relative, and adaptation to absolute, numbers). *Adaptivity* of a character therefore refers to its capacity to maintain a population under specified conditions notwithstanding the possibility that other characters might be adaptively and thus selectively superior. *Maladaptedness*, in the sense that a character endangers population persistence when prevalent, need not imply selective inferiority. In fact, *selection may proceed in an adaptively neutral manner*, as is well known to be the case in selective substitutions at the carrying capacity.

The primary identity to be preserved under Dobzhansky's concept of adaptation thus is genealogical continuity, while the adaptation itself is a means to guarantee achievement of this goal. Yet, extending this concept to general system identity implies that, once the identity of a population is specified, all selection changing this identity has to be considered as a counter-adaptive process. In a strict sense, even selectively neutral changes of the identity (as by drift, for example) would rather have to be considered as counter-adaptive than adaptively neutral.

On an ecosystem level, the system's identity is usually specified by the presence of certain "characteristic" (in the sense of ecologically dominant or key-stone) species. This differs fundamentally from the specification on the population or species level, in that the existence of genealogical relationships between species as elements of an ecosystem is not a constituent part of its identity. There seems to be no such thing as an ultimate identity of an ecosystem, the preservation of which is aided by the other component identities, as is the case with genealogical continuity and local adaptations for populations. Consequently, an ecosystem would terminate if one of its characteristic species were to become extinct, in spite of continuity in space and time of the remaining characteristics. The system can be conceived of as having entered a successional process, and the subsequent sequence of ecological communities (a sere) is by itself a new system with its identity defined by characteristics of this sere.

There are of course other characteristics of ecosystems that are used for their identification, one of which consists in attributes of the chemical budget or nutrient balance. Under this concept of ecosystem identity, the members of the biocoenosis are exchangeable as long as they are equivalent with respect to their position and function in the chemical or nutrient cycling. Replacement of one species by another, with the effect of maintaining some chemical balance that would otherwise not have persisted under changed external conditions, would then have to be classified as an adaptational process. At the extreme, where species composition is given no part in the system identity, even losses or replacements of characteristic species would not be considered a case of adaptive failure if they served the stabilization of chemical balances.

The last statement applies analogously to system identities defined by human benefits (concerning *e.g.* water supply, climatic regulation, availability of medicinal plants or other profitable biological matter including breeding material, etc.). However, it has to be taken into consideration that the underlying (frequently artificial or manipulated) ecological systems may not provide sufficient adaptive potential to secure the persistence of such identities. This refers particularly to the opportunities of sustainable management and use of ecological, agricultural, and silvicultural systems.

Regulatory and evolutionary adaptation

An important aspect of adaptation implicit in the above demonstrations consists in the two ways in which system identity can be preserved, namely without and with change in the state of the system (see *e.g.* Bergmann *et al.*, 1990). While gene frequency changes within populations or species are commonly accepted as a consequence of adaptation to changing environments, comparable changes in state on the ecosystem level, where populations or species are the system elements, could easily result in losses of system identity. Adaptive processes on the latter level may therefore be expected to take place frequently without changes in element composition or state. The following terminology reflects this:

> ***Regulatory adaptation*** *is a form of adaptation that takes place without changing the state of the system (in autecology frequently referred to as physiological adaptation).*
>
> ***Evolutionary adaptation*** *takes place by changing the state of the system.*

To elaborate the difference between the two forms of adaptation, it is helpful to restate the above heuristic definition of the adaptive system in the more precise terms of an input-output system. Adaptation to a specified condition (the system's input) requires that the pertaining response (*i.e.* the output) of the system be evaluated with respect to its adaptive value for this very condition. The adaptive value specifies whether the system's response to the condition can be regarded as a successful adaptation to the condition, *i.e.* whether the system's identity is preserved under this condition. The external conditions or influences thus act in two ways: (a) they initiate a response of the system depending on its state, and (b) they determine the reference for evaluation of the adaptedness of the response. It is therefore meaningful to divide the external influences into two components, one of which effects the system's responses while the other determines the adaptive value of these responses. In the event of a negative adaptive evaluation, a change in system state would be required in order to regain adaptedness (*i.e.* positive evaluation of the changed response), in which case evolutionary adaptation would have taken place.

Figure 1 illustrates this concept on the individual level, the adaptive system of which is characterized by the genotype-environment-phenotype relationship. The elements in the system are given by the genes. The genotype specifies the system's state in that it implies a rule of assignment of environmental conditions (as inputs) to phenotypes (as outputs). By definition, adaptation to an environmental change has taken place if the corresponding phenotypic modification, if there is any, receives a positive adaptive evaluation. Since this does not involve a change in genotype (and thus in system state), regulatory adaptation is said to have occurred. Negative evaluations, in turn, would require changes in genotype and thus evolutionary adaptation in order to restore adaptedness. Changes in genotype, however, can regularly be realized only within populations, so that one is forced to enter this higher level of biological organization in order to enable evolutionary adaptation. On this system level, genotypes take the role of system elements, and the respective genetic composition of the population forms the system's state.

Table 1 provides examples for the specification of system elements, states, and outputs for the three most important levels of biological organization in conservation biology. Concerning specification of the system output at the biocoenotic level in this table, the focus is here on characteristic species as defining system identity and on interspecific interactions as the major forces affecting the maintenance of this identity. Though intuitively obvious, it is important to realize that *only after the capacity for regulatory adaptation is exhausted does evolutionary adaptation come into play*. It is easily verified that this biological principle of adaptation, the validity of which was demonstrated for the individual level of organization, generalizes to the other two levels listed in Table 1.

The distinction between the two types of adaptation reflects the basic role played by biodiversity in evolution. Under non-cyclical environmental changes the limits of the regulatorily adaptive capacity are soon surpassed, and evolutionary adaptation is the only option for persistence. This, in turn, requires (intra- and/or

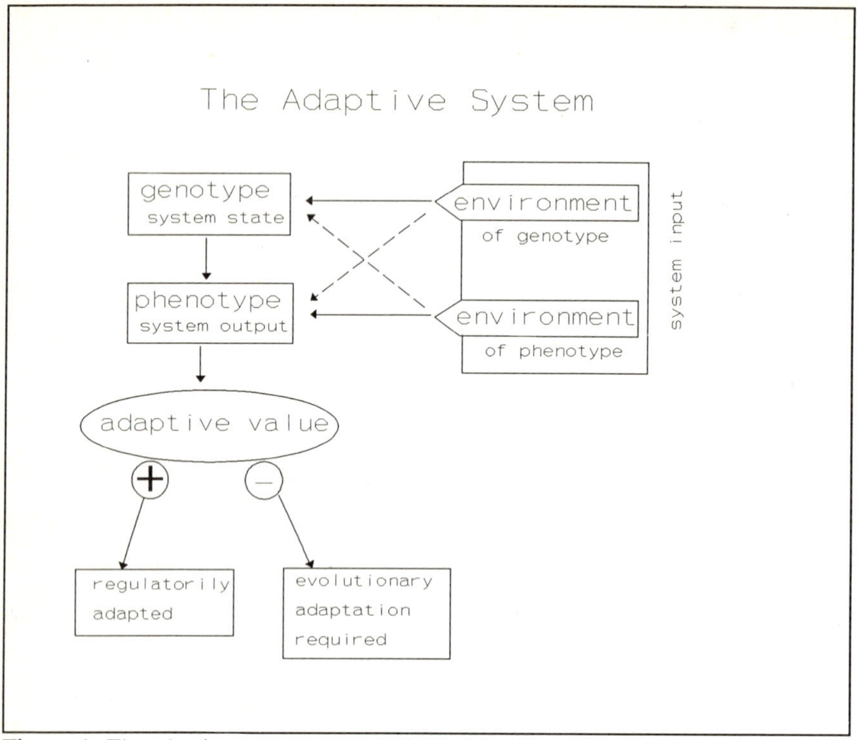

Figure 1. The adaptive system at the individual level.

Table 1. Examples for the specification of system elements, states, and outputs for the three most important levels of biological organization in conservation biology.

System level	Individual	Population	Biocoenosis
System elements	Genes	Genotypes	Populations or species
System states	Genotypes	Genetic Compositions	Species Compositions
System outputs	Phenotypes	Phenotypic Compositions	Modes of Interspecific Interactions

interspecific) genetic variation, and the higher the unpredictability of the changes, the more such variation is required in order to enable evolutionary adaptation and thus persistence of biological communities and their identities.

The genetic system as an adaptive system

In biology, usage of the term "evolutionary adaptation" is usually restricted to processes taking place within populations or species, and which depend basically on the respective genetic system as originally defined by Darlington (1939) and concisely formulated in Rieger *et al.* (1991) as "*any of the species specific ways of organization and transmission of the genetic material in pro- and eukaryotic organisms, which determine the balance between coherence and recombination of genes and control the amount and type of gene combinations*". While this definition is explicitly restricted to the mechanisms enabling genealogical continuity within species, it is desirable for the present purpose to stress those aspects of the genetic system that characterize its implications on persistence, preservation of identity, and those interactions within biological species that differ from those between biological species. This suggests the following generalizing definition:

> *The term* **genetic system** *is applied to systems characterized by the capacity of (l) extinction and identical self-replication of their elements, (2) recombination (combinations of elements behave temporarily as a single element with respect to self-replication), (3) modification of their elements, (4) regulatory and evolutionary adaptation. A genetic system is thus a special kind of adaptive system.*

Both in Darlington's and in the above sense, the genetic systems of individuals may consist of many elements (genes), the number of relations among which may be extremely large and may be of structural (arrangement on chromosomes and the possibility of recombination, degree of ploidy, assignment to cell compartments, etc.) and functional (regulation of gene activity, dominance, epistasy, polygeny, pleiotropy, etc.) kind. The forces combining the elements, such that their various relations can be realized act on the population level and comprise all phenomena which are at least directly connected with reproduction (sexual or asexual mode of reproduction, sexual system, mating system, etc.). Consequently, major components of the reproductive and migration system are part of the genetic system of a population or species.

Clearly, biological systems above the species level, such as biocoenoses, cannot be considered as genetic systems primarily because their elements, the biological species, lack the capacity for recombination among each other, as was emphasized in the Introduction. Moreover, concerning requirement (1) in connection with populations or species as system elements, the extinction-replication cycle has no analogue without either resorting to the individual members as elements or assuming a metapopulation type of reproduction. Yet such systems, including ecosystems, can be viewed as adaptive systems with self-replication restricted to their biotic elements, with modification of elements, and with regulatory and evolutionary adaptation. Changes in element composition caused by evolutionary adaptation may, for example, concern the companion or ecologically sub dominant species but must not concern the characteristic species of ecosystems. However, the significance of companions or subdominants for preservation of the system identity (*i.e.* for evolutionary adaptation) is limited to the degree to which they assist the characteristic species and not the other way round.

RELEVANT CHARACTERISTICS OF GLOBAL ENVIRONMENTAL CHANGES

Returning to the particular concern of the present topic, at least some of the most apparent characteristics that distinguish "global" from other kinds of environmental changes ought to be discussed with reference to implications for the conservation of genetic resources. At the outset, however, it might be useful to emphasize that under global changes the *opportunities for abiotic and biotic environmental regulation* (melioration, pest control, etc.) are very likely to decline drastically, so that habitat and ecosystem protection or the declaration of reserves alone may not suffice as measures for conservation of biological resources.

Global changes may result in *globally new conditions or spatial shifts of existing local conditions*, so that fundamental ecological niches of species may, for example, move towards higher latitudes or altitudes as average temperature increases globally. Genetic variation for the adaptation to globally new conditions can be expected to exist within species if, for example, the novelty of the conditions consists in a *new combination of pre-existing factors*. Adaptation to this situation could then occur through the selection of new combinations of pre-existing genes, provided the recombination system allows for such combinations. Since recombination is inhibited between species, potential for adaptation to such environmental conditions is unlikely to exist above the species level, such as in biocoenoses. It would in fact require introduction of a new species to allow for such adaptation, and this could lead to a loss of identity of this biocoenosis by initiating successional processes, as was argued above.

For example, one isoenzyme may show comparatively high sensitivity to temperature but little sensitivity to ultraviolet radiation, while another isoenzyme may show the opposite; both enzymes may or may not participate in the same metabolic process. If there existed allelic variation within a species at two corresponding isoenzyme loci, and if a change from originally predominant coolness and low radiation to warmth and high radiation or any other combination of conditions occurred, the genetic prerequisites for the required adaptation could be mobilized through recombination.

Previously disadvantageous and therefore rare allelic variants maintained by immigration or recurrent mutation might become important under *new conditions that cannot be derived by combination or modification of pre-existing conditions*. The same type of argument of course holds above the species level where, however, no equivalent to recurrent mutation exists, since this would require recurrent emergence of particular species from within the biocoenosis. Again, as compared to the situation within species, the evolutionarily adaptive potential above the species level appears to be reduced.

Even if complete reversal of a global change were possible, a species may find it difficult to recolonize its initial local habitats. The *spatial expanse* of the change implies that a species' populations previously distributed within regular migration distances no longer exist at these locations, so that successful replacement of local extinctions by neighbouring conspecific populations, as is otherwise a normal consequence of metapopulation structures, may not take place. Hence, even *in situ* methods of dynamic conservation, aiming at the retention of previous states of adaptedness, may not provide a basis for local recolonization and might even be dangerous.

Another aspect frequently associated with global change is the possibility of continuing local changes, *i.e.* the environmental dynamics is non-recurrent (non-cyclical, in particular) with respect to the original local conditions. In this case, static conservation of a collection could only make sense if the environmental change would cause a *spatial shift* rather than global extinction of this collection's niche. More generally, static conservation can therefore be useful if the endangered niche of a collection is or will be available at some other place or if it can be predicted to be restored within the time over which the method of conservation is effective. Since creation of a local niche, with all its relevant abiotic and biotic components, at some other place as well as restoration at the same place are highly unlikely under global changes, the utility of static conservation could be restricted to the situation described in the Introduction (emergency measure, providing the basis for subsequent steps of restoration of adaptability).

As current climate data confirm, at least the initial phases of the global change are very likely to be characterized by *enhanced local fluctuations in environmental conditions* (see *e.g.* Scholz 1993). Considering the discrepancy between the generation time of trees and the speed of these changes, the regulatory gain priority over the evolutionary adaptive capacities. Yet, without sufficient evolutionary potential, genotypes with large regulatory plasticity may be missing from the extant genotypic structure and may not be producible by recombination during sexual reproduction.

Summarizing the evidence presented so far it appears that:

under the limited opportunities for environmental regulation permitted by global changes, restoring, preserving and, if necessary, enlarging the adaptability within populations and species is the most efficient means of stabilizing identities both below and above the species level and thus in ensuring the success of measures of species, habitat, or ecosystem protection.

CONSEQUENCES FOR CONSERVATION PROGRAMMES

The reasoning presented so far demonstrates that, ideally, conservation programmes can be viewed to consist of two successive steps:

1. Collection of information about the evolutionarily adaptive potential inherent in the presumably endangered population, and
2. Based on this knowledge, application of measures for preservation, restoration, or enlargement of evolutionarily adaptive potentials, if necessary.

As to the first step, *conservation measures become urgent if non-regular population decline is observed in the absence of any apparent sign of evolutionary adaptation*. Surveys of the dynamics of population density or size and of genetic structures therefore provide the fundamental information on which conservation decisions are to be based. On the other hand, if evolutionary adaptation is in progress and is accompanied by sizable population decline, as might be necessitated by the replacement of prevalent by hitherto rare adaptations, the risk of loss of non-adaptive genetic variation and therewith adaptability as a consequence of small population size is immanent. In this case a meaningful second step of the conservation programme might consist in slowing down the adaptational process by environmental regulation (melioration) in order to enable recovery of population size. Yet, the applied melioration techniques should not hinder the current nor initiate new adaptational processes.

Moreover, since habitat characteristics are highly likely to be matched by corresponding adaptations, preservation of a population's identity may be effectively achieved by preservation of its habitat characteristics. This idea underlies the common practice of resource protection. Yet, since a population's habitat is always imbedded in an ecosystem, protection of the ecosystem may in many cases be imperative for protection of habitats and for preservation of their characteristics (Gregorius 1991). By this, adaptations to local environmental factors not affected by the global change can be protected, and further impairment of adaptability to the altered conditions is prevented.

Recognition of adaptive processes

The above reasoning behind the first step in a conservation programme proceeds from the idea that adapted genetic information must prevail in order to keep the current genetic load below the threshold endangering population persistence. Adapted genetic information can be referred to as the *operating genetic potential* (Bergmann *et al.* 1990), and it includes the major contributions of regulatory adaptation to the state of adaptedness of the population. Currently adaptively inferior genetic variants, frequently referred to as the *latent genetic potential* (see Stebbins and Hartl, 1988, or Bergmann *et al.*, 1990), constitute the current genetic load and are therefore rare. The latent genetic potential thus comprises the evolutionary potential for adaptation to new environmental conditions. As pointed out by Gregorius and Bergmann (1993), the separability between operating and latent genetic potential suggested by the genetic load argument is very likely to produce characteristic genetic frequency profiles at the concerned loci. These profiles were called *concentrated profiles* by the authors, and they are distinguished by the lack of genetic types with intermediate frequency.

Based on this idea, two kinds of adaptive environmental changes may be recognized, one leading to new conditions which call for the hitherto latent genetic potential, and another that extends or amplifies extant conditions and by this makes additional use of the currently operating genetic potential. Hence, evolutionary adaptation to a changing environment is expected to imply for the concerned trait that either:

- *prevalent genetic types decline in relative frequency, while rare genetic types (hitherto latent potential) increase in both relative and absolute frequency*, or

• *prevalent genetic types increase in both relative and absolute frequency while rare genetic types decline in relative frequency.*

The second situation should lead to even higher degrees of concentration of the genetic profile, while the opposite (lowered concentration, types at intermediate frequency) should result from the first situation. As pointed out earlier, a selective increase of genetic types in proportion but not in absolute number is by itself not a sufficient observation to infer adaptation, since it does not rule out the possibility of continuing population decline and thus of extinction. On the other hand, selection favouring rare types at the expense of prevalent types may be accompanied by population growth, in which case adaptation to conditions can be assumed that allow for expansion or higher density of the population. This is the type of observation on which experimental population genetic studies must focus when trying to reveal evolutionary adaptation, and it requires information on temporal dynamics in both relative frequencies of genetic types and population density or size.

Enlargement of evolutionarily adaptive potentials

In cases where the extant adaptive potential is suspected to be insufficient, measures of enlargement of the evolutionarily adaptive basis, by the introduction of new genetic variation, should consider the requirement for recombination between resident and immigrant genetic information in order to allow for the evolution of new co-adapted gene complexes and thus for integration into a population's genetic system. The emergence of large numbers of maladapted hybrid individuals could be prevented by selecting populations existing under similar ecological conditions but showing sufficient genetic differentiation.

Another opportunity for the recruitment of enlarged adaptive potentials might be found in the inclusion of populations from zones of introgression into conservation programmes. However, since this affects the identity with respect to both genetic composition and genealogical continuity, as demonstrated above, the loss of identity must be tolerable in order to secure persistence of the remaining genome or at least general species characteristics. Such a decision could be inevitable if the adaptive norm of a population (race, subspecies, etc.) is known to be exceeded by the environmental change and if the hybrid population can be expected to combine characteristics that aid the stability of the concerned ecosystem (as would apply to hybrids between ecologically dominant species).

Yet, in less extreme cases, methods of mobilization of a population's own adaptive potential should probably be given priority. These methods should aim at promoting and strengthening characteristics of the environment and the genetic system that may enhance evolutionary speed and genetic variation without implying drastic reduction in population size. Examples are provided by methods of increasing overall fertility, shortening generations by shifting the age distribution in iteroparous populations towards younger age classes, or increasing recombinant frequency by promoting cross-fertilization or mobility (including pollen and seed dispersal).

When applying undirected measures for enlargement of evolutionarily adaptive potentials, it is of utmost importance to consider the possibility of adaptive load resulting from the emergence of adaptively disadvantageous genetic variants. As was emphasized in the last subsection, an obvious minimum condition for keeping adaptive load within reasonable bounds consists in sufficient representation of the currently operating adaptive potential, which implies characteristic genetic frequency profiles (see Gregorius and Bergmann 1993). This puts emphasis on the fact that *evolutionary adaptability is determined by both the presence and the distribution of evolutionarily adaptive potential.*

Suitability of gene markers

Gene markers employed in the planning of conservation programmes need to be evaluated with respect to their adaptive significance. This became evident in the introductory discussion of the potential danger resulting from the inclusion of genetic information on adaptive versus adaptively neutral traits into the specification of a genetic resource's identity.

There is strong evidence that the genetic frequency profiles observable at a number of **isoenzyme loci** are the result of an evolutionary and functional optimization, as is suggested for example by the presence of the same prevalent allelic variants across environmentally strongly differentiated locations and even across species (Gregorius and Bergmann 1993). These profiles are signs of stable states of adaptedness. However, other enzyme loci frequently show clinal variation for their prevalent alleles, thus indicating their adaptivity (concerning forest trees, see *e.g.* the papers by Stutz and Mitton, 1988, Mitton *et al.*, 1989, or Bergmann and Gregorius, 1993).

On the other hand, such **DNA traits**, the variability of which cannot directly be associated with known function (RAPD's and RFLP's not defined by gene probes, for example), may be inappropriate, since they rarely allow assessment of their adaptive significance. The high degrees of polymorphism frequently observed for these traits better qualify them as useful tools for studies of reconstruction of descent. However, considering the possibility of moderate polymorphism and stochastic association with the operating genetic background, it cannot be ruled out that under special conditions and by application of more sophisticated methods of analysis, such traits (as well as other currently adaptively neutral traits) may be useful indicators of adaptation or the existence of adaptive potentials.

Gene loci involved in the regulation of gene activity (**regulatory genes**) are generally believed to form the class of genetic function which is most frequently engaged in adaptation. Yet, regulatory genes are difficult to identify both directly by detecting their regulating agents and indirectly by the response of a regulated gene (even methods of genetic analysis of regulatory genes inferred from observations on regulated genes are not well developed). Non-constitutive (adaptive) enzymes (such as certain peroxidases) could be candidates for either regulating or regulated gene markers.

Inferring adaptive potential from observable phenotypic and genetic variation

From a practical point of view, probably the most promising method of revealing evolutionarily adaptive potentials results from the observation of *traits attesting states of adaptedness*, such as stress response traits or symptoms. The emergence of stress symptoms, for example, can be considered as a reliable indicator of environmental changes which call for adaptation (consult Hoffmann and Parsons 1991, pp. 220ff, for example, concerning the significance of stress response traits for conservation in the face of global changes). Hence, the existence of evolutionarily adaptive potential for the stressful conditions (stressors) can be suspected if the group of individuals showing the stress symptoms (sensitive group) is genetically differentiated from the group showing no such symptoms (tolerant group). Conversely, the absence of genetic differences between the two groups signifies lack of adaptive potential, in which case regulatory adaptation remains as the last option, if there is any, for adjustment to the stressful conditions.

More precisely, assuming that stress reduces fitness, it can be expected that a genotype or gene occurring more frequently in the tolerant than in the sensitive group is likely to increase in overall relative frequency over the generations. Hence, if genes that are rare in the total population are more frequent in the tolerant than in the sensitive group, and if the reverse were true for prevalent genes, this could hint at adaptation taking place to the changed conditions causing the stress symptoms, as was argued above. A different type of information can be obtained from distributions of stress symptoms among carriers of a specified genetic type. The genetic type with the higher proportion of tolerant individuals way be addressed as "more tolerant" in the sense that the environment provides a higher proportion of individuals with conditions to which they react tolerantly. Yet, in order to derive this information from the distribution of genetic types within both phenotypic groups, one additionally requires knowledge on the proportions of the tolerant and sensitive groups in the total population.

Clearly, these considerations are based on the absence of genotype-environment associations, since these could feign genetic effects on stress responses even though the variation in these responses is solely determined by environmental variation. However, particularly for sessile organisms such as trees, the assumption of independent distribution of genotypes over environments is quite unrealistic in natural populations because of locally differential selection and non-random distribution of seeds over the habitat. To account for this possibility, methods of *structured sampling* of individuals such as the methods of "pairwise sampling"

(Gregorius 1989) need to be developed further. Moreover, the significance of genetic markers as tools for revealing adaptive processes and potentials need to be analyzed in more detail with emphasis on the mode according to which the above-described characteristics of evolutionary adaptation are transferred to adaptively neutral gene loci via stochastic association.

An application

Finally, to demonstrate the above considerations, a brief analysis of the data on genetic effects of air pollution in spruce populations summarized by Bergmann and Scholz (1989) shall be attempted. The authors investigated four stands of Norway spruce from different sites characterized by high deposition of air pollutants. The possibility of genetic effects on their sensitivity scores ("tolerant" and "sensitive" based on observations on needle damage) was analyzed with respect to the differences between tolerant and sensitive individuals found at four out of 14 isoenzyme gene loci known for their involvement in biochemical resistance mechanisms or pollutant-sensitive metabolic pathways. To account for the requirement of independent genotype-environment association, the authors applied the method of pairwise sampling of tolerant and sensitive individuals in each of the stands. Table 2 presents the observed allelic and heterozygote (for the major polymorphic locus) frequencies.

Even though the likelihood ratio test produced no statistically significant differences for the allelic frequencies between the groups of tolerant and sensitive individuals, there is consistency in ranking between the two groups for their prevalent allele across all stands (and thus across variable genetic backgrounds) at each of the minor polymorphic loci G6PDH-A, GRD-A, and PEPCA-A. This accords with the predicted sensitivity of the enzymes to the pollutants. At two of the loci (GRD-A, PEPCA-A) the respective prevalent allele has decreased in the tolerant as compared with the sensitive group, which, by the above reasoning, could hint at the presence of adaptive potential among the rare alleles.

Table 2. Genetic frequency distributions at four enzyme gene loci among trees sensitive (S) and tolerant (T) for air pollution in each of four stands of Norway spruce in the Harz and Fichtel Mts. (data taken from Tables 5 and 6 in Bergmann and Scholz, 1989)

Enzyme	Sonnenberg		Bruchberg		Farrenleite		Seehaus	
gene locus	S	T	S	T	S	T	S	T
GOT-B1	-	.02	.02	-	-	.02	-	.02
-B2	.64	.60	.46	.48	.60	.64	.51	.52
-B3	.36	.38	.52	.52	.40	.34	.49	.46
heterozygotes	.40	.56	.36	.56	.54	.64	.40	.69
G6PDH-A1	.10	.04	.16	.10	.14	.06	.14	.07
-A2	.90	.96	.84	.90	.86	.92	.84	.90
-A3	-	-	-	-	-	.02	.02	.03
GRD-A1	.04	.08	-	.02	.06	.03	.04	.01
-A2	.90	.82	.82	.80	.83	.74	.84	.73
-A3	.06	.10	.18	.18	.11	.23	.12	.26
PEPCA-A1	.90	.84	.84	.84	.85	.82	.86	.76
-A2	.10	.16	.16	.16	.15	.18	.14	.24

The opposite applies to the other minor polymorphic locus (G6PDH-A), where the previously prevalent allele becomes even more frequent, so that the latent adaptive potential inherent in the rare alleles could be rapidly lost. This could analogously be interpreted as a sign of adaptation without utilization of new potential, such as would occur if the present state of pollution merely amplified previously acting effects on this locus. However, since information on absolute frequencies is not available, one could also argue that the pollution constitutes an extreme environmental state, which merely intensifies previously existing disadvantages to the rare alleles. In the light of this intuitively appealing argument, the increase of rare alleles at the loci GRD-A and PEP CA-A becomes even more interesting, since it hints at the possibility that genetic types showing inferior adaptedness or maladaptation under normal conditions become adaptively superior under extreme conditions.

The fourth locus (GOT-B), which is major polymorphic for two alleles, does not show any consistent ranking in frequency of the two prevalent alleles between the sensitive and tolerant group. This is different for the heterozygotes, which are consistently more frequent among tolerant than among sensitive individuals. Since this implies that heterozygotes are more tolerant than homozygotes (see Appendix*), balancing selection with its associated persisting adaptive load could result, and this could, in turn, fix a state of reduced adaptedness for the population until the stressful conditions decline. Again, an explanation could be found in the intensification of previously existing disadvantages to both homozygotes caused by stress. The advantage, however, lies in the preservation of genic variation and thus of adaptive potential.

Probably the most interesting aspect of these results lies in the apparent diversity of adaptive reactions observable at only a few gene loci, which hints at the possibility of a highly complex action of air pollution. Whether the complexity is a primary consequence of the comparatively extensive distribution of this environmental change cannot be decided on the basis of the present data and the available methods of analysis. However, a better assessment of the portions of the genome indicating the existence of evolutionarily adaptive potential can be expected from information from increasing numbers of gene loci, the products of which are putatively affected by the stressful environmental changes (thus continuing the strategy followed in the above-cited work of Bergmann and Scholz).

---

* Note that the result in the Appendix proves superiority of the heterozygote over the set of all homozygotes, which does not in all cases imply superiority of the heterozygote to each of its two corresponding homozygotes.

REFERENCES

Bergmann F, Gregorius H-R, Larsen JB (1990) Levels of genetic variation in European silver fir (*Abies alba*) - Are they related to the species' decline? Genetica 82:1-10

Bergmann F, Gregorius H-R (1993) Ecogeographical distribution and thermostability of isocitrate dehydrogenase (IDH) alloenzymes in European silver fir (*Abies alba*). Biochemical Systematics and Ecology 21: 597-605

Bergmann F, Scholz F (1989) Selection effects of air pollution in Norway spruce (*Picea abies*) populations. Pages 143-160 *in* Scholz F, Gregorius H-R, Rudin D (eds). Genetic Effects of Air Pollutants in Forest Tree Populations. Springer-Verlag, Berlin

Darlington CD (1939) The Evolution of Genetic Systems. University Press, Cambridge

Dobzhansky T (1968) Adaptedness and fitness. Pages 109-121 *in* Lewontin RC (ed). Population Biology and Evolution. Syracuse University Press, Syracuse

Gregorius H-R (1989) The attribution of phenotypic variation to genetic or environmental variation in ecological studies. Pages 3-15 *in* Scholz F, Gregorius H-R, Rudin D (eds). Genetic Effects of Air Pollutants in Forest Tree Populations. Springer-Verlag, Berlin

Gregorius H-R (1991) Gene conservation and the preservation of adaptability. Pages 31-47 *in* Seitz A, Loeschcke V (eds). Species Conservation: A Population-Biological Approach. Birkhauser Verlag, Basel, Boston, Berlin

Gregorius H-R, Bergmann F (1993) Analysis of isoenzyme genetic profiles observed in forest tree populations. Submitted

Hoffmann AA, Parsons PA (1991) Evolutionary Genetics and Environmental Stress. Oxford University Press, Oxford

Lande R (1988) Genetics and demography in biological conservation. Science 241:1455-1460

Levins R (1968) Evolution in Changing Environments. Princeton University Press, Princeton

Miller PR (1993) Response of forests to ozone in a changing atmospheric environment. Angewandte Botanik 67: 42-46

Mitton JB, Stutz HP, Schuster WS, Shea KL (1989) Genotypic differentiation at PGM in Engelmann spruce from wet and dry sites. Silvae Genetica 38: 217-221

Rieger R, Michaelis A, Green MM (1991) Glossary of Genetics. 5th edition, Springer Verlag, Berlin

Scholz F (1993) Anforderungen an die forstliche Forschung aufgrund der prognostizierten Klimaveranderungen. Allgemeine Forst Zeitschrift 12: 592-595

Stutz HP, Mitton JB (1988) Genetic variation in Engelmann spruce associated with variation in soil moisture. Arctic and Alpine Research 20: 461-465

Stebbins GL, Hartl DL (1988) Comparative evolution: Latent potential for anagenetic advance. Proceedings of the National Academy of Sciences, USA, 85, 5141-5145

Templeton AR (1991) Genetics and conservation biology. Pages 15-29 *in* Seitz A, Loeschcke V (eds). Species Conservation: A Population-biological Approach. Birkhauser Verlag, Basel, Boston, Berlin

Wilkes G (1992) Strategies for Sustaining Crop Germplasm Preservation, Enhancement, and Use. Issues in Agriculture, No. 5, published by the CGIAR

Ziehe M, Gregorius H-R, Glock H, Hattemer HH, Herzog S (1989) Gene resources and gene conservation in forest trees: General concepts. Pages 173-185 *in* Scholz F, Gregorius H-R, Rudin D (eds.) Genetic Effects of Air Pollutants in Forest Tree Populations. Springer Verlag, Berlin

APPENDIX

To see that a higher frequency of heterozygotes among tolerant than among sensitive individuals implies that heterozygotes are more tolerant than homozygotes, let R denote a response variable with $R=t$ for tolerant and $R=s$ for sensitive individuals, and let G analogously denote a variable specifying the genotype so that $G=hom$, for example, refers to all homozygous individuals. With the usual notation $P(\ldots)$ for probabilities and $P(\ldots/\ldots)$ for conditional probabilities, the above verbally formulated question translates into "does $P(G=het/R=t) > P(G=het/R=t)$ imply $P(R=t/G=het) > P(R=t/C\text{-}hom)$?" In fact, since $P(G=het)$ is an average of $P(G=het/R=t)$ and $P(G=het/R=t)$, the first inequality implies $P(G=het/R=t) > P(G=het)$, so that after multiplication with $P(R=t)$ one obtains $P(G=het, R=t) > P(C=het) \cdot P(R=t)$, from which $P(R=t/C=het) > P(R=t)$ follows. In the same manner one derives $P(G=hom, R=t) = P(R=t) - P(G=het, R=t) < P(R=t) - P(G=het) \cdot P(R=t) = P(R=t) \;\; P(G=hom)$, from which $P(R=t/G=hom) < P(R=t)$ and thus $P(R=t/G=hom) < P(R=t/G=het)$ follows.

Chapter 10: The genetic response of plant populations to a changing environment: the case for non-Mendelian processes

Tore Skrøppa and Øystein Johnsen
Norwegian Forest Research Institute
1432 Ås
Norway

INTRODUCTION

Classical Mendelian inheritance assumes the existence of chromosomal genes which are transferred from the parents to the next generation in a random fashion. In a diploid plant species, the zygote is derived from the fusion of two haploid gametes, one contributed by its maternal and one by its paternal parent. These gametes were formed after a random segregation during meiosis in each parent. The fertilization of the female (egg) by the male gamete (pollen) is likewise thought to be random. Therefore, when no internal or external factors are operating, the genetic composition of the progeny population can be described with statistical precision by the laws of probability theory. Fundamental principles are regular segregation and independent assortment between different pairs of alleles (Grant 1975). Based on these assumptions a whole body of population and quantitative genetic theory has been developed for plant population changes under the evolutionary forces of natural selection, mutation, migration and drift (*i.e.* Falconer 1989; Hedrick 1985). The models have been verified in a large set of observational and experimental data.

Over the years, evidence has accumulated that gene transmission may not always follow this classical theory (Grant 1975). Maternal effects and cytoplasmatic inheritance are well-known examples. Recently, it has been suggested that selective events occurring during the sexual reproductive phase (gametophytic selection, preferential matings, postzygotic selection) may be important factors in the evolution of higher plants (Mulcahy 1979; Ottaviano *et al.* 1988). A number of reports of such

NATO ASI Series, Vol. I 20
Biodiversity, Temperate Ecosystems, and Global Change
Edited by T. J. B. Boyle and C. E. B. Boyle

effects and the possible influence of environmental factors, suggest that these mechanisms also should be considered among the possible evolutionary forces, see Hormaza and Herraro (1992) for a recent review.

Molecular characterizations of the plant genome have shown that rapid genomic changes can take place, sometimes influenced by environmental stress factors or after hybridizations, see Cullis (1991) for a recent review. These mechanisms are still poorly characterized, but should be considered when the genetic performance of plant populations does not follow the expectations from the classical inheritance models.

This paper was motivated by our findings in Norway spruce that adaptive properties of progenies after controlled crosses seem to depend on the climatic conditions at the site where the crosses are performed. As we are not able to explain the observed effects by temporary physiological changes or give more convential genetic explanations, the non-Mendelian processes have to be considered. We will first give an overview of our results with Norway spruce. Then short presentations will be given of results related to selection mechanisms during the reproductive phase and of rapid genomic changes. Finally, our results with Norway spruce will be discussed relative to the non-Mendelian mechanisms. It has to be remembered that little hard evidence is available of such effects and their initiating factors. Parts of the suggestions may therefore be provocative and should be regarded as a basis for further discussions and research.

A STORY OF NORWAY SPRUCE (*Picea abies*)

Norway spruce (*Picea abies* (L.) Karst.) is the most widely planted conifer tree species in Europe. Seeds for practical plantations are collected in forest stands or in seed orchards after matings between clones that have been selected for superior phenotypic performance. Within the natural range of distribution a great number of climatic races (provenances) have developed (Schmidt-Vogt 1978). They show clinal variation along latitudinal and altitudinal gradients for traits that describe the annual growth cycle, in particular growth start and cessation and development of autumn frost hardiness. This variability is thought to reflect climatic adaptation to photoperiod and temperature (Ekberg *et al.* 1979).

One of the constraints of seed production in Norway spruce orchards is a scarce flowering, in particular at northern latitudes and high altitudes. Seed orchards are therefore generally located in a warmer climate to enhance flowering. In Norway, seeds for the northern and central part of the country (lat. 63-67°N) are produced in an orchard located in the southern part at lat. 58°N, and seeds for high altudes are produced in orchards near the sea level. The clonal materials in these seed orchards consist of grafts of phenotypically selected trees from native stands in the same area where the seedlings from the orchard seed are to be planted.

When seedlings from seed produced in the southern seed orchard were cultivated in nurseries in the northern environment, it was discovered that they did not have the expected adaptive properties. In particular, they had a prolonged growth season in late summer and set terminal buds considerably later than seedlings from stand seed of the local provenance. Their growth rhythm was similar to that of a more southern Norway spruce provenance. These results were confirmed in experiments comparing controlled cross families in the seed orchard with their half-sibs after open pollination in the native stand, after cone collection on the selected plus trees (Bjørnstad 1981). Large differences were observed within half-sib family pairs for a number of traits, including autumn frost hardiness measured in artificial freezing tests. The differences persisted even after a clonal propagation with rooted cuttings (Johnsen 1989a). The hypothesis that the observed differences were an effect of the phenotypic plus tree selection had to be rejected as progenies of the plus trees did not perform differently from progenies of non-selected trees from the same stands (Johnsen and Østreng 1993).

In the first experiments, mixtures of pollen from seed orchard clones were used in the controlled crosses, while the northern families were from open pollation in the stand. The male parents were therefore different, even if they were of the same and northern provenance. To have better comparisons, the same controlled crosses were performed to produce identical full-sib families under different climatic conditions in Norway spruce seed orchards in southern and northern Finland. In eight out of nine full-sib family pairs, the seedlings from the southern cross were more damaged in freezing tests than their full-sibs from the southern locality (Skrøppa *et al.* submitted).

Identical single crosses were also performed on potted grafts in a greenhouse, and on the same clones in a nearby seed orchard (Johnsen *et al.* submitted). The progenies from the greenhouse crosses had significantly lower autumn frost hardiness in freezing tests than their full-sibs from the outdoor cross in the seed orchard.

Similar effects have been observed on progenies after crosses performed at low elevation sites between spruce clones originating from high altitudes (Johnsen 1988). Seedlings from controlled crosses had an extended shoot elongation period at ages nine and ten years compared with trees from high altitude provenances and had a lower autumn frost hardiness than comparable seedlings from natural stands (Skrøppa submitted).

Another study has shown that seedlings from bulked seedlots from the two years 1987 and 1989 in the southern seed orchard where these effects first were detected, differ considerably in their response to photoperiodic treatments (Kohmann and Johnsen submitted). The seedlings from the 1987 seed harvest responded faster to a reduced photoperiod and had an earlier bud-set than seedlings from 1989 seed. The climate in the winter and spring of these two years differed considerably. In 1987, the winter was cold, spring was late and flowering in spruce occurred 17 days later than in 1989, which had a very mild winter and an early flowering.

In Scots pine (*Pinus sylvestris* (L.)), similar results were observed, but were first thought to be temporary (Dormling and Johnsen 1992). Recent results, however, show that both mortality and growth in field tests seem to some extent to be dependent of the locality where crosses were performed, with the southern locality producing progenies with highest mortality and best growth performance (Lindgren and Wei submitted).

All the presented results show differences in phenotypic traits between identical or related controlled cross families after sexual reproduction at localities with apparently different climatic conditions. No relationships have been found to seed traits (weights) which may have a short-term physiological influence on early seedling performance. The observed differences are similar to the provenance variation in the same traits between northern and southern or high and low altitude provenances. They persist when seedlings of southern seed orchard seed are grown in their native northern environment. The effects are endurable (last for more than ten years) and persist under different environmental conditions. It is therefore less likely that they are non-genetic physiological effects.

A common factor of the observed results is that the sexual reproductive process has taken place under changed climatic conditions compared with the climates at the native origins of the parental genotypes. Both a north to south and a high to low altitude transfer imply an earlier flowering under changed climatic conditions. In the greenhouse, the pollinations were performed three weeks earlier than in the seed orchard, and the interval between pollination and fertilization was shorter. The differences in timing and climatic conditions relate to meiosis, pollination and fertilization. Is it possible that environmental influences on these processes may have genetic implications ? So far it has not been proven that the observed effects are due to genetic factors or that they are heritable. If they are, then clearly non-Mendelian processes must be involved. As many examples of such process recently have been reported in literature, this possibility should be investigated.

Selection during sexual reproduction

Flowering plants generally have two life phases, the gametophytic or haploid and the sporophytic or diploid phase. The haploid phase occurs during the reproductive cycle which starts with floral initiation and continues with the processes of meiosis, pollination, fertilization and seed development. At meiosis, the haploid male and female gametes are formed, and the haploid phase lasts until their fusion during fertilization. Its duration can vary considerably both between species and also between the two sexual types within the same species. As examples, the interval between pollination and fertilization can be as long as 14 months in some pine species (Owens and Blake 1985), while it lasts only 48 hours in tomato and 30 minutes in some species of chrysanthemum (Zamir 1983). These large differences are thought to be associated with the degree of maturity of the male gametes at the time of their release (Zamir 1983).

Disturbances during meiosis may result in unequal representation of gametic types from heterozygeous parents (preferential segregation or meiotic drive).

Although such cases are well known in plants (Grant 1975), little is known of environmental causes of the irregularities and their evolutionary significance. In most cases, deviations from Mendelian segregation ratios are not observable at this stage and can not be distinguished from segregation distortions due to gametic selection (Gillet and Gregorius 1992). However, the quantity of male gametes produced may to a large extent depend on environmental factors such as temperature and photoperiod. In a phytotron experiment, four Norway spruce clones produced pollen under four treatments, combining a low and a high temperature and short and long day (Johnsen and Skrøppa unpublished). The clones produced between 50 and 100% larger volumes of pollen at the low compared with the high temperature. A positive effect on pollen production was also observed from the long day treatment. Whether these differences also have genetic implications is not known, but is being tested.

Hormaza and Herrero (1992) have recently reviewed the subject of pollen selection and have discussed its consequences and possible utilization in plant breeding. Under intense pollen competition both non-Mendelian segregation ratios and a decrease in the variability of the progeny generation indicate that such selection indeed may take place, even if some data have been susceptible to alternative interpretations (Charlesworth 1988). The objections point to the difficulty of distinguishing between haploid gametophytic and post-zygotic diploid selection since observations in many cases are measures only of seed or progeny quality. Effects of seed size on progeny fitness may also be mistaken to be caused by pollen competition (Charlesworth 1988).

Pollen can be considered an independent organism expressing its own genetic information and is more than a transmission vector for the genome during the hapoid life cycle (Hormaza and Herrero 1992). Several studies have shown that a large portion of the genome of the male gametophyte is transcribed and translated during pollen development, germination and tube growth (Mascarenhas 1989). Many of these genes are expressed both in the gametophytic and sporophytic life stage. In tomato, more than 60% of isozymes studied in the sporophyte were also expressed in the male gametophyte (Tanksley *et al.* 1981). Similarly, in maize 72% and in barley 60% of the isozymes were expressed in both the male gametophytic and sporophytic tissue

(Sari-Gorla *et al.* 1986; Pedersen *et al.* 1987). Even larger percentages of overlap have been reported for three poplar species (Rajora and Zsuffa 1986) and in apple (Weeden 1986).

Two pollen traits are thought to be of major importance in pollen competition: its germinative ability and pollen tube growth. Variation between pollen donors in these traits has been demonstrated for several species (*e.g.* Marshall and Ellstrand 1986; Snow and Spira 1991). Several studies have also demonstrated a positive relationship between the pollen tube growth and sporophytic traits, *e.g.* in maize (Ottaviano *et al.* 1982), while no relationships have been found in other cases of suspected pollen selection, *e.g.* in *Silene diocia* and *Mimulus guttatus* (Searcy and Mulcahy 1985a). Two levels of pollen competition are thought operating (Hormaza and Herrero 1992); either as a direct competition among the haploid gametophytes or through interactions between the haploid male gametophyte and the diploid female tissue.

Selection at the haploid phase as a response to an environmental factor will influence the sporophytic population if there is a relationship between the phenotypic responses of the two phases. Several studies have shown that such relationships may exist, both under natural conditions and in controlled experiments. Table 1 lists a number of studies which have shown differential male gametophytic responses to environmental stress factors. In several of these examples a correlation has been observed between the behaviour of the gametophytic and sporophytic phases. In tomato, progeny from crosses made at low temperatures had a higher rate of root elongation than those from crosses at normal temperature (Zamir and Gadish 1987). Sacher *et al.* (1983) found that plants (*Lycopersicon* x *Solanum*) after fertilization under salt stress had both an increased resistance to salinity and a higher mean vigor. In walnut (*Juglans regia*), Polito *et al.* (1991) found that the optimum temperature for pollen germination increased with treatment temperatures given during pollen development, while Luza *et al.* (1987) showed in the same species that late-blooming clones showed higher temperature optima for pollen germination.

Table 1. Observed cases of male gametophytic responses to environmental factors

Environmental factor	Species	Reference
Temperature	*Lycopersicon* sp.	Zamir *et al.* 1981
		Zamir *et al.* 1982
		Zamir and Vallejos 1983
		Zamir and Gadish 1987
	Zea maize	Frova *et al.* 1986
	Hordeum bulbosum	Schön *et al.* 1991
	Pennisetum typhoides	Robert *et al.* 1989
	Medicago sativa	Mulinix and Iezzoni 1988
	Juglans regia	Luza *et al.* 1987
	Juglans regia	Polito *et al.* 1991
Nutrients	*Raphanus raphanistrum*	Young and Stanton 1990
Heavy metals	*Silene dioica*	Searcy and Mulcahy 1985a,
	Mimulus guttatus	b,c
	Lycopersicon sculentum	Searcy and Mulchay 1990
Salinity	*Armeria maritima*	Eisikowitch and Woodell 1975
	Lycopersicon x *Solanum*	Sacher *et al.* 1983
Ozone	*Nicotiana tabacum*	Feder 1986
	Petunia hybrida	
Air pollutants	*Populus* sp.	Venne *et al.* 1989
	Pinus sylvestris	
Acidity	*Betula* sp.	Cox 1989
	Pinus strobus	
Toxic componds	*Beta vulgaris*	Smith and Moser 1985
	Brassica napus	Hodgkin 1988
Herbicides	*Lolium rigidum*	Richter and Powles 1993

Characteristics of the male gametophytic phase are a very large population size and a direct expression of recessive traits. Selection during this phase will therefore theoretically be more efficient than sporophytic selection (Pfahler 1983). During meiosis in the ovule four megaspores are produced of which three degenerate. In conifer species, it is common that more than one egg per female gametophyte is fertilized (polyembryony) (Owens and Blake 1985), but only one will remain in the fully developed seed. Seed or fruit abortion is a common occurrence in many species (Stephenson 1981), in particular under resource limitation (Marshall and Ellstrand

1988). The mechanisms that regulate these processes are unknown, but selective influences of environmental stress factors are possible both during megaspore degeneration and embryo and seed abortion. It has been suggested that a possible selective maturation favouring higher-quality embryos might be more pronounced in unfavouble habitats (Casper 1988). One case of possible postzygotic selection was recently reported by Searcy and Macnair (1993), causing an increase in copper tolerant progeny when pollen recipients were grown with added copper. In conifers, the low seed set after self-pollinations is assumed to be caused by embryo abortions (Owens and Blake 1985). In several cases this has also been the explanation for observed differences in paternal success (*e.g.* El-Kassaby and Ritland 1992; Nakamura and Wheeler 1992). In other cases, it has not been possible to distinguish between pre- or postzygotic factors causing variations in male fertilities and segregation distortions after matings with mixtures of pollen from several paternal parents (Skrøppa and Lindgren, submitted). A conclusion is therefore that little information is available of selective effects of environmental factors at the postzygotic level during embryo and seed maturation.

The sexual reproductive process offers several possibilities for selective influences of external environmental factors. The observed relationships in genetic expressions between the gametophytic and sporophytic phases and their similar phenotypic responses to possible selective factors indicate that male gametophytic selection may be an important factor in the evolution of higher plants (Mulcahy 1979; Ottaviano and Mulcahy 1986; Ottaviano *et al.* 1988). The rate at which selection may proceeed at the haploid level makes this subject extremely interesting under rapidly changing environmental conditions, both for understanding the evolution of natural populations and for utilization in practical plant breeding. Similarly, possible selective environmental factors at the postzygotic level should be investigated.

Rapid genomic changes and epigenetic effects

It is well known that environmental stress factors generally will increase both recombination and mutation rates (Hoffmann and Parsons 1991). Even if most mutations are assumed to have deleterious effects, they may in one sense be called adaptive as they enable the organism to generate genetically variable offspring in response to environmental pressure (Hoffmann and Parsons 1991). Several authors have suggested that environmentally induced mutations may be directed, see references in Lenski and Mittler (1993). However, Lenski and Mittler (1993) find no support for the existence of directed mutations whose rates are dependent on their specific value to the organism.

Mendelian principles are violated in cases where the genome undergoes rapid reorganisations, as suggested by Walbot and Cullis (1985). Examples of such processes are chromosome rearrangements, chromosome imprinting, gene amplification and transposable elements (McClintock 1984; Walbot and Cullis 1985; Cullis 1990). Such processes may be induced by environmental stress, but it is unclear to what extent they are directed and repeatable or just random. The genomic changes occuring may be unstable and only produce a temporary phenotype without permanently affecting the genome. However, they may be transmitted to subsequent generations if occuring in generative cells or in somatic tissues which differentiate into generative cells. One example of genomic reorganiazation is the within-clone variability observed in tissue culture plants (Cullis 1991).

The most extensive studies of rapid DNA-changes have been made in flax (*Linum*) (Durrant 1958; Cullis 1987). In some varieties of this species, phenotypic and heritable changes were induced when plants had been growing one generation under some specific nutrient and temperature treatments. These changes can be stably inherited for several generations. They were observed for plant weight, height, DNA-content and isozymes. Both a nuclear and a cytoplasmatic component may control the response to the environment (Cullis 1987). The mechanisms responsible for thesc changes are unknown.

By genomic imprinting is meant that the gene can be reversably altered in its potential for expression. Two types of effects are considered (Matzke and Matzke 1993): i) the expression of a gene depends on the sex of the parent it was inherited from or ii) the gene expression is influenced by other genes as a result of an allelic or nonallelic gene interaction. Such effects are caused by mechanisms that modify DNA and/or chromatin structure. Little is known about such effects and the possible influences of environmental factors. Speculations can be made, however, about the possibility of certain genes being activated or deactivated, for shorter or longer periods of time, by environmental triggers during embryo development or earlier in the reproductive process. Such effects may have permanet influence in one generation, but the changed performance may not necessarily be transmitted to the next generation.

Genomic imprinting is an example of epigenetic effects which involve no changes in gene frequencies. In a rather speculative paper Jablonca *et al.* (1992) suggest that functional states in the parents can be transmitted to the offspring through the differentiation of generative cells into gametes. They base their arguments on evidence showing that chromatin structures, and not just DNA sequences are transmitted between generations, and on data indicating that some epigenetic variations can be passed from parents to offspring. The variations may be specifically induced by the environment and thus occur to a number of different individuals exposed to the same environmental condition, or they may be non-specific or random. However, as no conclusive evidence exists of the transmission of epigenetic variations between generations, this issue is still a speculative one.

The few examples presented above show that the molecular characterization of the plant genome and its responses to environmental changes are very complex. It is clear that both phenotypic and genetic variability may be generated by environmental stress factors, and that some forms of such variations may be inherited. The mechanisms involved, however, are not known.

DISCUSSION

In Norway spruce, the phenotypic performance of progenies depends to some extent on the climatic conditions at the site where reproduction has taken place. This phenomenon is observed for traits that characterize the annual growth cycle and therefore are strongly related to fitness. In particular the effects are pronounced in clones that are transferred from the climatic margins of the species. So far, we have too little information to draw conclusions about the underlying mechanisms.

Several characteristics of the northern spruce populations may be of importance for the observed effects. Norway spruce is in Norway a young species, on most localities less than 2,000 years old, and in the marginal areas considerably younger (Moe 1970). Sexual reproduction occurs very infrequently in populations at northern latitudes and high altitudes, mainly due to climatic stress factors. Anderssen (1965) found within population variability in tolerance to low temperatures and temperature changes during meiotic divisions. He therefore concluded that the northern Norway spruce populations are not well adapted in regard to the generative phase. Substantial genetic variability in adaptive traits is also present within populations (*e.g.* Johnsen and Østreng 1993). Mechanisms must therefore be present that keep up this variability. The species, like many other conifer species, seems to have the ability of fast adaptation to climatic conditions. Effects of both natural and artificial selection were observed in progenies among selected trees of southern origin in a Norway spruce provenance trial at a northern latitude (Johnsen and Skrøppa 1992; Skrøppa and Johnsen unpublished). Similar effects were observed in the same species after selection in a provenance trial in Canada (Boyle pers. comm.). In several species, seedlings from seeds harvested in first generation plantations of introduced provenances seem to be better adapted than direct introductions of the same provenances, *i.e.* Sitka spruce in Denmark (Nielsen 1993).

The sexual reproductive process in gymnosperms offers several possibilities for directed selection. However, the conditions for gametophytic selection are less favourable than in many angiosperm species as the limited size of the pollen chamber allows few pollen grains to germinate and compete during fertilization of the ovule. Pollen selection alone can therefore hardly account for the observed effects. Other possibilities are irregularities during male or female meiosis, initiated by climatic factors. However, as long as these processes are not well characterized, one can only speculate about their importance. It also remains to establish relationships between responses at the gametophytic and sporophytic level to the same selective factor.

Virtually no information is available about rapid genomic changes or epigenetic effects in spruce. It may be speculated that gene regulation or imprinting, caused by the activation of certain genes by environmental conditions during reproduction, regulate the phenotypic performance of certain traits. What is observed may therefore not be genetic differences in the Mendelian sense with changes in gene frequencies. Nevertheless, they are genetic or epigenetic effects which may last the whole life of the tree, and will have important implications for its survival and growth performance.

Nearly 30 years ago Rowe (1964) warned about environmental preconditioning of the succeeding generation caused by the ecological conditions experienced by the parent plants. In particular, he considered this problem in relation to seed production in grafted seed orchards. The observed effects in Norway spruce certainly have implications for the use of seed orchard seed and the future seed orchard programmes. However, if they are truely genetic, then they will also have large influences in the evolution of the species, in particular for the rate of evolutionary change. The consequences will also be large for gene conservation and for the expected performance of Norway spruce populations under climatic change.

This discussion shows that we know too little about the sexual reproductive process and how it is regulated by environmental factors. This may be the most sensible period in a plant's life and should therefore be investigated further. Our experimental species, Norway spruce, is an extremely difficult species for such studies due to its life length and long generation interval. The studies will be continued, however, due both to their practical and biological importance.

Acknowledgement

This research was supported by a grant from the Nordic Forest Research Co-operation Committee.

REFERENCES

Anderssen E (1965) Cone and seed studies in Norway spruce (*Picea abies* (L.) Karst.). Stud For Suec 23: 1-214

Bjørnstad Å (1981) Photoperiodical after-effect of parent plant environment in Norway spruce (*Picea abies* (L.) Karst.) seedlings. Medd Nor Inst Skogforsk 36: 1-30

Casper BB (1988) Evidence for selective embryo abortion in *Crytantha flava*. Am Nat 132: 318-326

Charlesworth D (1988) Evidence for pollen competition in plants and its relationship to progeny fitness: a comment. Am Nat 132: 298-302

Cox, RM (1989) Natural variation in sensitivity of reproductive processes in some boreal forest trees to acidity. Pages 77-88 *in* Scholz F, Gregorius HR, Rudin D (eds) Genetic effects of air pollutants in forest tree populations. Springer-Verlag, Berlin, Heidelberg, New York

Cullis CA (1987) The generation of somatic and heritable variation in response to stress Am Nat 130: S62-S73

Cullis CA (1990) DNA rearrangements in response to environmental stress. Advances in Genetics 28: 73-97

Cullis CA (1991) Molcular characterization of plant reponses to stress. Pages 245-264 *in* Taylor GE, Pitelka LF and Clegg MT (eds) Ecological genetics and air pollution. Springer-Verlag, New York, Berlin, Heidelberg

Dormling I, Johnsen Ø (1992) Effects of the parental environment on full-sib families of *Pinus sylvestris*. Can. J. For. Res. 22: 88-100

Durrant A (1958) Environmental conditioning of flax. Nature 181: 928-929

Eisikowitch D, Woodell SRJ (1975) Some aspects of pollination ecology of *Armeria maritima* in Britain. New Phyt 74: 307 -322

Ekberg I, Eriksson G, Dormling X (1979) Photoperiodic reactions in conifer species. Holarct Ecol 2: 255-263

El-Kassaby YA, Ritland K (1992) Frequency-dependent male reproductive success in a polycross of Douglas fir. Theor Appl Genet 83: 752-758

Falconer DS (1989) Introduction to quantitative gentics. 3rd ed. Longman Group

Feder WA (1986) Predicting species response to ozone using a pollen screen. Pages 89-94 *in* Mulcahy DL, Mulcahy GB, Ottaviano E (eds) Biotechnology and ecology of pollen. Springer-Verlag, New York

Frova C, Binelli G, Ottaviano E (1986) Male gametophyte response to high temperatures in maize. Pages 33-38 *in* Mulcahy DL, Mulcahy GB, Ottaviano E (eds) Biotechnology and ecology of pollen. Springer-Verlag, New York

Gillet E, Gregorius HJ (1992) What can be inferred from open-pollinated progenies about the source of observed segregation distortion? - A case study in *Castanea sativa* Mill. Silvae Genet 41:82-87

Grant V (1975) Genetics of flowering plants. Columbia University Press New York

Hedrick PW 1985 Genetics of populations. Jones and Bartlett Publishers Boston

Hodgkin T (1988) In vitro pollen selection in *Brassica napus* L. Pages 57-62 *in* Cresti M, Goro P, Pacini E (eds) Sexual reproduction in higher plants. Springer-Verlag, Berlin, Heidelberg, New York

Hoffman AA, Parsons PA (1991) Evolutionary Genetics and Environmental Stress. Oxford University Press, Oxford

Hormaza JI, Herrero M. (1992) Pollen selection. Theor Appl Genet 83: 663-672

Jablonka E, Lachmann M, Lamb MJ (1992) Evidence, mechanisms and models for the inheritance of acquired characters. J Theor Biol 158: 245-268

Johnsen Ø (1988) After-effects on progenies from seed orchard clones moved to non-native environments. Pages 1-11 *in* Worral J, Loo- Dinkins J, Lester DT: Physiology and genetics of reforestation. Proc of the Tenth North American Forest Biology Workshop, Vancouver BC

Johnsen Ø (1989a) Phenotypic changes in progenies of northern clones of *Picea abies* (L.) Karst. grown in a southern seed orchard. I. Frost hardiness in a phytotron environment. Scand J For Res 4: 317-330

Johnsen Ø, Skrøppa T (1992) Possible influence of natural and artificial selection on autumn frost hardiness in *Picea abies*. Medd Skogforsk 45.3

Johnsen Ø, Østreng G (1993) Effects of plus tree selection and seed orchard environment on progeies of *Picea abies*. Can J For Res. In press

Lenski RE, Mittler JE (1993) The directed mutation controversy and neo-Darwinism. Science 259: 188-194

Luza JG, Polito VS, Weinbaum SA (1987) Staminate bloom date and temperature responses of pollen germination and tube growth in two walnut (*Juglans*) species. Amer J Bot 74: 1898-1903

Matzke M, Matzke AJM (1993) Genomic inprinting in plants: Parental effects and trans-inactivation phenomena. Ann Rev Plant Physiol Plant Mol Biol 44: 53-76

Marshall, DL, Ellstrand NC (1988) Effective mate choice in wild raddish: evidence for selective seed abortion and its mechanism. Am Nat 131: 739-756

Mascarenas JP (1989) The male gametophyte of flowering plants. The Plant Cell 1:657-664

McClintock B (1984) The significance of reponses of the genome to challenge. Science 226: 792-801

Moe D (1970) The post-glacial immigration of *Picea abies* into Fennoscandia. Bot Notiser 123: 61-66

Mulcahy DL (1979) The rise of the angiosperms: A genecological factor. Science 206:20-23

Mulinix CA, Iezzoni AF (1988) Microgametic selelction in two alfalfa (*Medocago sativa* L.) clones. Theor Appl Genet 75: 917-922

Nakamura RK, Wheeler, NC (1992) Pollen competition and paternal success in Douglas-fir. Evolution 46: 846-851

Nielsen UB (1993) Genetic variation in sitka spruce (*Picea sitchensis* (Bong.) Carr.) regarding height growth, stem form and frost hardiness at the provenance, progeny and clonal level based on Danish field trials. Unpublished PhD. dissert. The Royal Veterinary and Agricultural University, Copenhagen

Ottaviano E, Mulcahy DL (1986) Pages 101-120 *in* Barigozzy C (ed) The origin and domestication of cultivated plants. Elsevier, Amsterdam

Ottaviano E, Sari-Gorla M, Frova C, Pè E (1988) Male gametophytic selection in higher plants. Pages 35-42 *in* Cresti M, Gori P, Pacini E (eds) Sexual reproduction in higher plants. Springer-Verlag, Berlin, Heidelberg, New York

Ottaviano E, Sari-Gorla M, Pè E (1982) Male gametophytic selection in maize. Theor Appl Genet 76:601-608

Ottaviano E, Sari-Gorla M, Villa M (1988) Pollen competitive ability in maize: within population variability and response to selection. Theor Appl Genet 76:601-608

Owens JN, Blake MD (1985) Forest tree seed production. Information Report PI-X-53 Petawawa National Forestry Institute Chalk River Ontario

Pedersen S, Simonsen V, Loeschcke V (1987) Overlap of gametophytic and sporophytic gene expression in barley. Theor Appl Genet 75: 200-206

Pfahler PL(1983) Comparative effectiveness of pollen genotype selection in higher plants. Pages 361-366 *in* Mulcahy DL, Ottaviano E (eds). Pollen: biology and implications for plant breeding. Elsevier, New York, Amsterdam, Oxford

Polito VS, Weinbaum SA, Muraoka TT (1991) Adaptive responses of walnut pollen germination to temperature during pollen development. J Amer Soc Hort Sci 116: 552-554

Rajora OP, Zsuffa L (1986) Sporophytic and gametophytic gene expression in *Populus deltoides* Marsh., *P. nigra* L., and *P. maximowiczii* Henry. Can J Gent Cytol 28: 476-482

Richter J, Powles SB (1993) Pollen expression of herbicide targe site resistance genes in annual ryegrass (*Lolium rigidum*). Plant Physiol 102: 1037-1041

Robert T, Sarr A, Pernès (1989) Sélections sur la phase haploide chez le Mil (*Pennisetum typhoides* (Burm.) Stapf et Hubb.): effet de la température. Genome 32: 946-952

Rowe, JS (1964) Environmental preconditioning with special reference ot forestry. Ecology 45: 399-403

Sacher RF, Mulcahy DL, Staples RC (1983) Developmental selection during self pollination of *Lycopersicon* x *Solanum* F_1 for salt tolerance to F_2. Pages 335-342 *in* Mulcahy DL, Ottaviano E (eds). Pollen: biology and implications for plant breeding. Elsevier, New York, Amsterdam, Oxford

Sari-Gorla M, Frova C, Binelli G, Ottaviano E (1986) The extent of gametophytic-sporophytic gene expression in maize. Theor Appl Genet 72:42-47

Schmidt-Vogt H (1978) Genetics of *Picea abies* (L.) Karst. Annales Forestales 7: 145-186

Schön CC, Hayes PM, Blake TK, Knapp SJ (1991) Gametophytic selection in a winter x spring barley cross. Genome 34: 918-922

Searcy KB, Macnair MR (1993) Developmental selection in response to the environmental conditions of the maternal parent in *Mimulus guttatus*. Evolution 47: 13-24

Searcy KB, Mulcahy DL (1985a) The parallel expression of metal tolerance in pollen and sporophytes of *Silene dioica* (L.) Clairv., *Silene alba* (Mill.) Krause and *Mimulus guttatus* DC. Theor Appl Genet 69:597-602

Searcy KB, Mulcahy DL (1985b) Pollen tube competition and selection for metal tolerance in *Silene dioica* (Caryophyllaceae) and *Mimulus guttatus* (Scrophulariaceae). Amer J Bot 72: 1695-1699

Searcy KB, Mulcahy DL (1985c) Pollen selection and the gametophytic expression of metal tolerance in *Silene dioica* (Caryophyllaceae) and *Mimulus guttatus* (Scrophulariaceae). Amer J Bot 72: 1700-1706

Searcy KB, Mulcahy DL (1990) Comparison of the response to aluminium toxicity in gametophyte and sporophyte of four tomato (*Lycopersicon esculentum* Mill.) cultivars. Theor Appl Genet 80: 289-295

Snow AA, Spira TP (1991) Pollen vigour and the potential for sexual selection in plants. Nature 352: 796-797

Smith GA, Moser HS (1985) Sporophytic-gametophytic herbicide tolerance in sugarbeet. Theor Appl Genet 71: 231-237

Stephenson AG (1981) Flower and fruit abortion: proximate causes and ultimate functions. Ann Rev Ecol Syst 12: 253-279

Tanksley SD, Zamir D, Rick CM (1981) Evidence for extensive overlap of sporophytic and gametophytic gene expression in *Lycopersicon esculentum*. Science 213: 453-455

Venne H, Scholz F, Vornweg A (1989) Effects of air pollutants on reproductive processes of poplar (*Populus* ssp.) and Scots pine (*Pinus sylvestris* L.). Pages 89-103 *in* Scholz F, Gregorius HR, Rudin D (eds) Genetic effects of air pollutants in forest tree populations. Springer-Verlag, Berlin, Heidelberg, New York

Walbot V, Cullis CA (1985) Rapid genomic changes in higher plants. Ann Rev Plant Phys 36: 367-396

Weeden NF (1986) Identification of duplicate loci and evidence for postmeiotic gene expression in pollen. Pages 9-14 *in* Mulcahy DL, Mulcahy GB, Ottaviano E (eds) Biotechnology and ecology of pollen. Springer-Verlag, New York

Young HJ, Stanton ML (1990) Influence of environmental quality on pollen competitive ability in wild radish. Science 248: 1631-1633

Zamir D (1983) Pollen gene expression and selection: applications in plant breeding. Pages 313-329 *in* Tanksley SD, Orton TS (eds) Isozymes in plant genetics and breeding, Part A

Zamir D, Gadish I (1987) Pollen selection for low temperature adaptation in tomato. Theor Appl Genet 74: 545-548

Zamir D, Thanksley SD, Jones RA (1981) Low temperature effect and selective fertilization by pollen mixtures of wild and cultivated tomato species. Theor Appl Genet 59: 235-238

Zamir D, Thanksley SD, Jones RA (1982) Haploid selection for low temperature tolerance of tomato pollen. Genetics 101: 129-137

Zamir D, Vallejos EC (1983) Temperature effects on haploid selection of tomato microspores and pollen grain. Pages 335-342 *in* Mulcahy DL, Ottaviano E (eds). Pollen: biology and implications for plant breeding. Elsevier, New York, Amsterdam, Oxford

Chapter 11: The Effects of Natural Disturbance on Plant Communities in Temperate Ecosystems

D.C. Malcolm
School of Forestry
Institute of Ecology and Resource Management
University of Edinburgh
Mayfield Road
Edinburgh EH9 3JU

INTRODUCTION

The purpose of this meeting is to consider the research required to predict potential shifts in ecological systems and their functioning as a result of predicted climatic changes. The rates of change predicted are a global surface warming of about 0.3°C per decade with a variation of about the same amount (Houghton *et al.* 1992). Regional effects remain speculative at the scales of resolution yet possible, while there are no clear indications about the impact of climate change on storm occurrence, general cloudiness or extremes of climatic variation.

There has been a general warming over the last century but it is not yet possible to attribute this to increased greenhouse gas concentration. Sensing changes at these proposed rates against the background variability will not be easy, particularly in vegetation influenced by man. Insect populations, which are known to be temperature sensitive, may provide early markers of change, as may changes in the relative nuclear DNA amounts in phenologically distinct types within plant populations (Grime 1993). The rate of change is greater than any experienced by vegetation in post-glacial time and will occur well within the life-span of existing trees.

To understand potential change in vegetation demands a knowledge of existing relationships between climate and vegetation and the factors that drive its development. Growth cycles and the structure of vegetation provide the framework for regeneration

NATO ASI Series, Vol. I 20
Biodiversity, Temperate Ecosystems, and Global Change
Edited by T. J. B. Boyle and C. E. B. Boyle

(Whitmore 1982), the opportunity for invasion and thus migration. The disturbance regime is critical for creating space and opportunity for invasion, the initiation of "secondary" succession and the distribution of communities in the landscape.

In recent years there have been several major reviews of both disturbance and succession (*e.g.* White 1979, Sousa 1984, Pickett and White 1985, Miles 1987). The complex of spatial scales, environmental heterogeneity and temporal variability involved have encouraged many attempts to model vegetation relationships from ecophysiological responses to regional and global perspectives (*e.g.* Running and Coughlan 1988, Shugart *et al.* 1981). This paper draws on these reviews and attempts to highlight those aspects of disturbance that may become important if the predicted changes in climate take place.

NATURAL DISTURBANCE

Defining natural disturbance is difficult. The variation in current vegetation reflects the variation in environmental conditions, geology, soil development, and topography all interacting with current climate. This variation in species assemblages and ecosystem properties (*e.g.* productivity) is the basis for various classifications designed to achieve a better understanding of the relationships between different ecosystems *and* their management. All systems are an expression of past as well as current environments. Even systems supposed to be at equilibrium have developed through succession from some previous condition and can be expected to change in the future. Some authors have suggested that within a region a "*shifting mosaic steady state*" might exist (Bormann and Likens 1979) and have sought to separate endogenous disturbances, required to maintain the system, from exogenous, unpredictable disturbing effects (*e.g.* Marrs 1986). This separation is difficult to apply as the scale of disturbance can vary from the disruption of a single plant to the destruction of considerable areas at once. Levin (1976) recognises "*endogenous microcycles*" which are age-dependent extinctions, but here again there is interaction between the stage of development of the plant (or ecosystem) in size, accumulation of necromass and the disrupting forces.

An encompassing definition (quoted by Sousa 1984) is:

"disturbance is a discrete, punctuated killing, displacement or damaging of one or more individuals (or colonies) that directly or indirectly creates an opportunity for new individuals (or colonies) to become established."

The opportunities created for new individuals depend on the availability of propagules and their ability to invade, migrate or enter the new succession, and thus their dispersal mechanisms interact with the magnitude of the disturbance and the heterogeneity of the surrounding ecosystems.

Natural disturbances are clearly those not planned or mediated by man. However, very few ecosystems have not been directly or indirectly influenced by human activities. Direct influences, say by the use of fire from prehistoric times (Day 1990), are obvious enough but control of fire, increases in herbivore populations, and shifting agriculture are also examples where successional development may lead to the predisposition of communities to other disturbances. Global climate change has been a "natural" phenomenon in the past but now, through emissions and their effects, is distinctly anthropogenic. Studying the effects of specific disturbance factors without a consideration of human influences may be useful but a general consideration without them is now hardly valid.

Disturbances can be conveniently divided into physical and biological, which may act independently or together in an episodic or continuous manner. In order to compare disturbance effects, it is useful to be able to describe their attributes unambiguously (Sousa 1984, Table 1).

PHYSICAL DISTURBANCES

Fire

Fire has been the most commonly studied disturbance. There is ample evidence of its occurrence throughout post-glacial times in the temperate zones as well as of its use by man to alter vegetation patterns,making it difficult to separate natural from induced occurrences.

Table 1 Descriptors of Disturbances (after Sousa 1984)

Extent	-	area affected
Magnitude	-	intensity (*e.g.* windspeed, fire temperature)
	-	severity (amount of damage)
Frequency	-	number unit time^{-1}
	-	random point frequency (return interval)
	-	regional frequency (total no. in geographic area)
Predictability	-	variance in the mean time between disturbances
Turn-over rate	-	mean time to disturb whole area

The incidence and severity of fire depends on coincidence of ignition sources, adequate fuel and suitable weather conditions. These tend to occur together in continental climates in older stands of tree vegetation and in grasslands. Heinselman (1973) showed that over the last 400 years an average of about 1 percent of the area was burned each year in north Minnesota. Fires were relatively infrequent (50-100 year intervals) but were large scale (10^2 - 10^5 hectares). Recovery was dominated by broadleaved species, notably *Betula papyrifera* and *Populus tremuloides* which suckered or sprouted from residual organs. Succession leads to a forest with *Abies balsamea, Thuja occidentalis* and *Picea* spp. which together with dying pioneers rebuild fuel supply in about 300 years. Detailed studies dating charcoal deposits and corresponding pollen from lake sediments in other parts of the conifer-hardwood forests at this latitude (42° - 50° N) show variable fire return periods with large and severe fires at about 300 year intervals before 6000 BP and at much larger intervals after that, although less severe fires continued at about 350 year intervals (Walker 1982).

Drawing on this North American data and an exceedingly long record (400,000 years) from Lake George, New South Wales, Walker (1982) argued that repeated fire allows the immigration of a range of species with different responses. These lead to a diverse assemblage and mosaic of forest types that may eventually be less damaged by repeated disturbance. Although more frequent or intense fires could disrupt these systems, lack of fire can be equally deleterious as a few species become dominant, excluding pioneers too long for them to respond when disturbance comes. For example, it took *Picea abies* 400 years after fire to reform on open stand, close to its limit in north Finland, where the pioneer *Betula tortuosa* was subject to cyclic fungal attack, and *Pinus sylvestris* had long intervals between seed years (Siren 1955). Dominance of a few species may be a response to strong environmental stresses (Walker 1982) whereas more stable circumstances with recurrent perturbations favour a more diverse assemblage of species.

In less continental climates fire may be less frequent but still the most important means of stimulating forest regeneration. For example, on Mt Rainier in Washington (Franklin *et al.* 1988), 90 percent of the forests were of fire origin with an estimated natural return interval of 460 years. Moisture, temperature and snow pack then conditioned the direction of succession.

Some shrub dominated eocsystems are maintained solely by fire, often man induced, such as the ericaceous heathlands of western Europe managed for sheep and grouse. If fire is withheld, the sward of *Calluna vulgaris* gradually breaks up due to senescence of individuals, creating gaps that can be invaded by other species. Miles (1987) described the cyclic tendency for these heathlands to develop into *Betula* woodland with suppression of the heathland, until light conditions again allow re-invasion of *Calluna*, initiating a new heathland phase. Such a cycle however, depends on the lack of propagules of successionary species in a largely treeless landscape. With repeated burning Calluna retains dominance, sprouting from rootstocks and a seed bank of considerable longevity. Severe fires may destroy the raw humus, together with rootstocks and seed bank. Regeneration may thus be precluded, and erosion may follow. Dimbleby (1962) considered these systems inherently unstable. The establishment of similar heathlands from fire-destroyed forest has been reported from Newfoundland (Richardson 1975) where a *Kalmia* spp becomes dominant.

The effects of fire can be considered to result in a period of "*extremely negative ecosystem productivity*" (Sprugel 1985), because much of the biomass and accumulated litter is destroyed. However, root systems and some dead wood are added to the soil organic matter. Recovery time depends on the nature of the soil and the returning vegetation. As in the heathlands described above, on some soils loss of organic matter and a proportion of the site nutrients can alter soil moisture regimes, reducing future productivity and enhancing the potential for erosion. On the other hand, in cold climates decomposition is so slow that mor humus gradually accumulates most of the nitrogen and phosphorus, resulting in stagnation of growth in older stands. Disturbance by fire then provides a necessary stimulus to restore active mineralization and nutrient cycling. Carbon, nitrogen and sulphur are dissipated by fire but most of the phosphorus, potassium and other bases are released to the soil, stimulating biological activity (Vitousek 1985).

The effects of fire disturbance clearly vary from site to site according to general climate, soil and vegetation type (Clark 1990), and its frequency is critical to the direction of succession. An increase in the frequency could result in increases in early successional species or on some sites loss of forest vegetation, as has been predicted for parts of Canada (Flannigan and Van Wagner 1991).

Wind

In oceanic and montane climates, where fire is relatively infrequent, wind plays a major role in forest rejuvenation. Wind affects vegetation in several ways, from cooling foliage to boundary layer effects and mechanical damage, but of interest here is its ability to uproot trees. Several forces are involved which relate to soil strength, root distribution and depth and, most importantly, the size and shape of the tree (Coutts 1986). Essentially taller and denser stands are more prone to disruption particularly on shallowly-rooted soils (Foster 1988). Because wind is the dominating factor in British silviculture, it has attracted much attention. As a result it is now possible to predict empirically the height at which stands (plantations) will begin to break up. The factors used are the windiness of the area, elevation of the site, tree rooting depth, and an estimate of geomorphic shelter (Miller *et al.* 1987). This relationship has been developed in a climate with perhaps 10 or more gales (gusting to > 30 m s^{-1}) annually. Such winds give rise to so-called "endemic" windthrow which is expected and predictable. On the most susceptible sites the critical height is about 16 m for spruce stands. A similar height (18 m) is quoted by Jane (1986) for *Nothofagus solandri* in New Zealand. However, massive disruption arises from "catastrophic" winds maintaining speeds in excess of about 40 m s^{-1}(gusting to 50 m s^{-1}) over several hours. These winds are stochastic in occurrence and direction, have a front up to 100 km and blow down virtually all mature or semi-mature stands in their path (Table 2). Return periods in any location are estimated at 150-200 years although the records are rather shorter than that.

Windthrow can thus create openings from the single-tree gap to tree-fall over extensive areas. Smaller gaps may be rapidly occupied by shade-tolerant species in the understorey or in larger gaps regeneration of surviving individuals or neighbouring stands. Canham and Marks (1985) suggest that, for slower growing species in the lower canopy, multiple gap occurrences may be needed for them to reach the canopy.

Table 2. Catastrophic storms affecting Britain since 1945 (after Quine 1990)

Date	Area affected by 36 m/s gusts km^2	Max. gust recorded ms^{-1}	Mean of max. gusts within zone 36 m s^{-1}	Vol. timber blown down Mm^3
31 Jan 1953	370	50	43	1.80
15 Jan 1968	510	52	43	1.64
2 Jan 1976	890	46	39	0.96
16 Oct 1987	220	51	41	3.91
25 Jan 1990	690	47	39	1.26

The effects of wind disturbance differ from fire because the material blown down is not consumed but added to the soil, where decomposition may take many years. At the same time, the soil is disturbed, much as in cultivation. Organic layers are loosened and aerated, enhancing mineralization by increased temperature and moisture although this appears not to have been much studied (Vitousek 1985). Certainly on soils with high winter water tables or poor drainage, the hummocks formed allow more exacting species or relatively shade intolerant species to enter the canopy. This is the case for *Picea sitchensis* in the rainforests of the Queen Charlotte Islands (Day 1957). In catastrophic winds well-rooted trees may be broken rather than uprooted but generally these form a minority of damaged stems and in temperate conditions rarely contribute to the succeeding stand by sprouting epicormic branches (Peterson and Pickett 1991).

Geological activity

Landslips are frequently initiated by earthquake or excessive rainfall on unstable slopes in juvenile mountain ranges. Although relatively small in extent they may occur frequently in space and time and be locally important disturbances. Succession is restarted and the bare areas are colonised by tree or other vegetation depending on substrate and the surrounding stands, eventually returning to the mature vegetation (Merk *et al.* 1989).

Vulcanism can destroy vegetation over large areas by massive eruption (*e.g.* Mt St Helens) or by initiating lava flows and ash deposits. Tree vegetation seems capable of surviving ash deposition even of several metres depth (Eggler 1948) and shallow deposits may act as mild fertilisation or create new seedbeds. On the other hand herbaceous species appear to be most susceptible and are differentially killed. In an area of repeated activity in New Zealand Clarkson (1990) estimated that 20,000 ha had been affected over a 450 year period influencing both primary and secondary successions. Colonisation of these "new" soils is effected more rapidly if nitrogen-fixing species are present.

Volcanic activity influences vegetation indirectly, at least on a regional basis, through dust emissions that reduce surface irradiation, increase cloudiness generally and thus lead to cooler surface temperatures. Such effects are unlikely to alter major vegetation types but rather to reduce growth rates. Abnormal growth rings in *Picea glauca* in Alaska were attributed by Oswalt (1957) to eruptions in Iceland and Japan.

Volcanoes contribute large emissions of sulphur to the atmosphere, which apart from their effects on deposition have recently been shown to have a cooling effect on the earth's surface (Houghton *et al.* 1992). Emission effects are another instance where it is difficult to disentangle anthropogenic and natural effects.

Extreme climatic events

Natural vegetation is adapted to its local climate, but climates have fluctuated in post-glacial times sufficiently to alter species distributions. Response to climatic change has to accommodate the distribution pattern and potential for migration of individual species. In some cases, geographical barriers may prevent or slow species migration to areas to which they are well adapted. For example, Norway spruce, known to be in Finland at 5000 BP, only reached western Norway in 1000 BP, after migrating around the Gulf of Bothnia (Moe 1970 quoted by Miles 1979).

Once long-lived vegetation is in place, seasonal and annual variation in climatic factors may be reflected in growth rates and periodicity in flowering. The lower canopy and ground vegetation may be largely protected from these variations. It is a moot point when, or whether, such small scale variation becomes disturbance; however, periodically extreme values of climatic variables occur with important effects on vegetation. These events perhaps may be best expressed in probabilistic terms as potential return periods. The stress induced by these events may be seen most clearly at the latitudinal or altitudinal limits of the species or vegetation type.

Low temperatures tend to be most disruptive when they recur in unusual circumstances. For example the -40 °C temperatures recorded in winter 1984/85 in northern Fennoscandia froze the soil before the usual insulation of snowpack arrived. This is thought to have severely damaged fine roots with resulting foliage loss in the next season with consequent reduced growth. Damage was also recorded for the ericaceous ground vegetation important for berry crops (Raatikainen and Vanninen 1988). Similarly *Calluna* heaths can be severely damaged by winter cold (Marrs 1986). Sudden falls in air temperature, such as those which occur in radiation frosts, can destroy unhardened tissues, or periods of freezing fog (ice storms) mechanically damage trees rendering them prone to fungal attack (Sanzen- Baker and Nimmo 1941). These forms of damage are particularly important at higher elevations where loss of foliage causes a negative carbon balance (Grace 1989).

Distinguishing between occasional, extreme events and frequently repeated occurrences of a less extreme nature is difficult. It is also a problem with periods of high temperature, usually associated with drought. Here again effects are usually noted as reductions in growth rate but sometimes in death of susceptible species or age-classes. Topography and soil type are often critical. Studies of annual growth rings in the forests of the Jura and Vosges mountains in France showed strong correlations with periods of drought this century (Bert and Becker 1990) in *Abies alba* stands. Serious mortality of older age-classes of *Nothofagus* spp in New Zealand (Hosking and Hutcheson 1988) is thought to be part of the natural disturbance order leading to forest renewal. The effects of a drought may extend over several years possibly due to death of fine roots and the inability of older trees to restore them so that debilitated trees are then prone to other disturbance.

Biotic disturbances

Plant communities are only part of the ecosystem to which they contribute and vertebrates, invertebrates and microbes all interact with the plants which provide their resources and habitat. It is not possible to summarise here these relationships or the effects of disturbance but simply to draw attention to major types of biotic disturbances.

Browsing animals are the most obvious groups of vertebrates to influence succession in vegetation. Large scale disturbance provides a rapid increase in resources, and where predation is inadequate, they can restrict regeneration of species of high palatability. Shade tolerant late seral species are often the most affected. Human involvement in driving or increasing game populations is of long standing. Burrowing animals (*e.g.* moles, badgers) create opportunities for invasion of existing vegetation which otherwise may competitively exclude immigration (Watt 1974).

Birds are involved in seed dispersal, particularly of heavy seeded species, the spread of which would otherwise be restricted to much slower migration rates. Migratory bird populations may be affected by disturbance of either their winter or breeding habitats with feedback effects on the alternative habitat.

Until recently insect interactions with plant communities were usually described in terms of damage limitation. Now they are generally considered as integral to ecosystem functioning and interacting with the disturbance regime (Schowalter 1985). Population increases follow the development of stress in the vegetation already affected by fire, drought, etc. and can predispose communities to disturbance from other sources, *e.g.*, by the accumulation of fuel. Stress conditions in plants that encourage insect attack include any factor limiting photosynthetic efficiency and the production of secondary metabolites. Schowalter (1985) considers that insects may increase disturbance reliability and hence reduce its severity. In short, insect outbreaks are a function of stand structure spatially, the history of disturbance and other influences such as parasitism (Sousa 1984).

Microbial contributions to disturbance regimes are, like insect damage, often related to the structure and age of the community. Root rots develop in old trees subject to stress, predisposing them to windblow, often creating the gap disturbances required by successional species (Worrall and Harrington 1988).

Massive disturbance, particularly from fire, may seriously reduce populations of mycorrhizal fungi essential to healthy growth of plants on nutrient poor sites. The importance of this effect has not often been considered but there is increasing evidence (Perry *et al.* 1992) that it may influence competitive abilities in regeneration.

Finally in considering the effects of disturbance, the role of genetic differentiation in populations, caused both by small scale disturbances over long periods, and by the effects of catastrophic events, needs to be considered. In some instances inbreeding may be common, for example where few individuals are widely dispersed in migration to new environments or, in other cases, potential for differentiation may be high following explosive population increase. To these possibilities must be added the range of natural selection effects including competitive removal of less fit individuals in successional stages, the past history of species distributions and disturbances. Although not much unifying theory is available, Rice and Jain (1985) suggest that selective processes may be easier to measure in patchy vegetation because of the often abrupt boundaries between different conditions.

DISCUSSION AND CONCLUSIONS

Although most ecologists now accept that the concept of succession leading to some hypothetical climax vegetation type is not supported by empirical evidence, there are still many attempts to construct models of so-called equilibrium states. These analyses of the frequency and distribution of disturbance make use of the age determination of cohorts of regeneration. For example, Frelich and Lorimer (1991) studied 23,000 ha of forest dominated by sugar maple (*Acer saccharum*) and eastern hemlock (*Tsuga canadensis*) and found that canopy residence time was about 145-175 years with a return period of 69 years or 1,920 years for canopy disturbances of 10 or 60 percent respectively. This regime gives rise to a uniform age class distribution leading to the conclusion that the landscape generally appeared to be in equilibrium.

Many studies of this nature that analyse disturbance on the landscape scale find, as one could expect, that agents of disturbance vary, certainly in time and often topographically. In the Appalachian mountains, White *et al.* (1985) found small windthrows (< 200 m^2) most common but larger gaps were due to large windthrows, lightning fires and debris avalanches. The 1-10 year age class occupied from 6-17 percent of the stand studied and the authors concluded from a study of the capture of the smaller gaps by the three dominant species (*Abies fraseri, Picea rubens* and *Betula lenta*) that these species were in equilibrium.

The wave regeneration systems described by Sprugel and Bormann (1981) in *Abies balsamea* and Matson and Boone (1984) in *Tsuga mertensiana* perhaps provide good examples of an equilibrium involving disturbance and continuous rejuvenation of the vegetation. In the former case disturbance is apparently climatically initiated whereas in the mountain hemlock case the fungus *Phellinus weirii* is thought to be responsible. The fungal attack is related to low nitrogen status in older stands but nitrogen becomes available to regeneration after the disturbance.

These examples of the supposed dynamic equilibria of vegetation are all heavily dependent on the spatial and temporal scale, as well as the species involved. The occupation of the gaps created by disturbance is conditioned by the morphology and reproductive biology of the species being disturbed or those of species within dispersal distance. The success of any species depends on the nature of the disturbance and its interaction with the reproductive strategy it adopts. The existence of seed banks, undamaged tissues which can sprout, or anemophilous dispersal will dictate which species reclaims the disturbance site. Given the variability of size, shape and microclimate within disturbed areas the direction of succession is likely to be unpredictable. Add the conjectural nature of the timing, frequency and distribution of disrupting events and it is hard to accept that any equilibrium exists.

A further problem in the temperate zone is that general climatic conditions are known to have fluctuated markedly in post-glacial times so that any apparent equilibrium may be illusory. Because of the time scales of adjustment of long-lived vegetation, the gradual change of environmental influences may be buffered by the vegetation itself although its reproductive capacity is reduced.

Although it seems probable that only a few plant communities anywhere have been devoid of anthropogenic effects in the past this is clearly now not the case. Global emission of "greenhouse" gases and other atmospheric pollutants undoubtedly already influence all temperate ecosystems. So far predictions of climatic change are somewhat vague in terms of regional effects and, more importantly in this context, the variability of climatic factors. The predicted rates of change in temperature alone are greater than previously experienced by temperate biota and if substantiated will amount to a major perturbation in existing plant communities. However it is the variability and the frequency of extreme events that drive disturbance regimes. It might be already too late to further determine the effects of natural disturbance on plant communities.

REFERENCES

Bert GP, Becker HM (1990) Present and past vitality of silver fir (*Abies alba*) in the Jura Mountains. Ann. Sci. For. 47: 395-412

Bormann FH, Likens GE (1979) Pattern and Process in a Forested Ecosystem. Springer-Verlag, New York

Canham CD, Marks PL (1985) The response of woody plants to disturbance: Patterns of establishment and growth: Pages 197-216 *in* Pickett STA, White PS (eds) The Ecology of Natural Disturbance and Patch Dynamics. Academic Press, London

Clark JS (1990) Fire and climate change during the last 750 years in northwestern Minnesota. Ecological Monographs 60: 135-159

Clarkson BD (1990) A review of vegetation development following recent (< 450 years) volcanic disturbance in North Island, New Zealand. N.Z. J. Ecol. 14: 59-71

Coutts MP (1986) Components of tree stability in Sitka spruce on peaty gley soil. Forestry 59: 173-197

Day R (1990) The dynamic nature of Ontario's forests from post-glacial times to the present day. Pages 63-122 *in* Old Growth Forests. Canadian Scholars Press Inc, Toronto

Day WR (1957) Sitka spruce in British Columbia. For. Comm. Bull. 28, HMSO London, Pages 110

Dimbleby GW (1962) The development of British heathlands and their soils. Oxford Forestry Memoirs 23

Eggler WA (1948) Plant communities in the vicinity of the volcano El Paricutin, Mexico after two and a half years of eruption. Ecology 29: 415-436

Flannigan MD, Van-Wagner CE (1991) Climate change and wildfire in Canada. Can. J. For. Res. 21: 66-72

Foster DR (1988) Species and stand response to catastrophic wind in central New England, USA. J. Ecol. 76: 135-151

Franklin JF Moir WH, Hemstrom MA, Greene SE, Smith BG (1988) The forest communities of Mount Rainier National Park. Scientific Monograph Series, National Park Service, US Dept. of the Interior, 19: 1-194 (seen in abstract)

Frelich LE, Lorimer CG (1991) Natural disturbance regimes in hemlock-hardwood forests of the Upper Great Lakes Region. Ecol. Mono. 61: 145-164

Grace J (1989) Tree lines. Pages 233-245 *in* Forests, Weather and Climate. Phil. Trans. Roy. Soc. Lond. B 324

Grime JP (1993) Vegetation functional classification systems as approaches to predicting and quantifying global vegetation change. Pages 293-305 *in* Solomon AM, Shugart HH (eds) Vegetation Dynamics and Global Change. Chapman and Hall, New York

Heinselman ML (1973) Fire in the virgin forests of the Boundary Waters Canoe Area, Minnesota. Quaternary Research 3: 329-382

Hosking GP, Hutcheson JA (1988) Mountain beech (*Nothofagus solandri*) decline in the Kaweka Range, North Island, New Zealand. N.Z. J. Bot. 26: 393-400

Houghton JT, Callander BA, Varney SK (eds) (1992) Intergovernmental Panel on Climate Change 1992. Cambridge University Press

Jane GT (1986) Wind damage as an ecological process in mountain beech forests of Canterbury, New Zealand. N.Z. Journal of Ecology 9: 25-39 (seen in abstract)

Levin SA (1976) Population dynamic models in heterogeneous environments. Ann. Rev. Ecol. Syst. 7: 287-310

Marrs RH (1986) The role of catastrophic death of *Calluna* in heathland dynamics. Vegetatio 66: 109-115

Matson PA, Boone R (1984) Natural disturbance and nitrogen mineralization: wave form dieback of mountain hemlock in the Oregon Cascades. Ecology 65: 1511-1516

Merk AF, Dickinson KJM, Fife AJ (1989) Forest succession on landslides in the Florid Ecological Region, Southwestern New Zealand. N.Z. J. Bot. 27: 369-390

Miles J (1979) Vegetation Dynamics. Outline Studies in Ecology, Chapman and Hall, London

Miles J (1987) Vegetation succession: past and present perceptions. Pages 1-30 *in* Gray AJ, Crawley MJ, Edwards PJ (eds) Colonization, Succession and Stability. Blackwell Scientific Publications, Oxford

Miller KF, Quine CP, Hunt J (1987) The assessment of wind exposure for forestry in upland Britain. Forestry 60: 179-192

Oswalt WH (1957) Volcanic activity and Alaskan spruce growth in AD 1783. Science 126: 928-929 (seen in abstract)

Perry DA, Bell T, Amaranthus MP (1992) Mycorrhizal fungi in mixed species forests and other tales of positive feedback, redundancy and stability. Pages 151-180 *in* Cannell MGR, Malcolm DC, Robertson P (eds) The Ecology of Mixed Species Stands of Trees. Blackwell Scientific Publications, Oxford

Peterson CJ, Pickett STA (1991) Tree fall and resprouting following catastrophic windthrow in old growth hemlock-hardwoods forest. For. Ecol. Mgmt. 42: 205-217

Pickett STA, White PS (1985) The Ecology of Natural Disturbance and Patch Dynamics. Academic Press, London

Quine CP (1991) Recent storm damage to trees and woodlands in southern Britain. Pages 83-89 *in* Hodge SJ (ed) Research for Practical Arboriculture. Forestry Commission Bulletin 97, HMSO, London

Raatikainen M, Vanninen I (1988) The effect of the 1984-1985 cold winter on the bilberry and lingonberry yield in Finland. Acta Bot. Fenn. 136: 43-47

Rice K, Jain S (1985) Plant population genetics and evolution in disturbed environments. Pages 287-303 *in* Pickett STA, White PS (eds) The Ecology of Natural Disturbance and Patch Dynamics. Academic Press, London

Richardson J (1975) Regeneration after disturbance in Newfoundland forests: patterns, problems and prescription. Information Report NX 130. Newfoundland Forest Research Centre

Running SW, Coughlan JC (1988) A general model of forest ecosystem processes for regional applications. I. Hydrological balance, canopy gas exchange and primary production processes. Ecological Modelling 42: 125-154

Sanzen-Baker PG, Nimmo M (1941) Glazed frost 1940 - damage to forest trees in England and Wales. Forestry 15: 37-54

Schowalter TD (1985) Adaptations of insects to disturbance. Pages 235-252 *in* Pickett STA, White PS (eds) The Ecology of Natural Disturbance and Patch Dynamics. Academic Press, London

Shugart HH, West DC, Emanuel, WR (1981) Patterns and dynamics of forests: an application of simulation models. Pages 74-94 *in* West DC, Shugart HH, Botkin DB (eds) Forest Succession: Concepts and Applications. Springer-Verlag, Berlin

Siren G (1955) The development of spruce forest on raw humus sites in northern Finland. Acta For. Fenn. 62

Sousa WO (1984) The role of disturbance in natural communities. Ann. Rev. Ecol. Syst. 15: 353-392

Sprugel DG (1985) Natural disturbance and ecosystem energetics. Pages 335-352 *in* Pickett STA, White PS (eds) The Ecology of Natural Disturbance and Patch Dynamics. Academic Press, London

Sprugel DG, Bormann FH (1981) Natural disturbance and the steady state in high-altitude balsam fir forests. Science 211: 390-393

Vitousek PM (1985) Community turnover and ecosystem nutrient dynamics. Pages 325-332 *in* Pickett STA, White PS (eds) The Ecology of Natural Disturbance and Patch Dynamics. Academic Press, London

Walker D (1982) The development of resilience in burned vegetation. Pages 27-43 *in* Newman EI (ed) The Plant Community as a Working Mechanism. Blackwell Scientific Publications, Oxford

Watt AS (1974) Sensescence and rejuvenation in ungrazed chalk grassland (Grassland B) in Breckland: The significance of litter and of moles. J. Appl. Ecol. 11: 1157-1171

White PS (1979) Pattern, process and natural disturbance in vegetation. Bot. Rev. 45: 229-299

White PS, Mackenzie MD, Busing RT (1985) Natural disturbance and gap phase dynamics in Southern Appalachian spruce-fir forests. Can. J. For. Res. 15: 233-240

Whitmore TC (1982) On pattern and process in forests. Pages 45-59 *in* Newman EI (ed) The Plant Community as a Working Mechanism. Blackwell Scientific Publications, Oxford

Worrall JJ, Harrington TC (1988) Etiology of canopy gaps in spruce-fir forests at Crawford Notch, New Hampshire. Can. J. For. Res. 18: 1463-1469

Chapter 12: Ecological Processes Linking Forest Structure and Avian Diversity in Western Oregon

Andrew J. Hansen, Robyn M. Vega[a], Arthur W. McKee[a], and Andrew Moldenke[b]
Biology Department
Montana State University
Bozeman, MT 59717 USA

INTRODUCTION

Ecologists have long debated the factors that may influence patterns of faunal diversity. Recent studies have found that landscape, regional, and continental-scale features explain some of the variation in bird diversity (Terborgh 1989, Huff and Raley 1991, Currie 1991). Yet, studies considering multiple spatial scales have found that many bird species are most strongly correlated with habitat features within forest stands (e.g., Lehmkuhl et al. 1991, McGarrigal, personal communication), especially forest structural complexity (see Cody 1985). Structurally complex forests are thought to provide relatively more habitat niches and, thus, support high animal species diversity (James 1971, Urban and Smith 1989).

In natural coniferous forests of the Pacific Northwest (PNW) United States, structural complexity is bimodally distributed over the course of succession, being relatively high in the early and late-seral stages (Spies et al. 1988). This is because structural features such as large trees, snags, and fallen logs often survive canopy-replacement disturbances, and hence are incorporated into early successional forests. These features then decay and structural complexity decreases, until large trees are recruited in the old-growth stage. Animal diversity in PNW forests is also thought to

[a] Dept. of Forest Science, Oregon State University, Corvallis, OR 97331, USA
[b] Entomology Dept., Oregon State University, Corvallis, OR 97331, USA

NATO ASI Series, Vol. I 20
Biodiversity, Temperate Ecosystems, and Global Change
Edited by T. J. B. Boyle and C. E. B. Boyle

be relatively high both early and late in succession (Harris 1984). Ecologists generally assume this follows from the patterns of structural complexity and habitat niche diversity (Hansen et al. 1991).

Forest management practices have provided a test of this hypothesis. Traditional clearcutting removes structural legacy, and recently harvested plantations are relatively homogeneous in structure. As predicted, bird diversity has been found to be lower in closed-canopy plantations than in natural closed-canopy forests (Hansen et al. in review). Consequently, many forest managers are now retaining canopy trees, snags, and fallen trees in harvest units to maintain suitable habitats for native organisms. Initial studies show a positive relation between animal richness and structural retention (Hansen and Hounihan in press). This provides support for the "Habitat Niche" hypothesis that forest structural complexity strongly influences vertebrate species diversity.

Studies in recent clearcuts, however, suggest factors in addition to habitat niche diversity may influence animal diversity. Hansen et al. (in review) found that clearcut sites with no large live or dead trees supported levels of bird diversity comparable to those in mature forests. This observation led us to a novel hypothesis that the level of energy available to consumers is higher in open-canopy than in closed-canopy stands. We predict that relatively more of the net primary production (NPP) in open-canopy, hardwood-dominated stands is fixed as palatable leaves, fruits, and seeds, and that this higher level of energy available to consumers supports higher levels of animal diversity. Closed-canopy, conifer stands, in contrast, are predicted to fix relatively more energy as wood and unpalatable conifer needles, have less energy available to consumers, and thus support lower levels of animal diversity. In support of this hypothesis, Turner and Long (1975) found that the production of understorey plants and leaves of trees was about 48% higher in 20- to 30-year-old (relatively open-canopy) stands than in 40- to 70-year-old stands (closed-canopy).

A third hypothesis on factors controlling bird diversity involves nest predation. Previous studies of bird response to canopy density indicated that the abundances of some ground and shrub-nesting birds are dramatically lower in harvest units with a few canopy trees (1-5 per ha) than in units with no canopy trees (Hansen and

Hounihan in press). These same species are also more abundant in the centres of clearcuts than near forest edges (Hansen et al. in prep.). These findings raise the hypothesis that nest predation is higher in sites where predators can use canopy trees as perches for locating the nests of prey species (Vega 1993).

Here we report the results of a coordinated set of studies on the effects of stand energetics, structural complexity, and predation rates on bird diversity (Figure 1). Specific hypotheses were:

H_0: Taxonomic diversity of birds is not influenced by the level of net primary productivity consumable by animals, by the structural complexity of the vegetation, nor by nest predation rates.

H_1: In PNW coniferous forests, the energy available to consumers is inversely related to canopy density (Figure 2a) because relatively more energy is fixed as wood and unpalatable conifer needles in closed-canopy stands than in open-canopy stands. Taxonomic diversity of birds is positively related to energy availability because greater resource partitioning is possible (Figure 2b). Consequently, bird diversity will be negatively related to canopy tree density (Figure 2c).

H_2: The vertical and horizontal complexity of vegetation structure is highest at some intermediate canopy density (Figure 3a) because more canopy layers and variation in canopy layers are possible. Animal diversity increases with structural complexity (Figure 3b) because of the greater diversity of habitat niches. Consequently, animal diversity will be highest under intermediate canopy densities (Figure 3c).

H_3: Detection of bird nests by predators is highest under intermediate canopy densities (Figure 4a). Bird reproduction and diversity are inversely related to rates of nest predation (Figure 4b), hence bird reproduction and diversity show the response to canopy density depicted in Figure 4c.

Primary results on nest predation rates and bird response to habitat features of stands are reported in Vega (1993). Here we summarize those results, and also examine the Energy and Habitat Niche hypotheses.

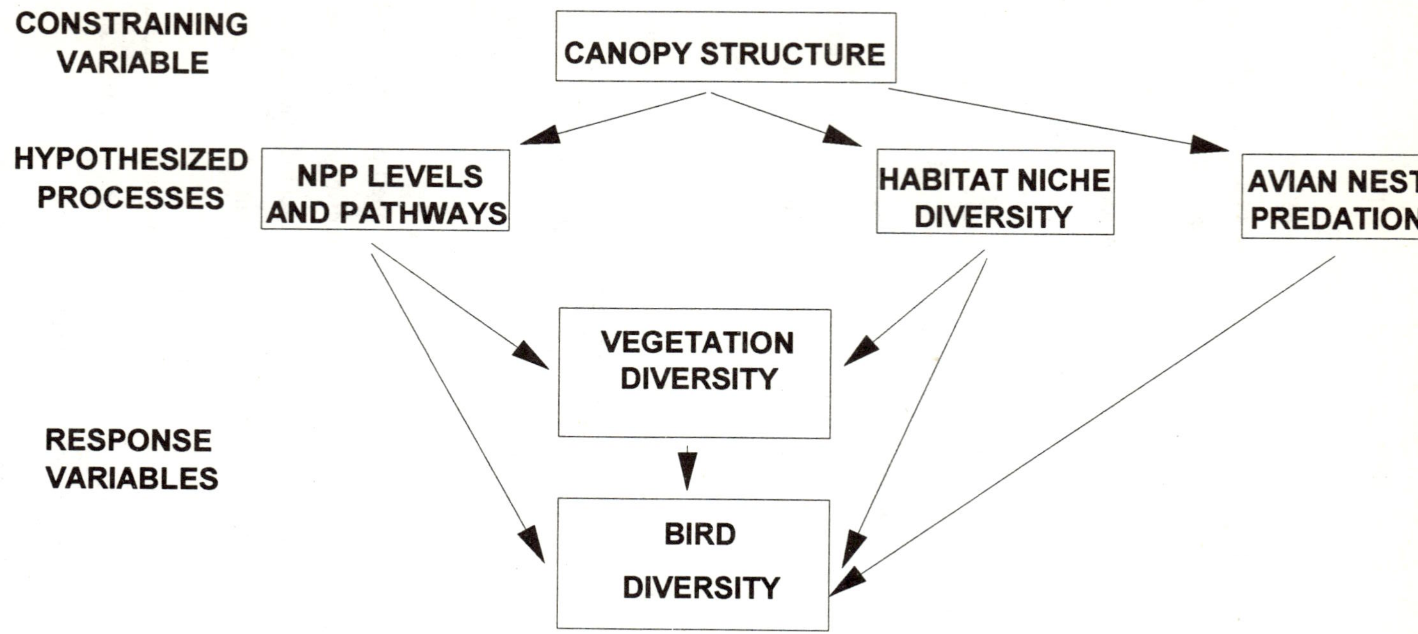

Figure 1. Driving variables, hypothesized pathways, and response variables examined in this study.

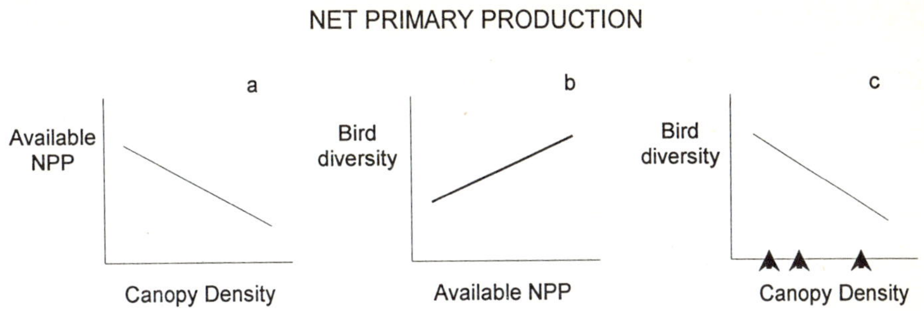

Figure 2. Hypothesized relationships between: (a) net primary productivity that is consumable by animals and canopy density; (b) consumable energy and animal community characteristics; and (c) canopy density and animal diversity. Vertical arrows denote the canopy densities for the three stand types studied.

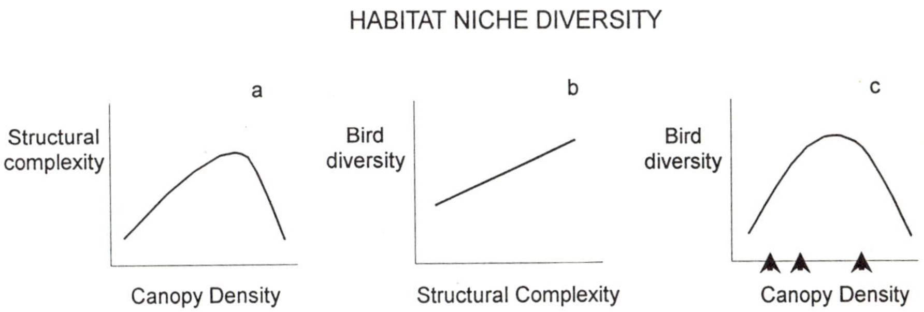

Figure 3. Hypothesized relationships between: (a) canopy density and stand structural complexity; (b) stand structural complexity and animal diversity; and (c) canopy density and animal diversity. Vertical arrows denote the canopy densities for the three stand types studied.

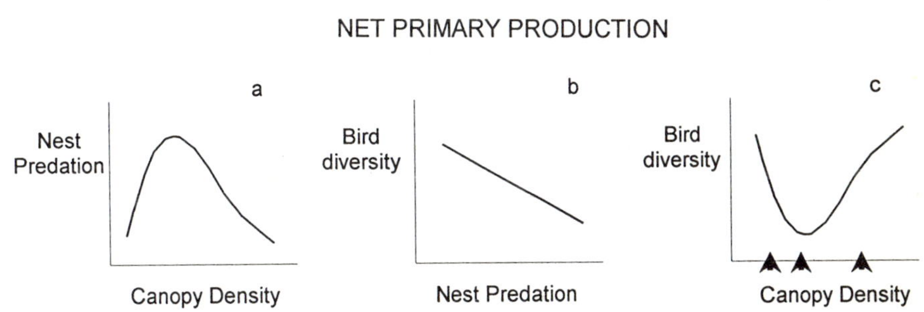

Figure 4. Hypothesized relationships between: (a) canopy density and nest predation rates; (b) nest predation rates and bird reproduction and diversity; and (c) canopy density and bird reproduction and diversity. Vertical arrows denote the canopy densities for the three stand types studied.

METHODS

Experimental Design

Three stand types were studied: structurally-simple, open-canopy stands (clearcuts with no canopy retention) (referred to hereafter as OCSS); structurally-complex, open-canopy stands (harvest units retaining about 4-13 canopy and subcanopy trees per ha) (OCSC); and structurally-complex, closed-canopy stands (natural mature forest) (CCSC). Four replicates of each canopy configuration were established for a total of 12 study sites. The stands were located in western Oregon in the vicinity of the H.J. Andrews Experimental Forest (Figure 5). Attributes of the stands are described in Table 1. Two parallel transects were placed 100 m apart at each site, and 3 plots were spaced at 100-m intervals along each transect. All plots were more than 75 m from the stand edge. Vegetation structure, biomass, nest predation, and bird densities were sampled at each of the six plots in a stand during spring and summer of 1992.

Forest Structure

Habitat attributes measured at each sample plot are listed in Table 2 and the sampling strategy depicted in Figure 6. Forest structural complexity was quantified based on and tree density by 10-cm dbh class and shrub density. Vertical habitat complexity was calculated as tree and shrub size-class diversity (mean Shannon's index among the 24 subplots within a stand). Horizontal habitat heterogeneity was quantified as the standard deviation in tree and shrub size-class diversity within a stand.

Forest Productivity

Vegetation biomass was estimated by life form (tree, shrub, subshrub, forb), plant part (wood, leaf), and palatability, using allometric relationships from the software package BIOPAK (Means, personal communication). Tree biomass was based on dbh, shrub biomass on diameter at the base of the stem, and subshrub biomass was based on percent cover. Herb and forb biomass (hereafter called herb biomass) was measured directly by clipping and measuring dry weights (see Table 2).

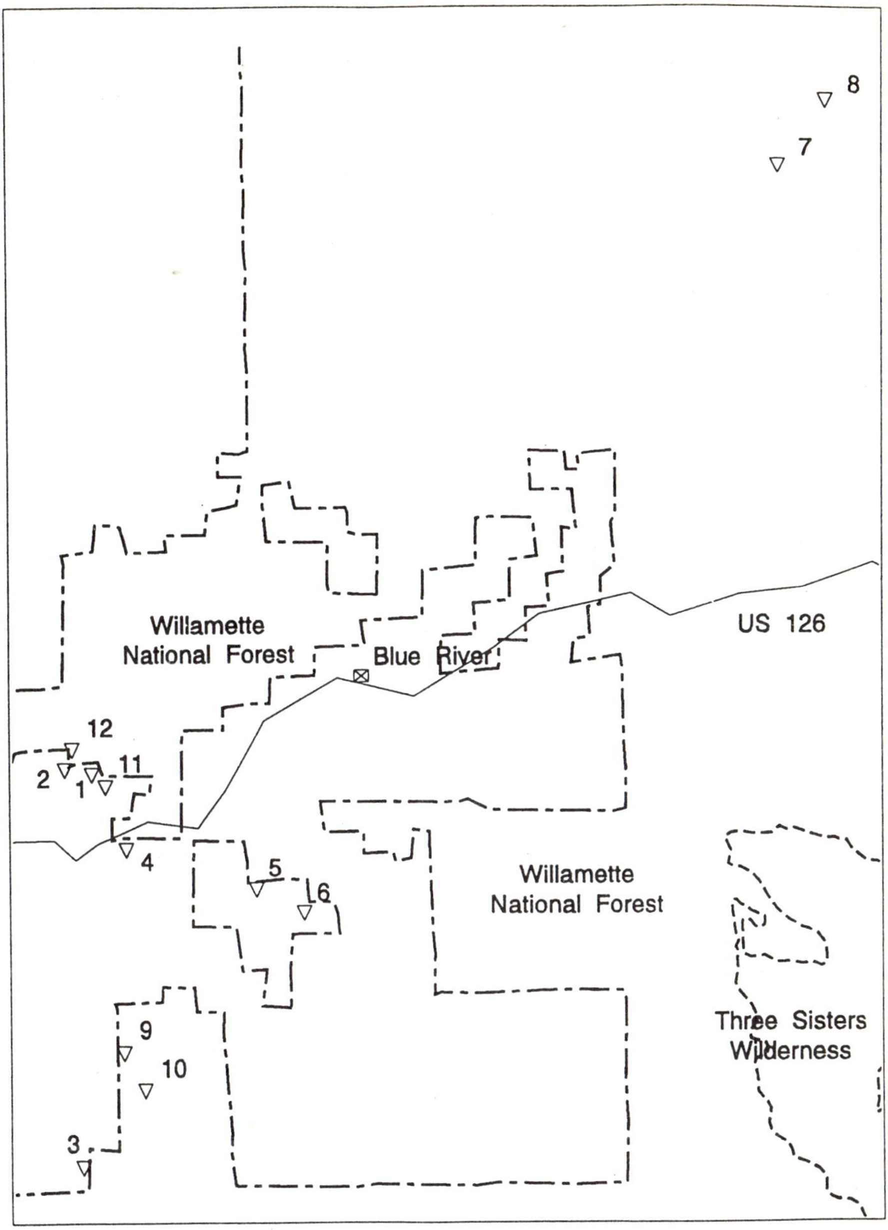

Figure 5. Location of study sites in Western Oregon. From Vega (1993).

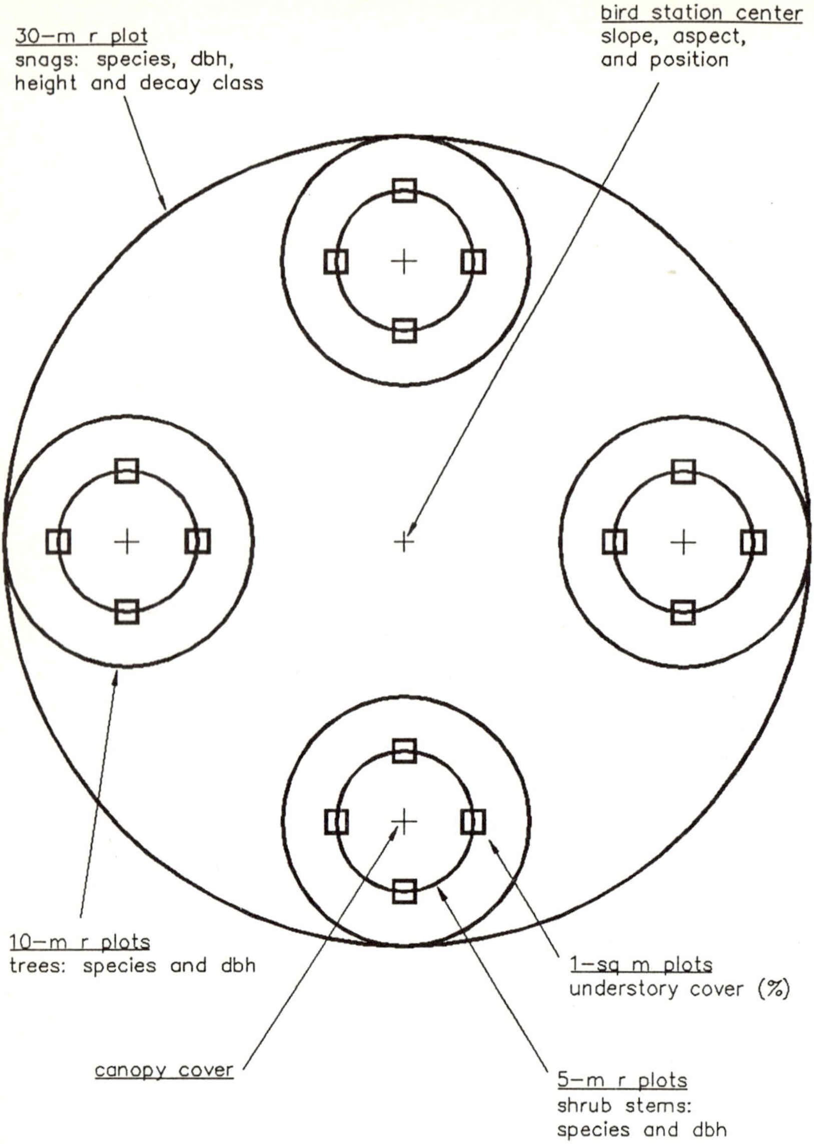

Figure 6. Sampling scheme for habitat attributes. From Vega (1993).

Table 1. Attributes of the 12 stands examined in this study. Modified from Vega (1993).

	CLEARCUT				RETENTION				MATURE FOREST			
STAND[1]	1	2	3	4	5	6	7	8	9	10	11	12
SIZE (ha)	16	17	25	51	28	33	23	19	N/A	N/A	N/A	N/A
ELEVATION (m)	870	940	970	320	590	630	670	780	560	540	620	880
YEAR OF HARVEST	1985	1986	1986 -87	1988-89	1989	1989	1990	1990	N/A	N/A	N/A	N/A
SITE PREPARATION[2]	BCB	BCB	N/A	BCB	PTB	BCB	PAB	PTB	N/A	N/A	N/A	N/A
YEAR OF PLANTING	1987	1987 -88	1990	1989-90	1991	1990	1991	1991	N/A	N/A	N/A	N/A

[1] Numbers correspond with study site locations shown in Figure 1.

[2] BCB - Broadcast burn, PTB - Partial burn, PAB - Pile and burn.

Table 2. Habitat variables measured in this study.

Attribute	Source	Plot-Level Habitat Attributes Definition and Collection Method
Elevation	Map	Elevation above sea level (m) at plot centre
Slope	Field	Average slope (%) within 20 m of plot centre measured with a clinometer
Aspect	Field	Average aspect (degrees) within 20 m of plot centre measured with a compass
Slope Position	Field	Position within 20 m of plot centre by 4 classes
Tree Density	Field	Density of stems (> =2 cm dbh) (number/ha) by species by 10 cm dbh classes within 4 10-m radius subplots placed 20 m from plot centre. Also, within 30 m radius in retention units. This larger area was sampled because of the highly variable distribution of retention trees.
Shrub Density	Field	Density of shrubs (> =0.5 cm dbh) (number/ha) by species by dbh within 4 5-m radius subplots placed 20 m from plot centre
Snag Density	Field	Density of snags (> =2 cm dbh) (number/ha) by species by 5 size classes within 30 m of plot centre
Snag Height	Field	Height of snags measured by clinometer
Snag Decay Class	Field	Decay class of snags in 5 classes of Cline
Canopy Cover	Field	Percent cover of all vegetation above 2 m in height as determined by the moosehorn technique at 4 points 220 m from plot centres
Understory Cover	Field	Percent cover of understory vegetation in the 0-2m height class within 4 0.5-m subplots distributed around each of 4 points 20 m from plot centres

Table 2 (continued)

Understory Habitat	Field	% cover of understory vegetation in each 1-m height class from 0-5 m, within 4 1-m^2 subplots distributed around each of 4 points 20 m from plot centres.
Shrub Basal	Field	Number of shrubs with basal diameters > =0.5 cm by species within 1 5-m radius subplot 20 m south of plot centre.
Understory Biomass	Field	% cover of vegetation in 1 m^2 subplots located 25m south of plot centres. In following categories: all subshrubs, herbs, forbs; all herbaceous plants; and special plant species.
Herbaceous Biomass	Field/lab	Herbaceous material in 1 m^2 subplots located 25m south of plot centres that did not fall into one of the above categories was clipped at ground level, dried, and weighed.

All deciduous tree and shrub leaves, subshrubs, and herbs were considered highly palatable to consumers. All coniferous leaves were considered moderately palatable to consumers.

A first approximation of average annual NPP of wood was made by dividing tree wood biomass by approximate time since last disturbance (100 years). Shrub wood biomass was divided by 5, based on the estimated average age of shrubs across the sites. Leaf NPP for conifers was calculated by dividing foliar biomass by 5, the average life span (in years) of conifer needles. The annual NPP of deciduous tree leaves, subshrubs, and herbs was assumed to be equal to the biomass estimates.

Nest Predation

Relative rates of nest predation among stand types were estimated by an artificial nest experiment (Vega 1993). In June 1993, 12 artificial cup nests constructed from grass were placed at shrub height in each study site. One nest was placed in each of two directions from each bird census point. Nests were located approximately 25 meters from each point along bearings that were roughly perpendicular to one another. The nests were wired onto a sturdy shrub stem approximately 0.5-1.0 m from the ground. Two Japanese quail eggs were placed in each nest. In order to minimize the extent to which human scent influenced the activity of predators, rubber gloves and boots were worn while placing nests and eggs in the study sites. After six days, observers checked the nests and recorded whether or not they were depredated. Signs of depredation included at least one broken or missing egg, or a missing nest.

Bird Censusing

The Variable Centred Plot method (Reynolds et al. 1980) was used to sample birds during the breeding season. Observers walked to a sample point, waited two minutes, then recorded all birds seen or heard for an eight minute period. The bird species, means of detection, distance class from census point, distance from stand edge and type of stand edge were recorded. Flagging was placed 40 m from each census point along the four cardinal directions. These markers helped to determine distance of an observed bird from the census point. The observers, order of censusing stands, and order of visiting census points were systematically altered to minimize sampling biases. Censusing was done each day during the four hours following dawn. Four or five censuses were conduced at each site during May 15-June 20 1993.

Only birds registered within 50 m of plot centres were included in the analyses. This prevented overlap in the areas covered from adjacent plots. Plots of bird detection distance indicated that the songs of all bird species in the study area can be detected within 50 m (Vega 1993); thus we calculated the relative abundance of each species as the number of individual registered within the 50-m radius plot surrounding the plot centre. The results for each species were averaged over plots within a stand and across censuses and are reported as mean number of registrations/ha/census. Only species registered more than 12 times were included in the analyses.

Analyses were done on individual bird species, the understorey bird community (25 species that nest and/or forage primarily on the ground or in the shrub layer), the overstorey bird community (23 species that nest and forage primarily in the overstorey, and the entire bird community. Variables examined included relative bird abundance, bird species richness, and bird diversity (Shannon's index).

Data Analyses

The data were analyzed both to determine the extent to which the faunal communities differed among the three stand types and to test the three hypotheses on the processes underlying these responses. Differences in bird communities among stand types were evaluated using Analysis of Variance. The hypotheses were examined by plotting data as depicted in Figs. 2-4 and analyzing with linear regression.

RESULTS

Bird Community Attributes

Fifty species of birds were registered across the 12 stands. Most of these were uncommon; only 20 species had more than 12 registrations (Table 3). Eleven of these birds differed significantly in abundance among the stand types (Table 4). Lazuli bunting and Willow flycatcher were more abundant in OCSS than in the other stand types (Figure 7). Dark-eyed junco was more abundant in OCSS than CCSC. Steller's jay, a nest predator, reached highest abundance in OCSC. MacGillivray's warbler and Song sparrow were more abundant in both open-canopy stand types than in closed forest. Species significantly associated with closed-canopy stands were Brown creeper, Golden-crowned kinglet, Hammond's flycatcher, Hermit/Townsend's warbler, and Winter wren. Chestnut-backed chickadee was significantly more abundant in CCSC than in OCSS.

Total bird abundance and species richness were slightly higher in OCSS and OCSC than in OCSS, but these differences were not significant (Table 5). Shannon's diversity index was significantly higher in OCSC than in OCSS, and intermediate in CCSC. Understorey bird abundance, richness, and diversity differed significantly among treatments, being higher in the open-canopy stand types. The overstorey bird community was significantly associated with closed-canopy stands.

Table 3. Bird species with greater than 12 registrations in this study.

COMMON NAME	LATIN NAME	CODE	REGISTRATIONS[1]
Brown creeper	*Certhia americana*	BRCR	22
Chestnut-backed chickadee	*Parus rufescens*	CBCH	86
Dark-eyed junco	*Junco hyemalis*	DEJU	129
Evening grosbeak	*Coccothraustes vespertinus*	EVGR	13
Golden-crowned kinglet	*Regulus satrapa*	GCKI	24
Hammond's flycatcher	*Empidonax hammondii*	HAFL	60
Hairy woodpecker	*Picoides villosus*	HAWO	17
Hermit/Townsend's warbler[2]	*Dendroica occidentalis and D. townsendi*	HTWA	63
House wren	*Troglodytes aedon*	HOWR	18
Lazuli bunting	*Passerina amoena*	LABU	14
MacGillivray's warbler	*Oporornis tolmiei*	MGWA	106
Pine siskin	*Carduelis pinus*	PISI	25
Rufus-sided towhee	*Pipilo erythrophthalmus*	RSTO	57
Rufous hummingbird	*Selasphorus rufus*	RUHU	25
Song sparrow	*Melospiza melodia*	SOSP	25
Steller's jay	*Cyanocitta stelleri*	STJA	19
Swainson's thrush	*Catharus ustulatus*	SWTH	15
White-crowned sparrow	*Zonotrichia leucophrys*	WCSP	60
Willow flycatcher	*Empidonax traillii*	WIFL	39
Winter wren	*Troglodytes troglodytes*	WIWR	53

[1] Number of registrations within 50 m of a plot centre across all stands and censuses.
[2] These two species hybridize in the study area and are difficult to distinguish by song. Consequently, they were lumped in this study.

Table 4. Results of ANOVA and Multiple Range Tests for bird species across the three stand types. Stand types are: OCSS - open-canopy, structurally-simple; OCSC - open-canopy, structurally-complex; CCSC - closed-canopy, structurally-complex. Density is mean registrations/census/ha across the 12 stands. Data values with similar superscripts (e.g., ^{a}) differ significantly ($P <= 0.05$). Codes for bird species are defined in Table 3.

	BIRD ABUNDANCE					
CODE	OCSS	OCSC	CCSC	R^2	F	P-VALUE
BRCR	0.0^{a}	0.0^{b}	0.25^{ab}	.71	10.9	.0040
CBCH	0.0^{a}	0.32	0.62^{a}	.45	3.7	.0700
DEJU	0.88^{a}	0.46	0.03^{a}	.54	5.2	.0300
EVGR	0.0	0.0	0.03	.27	1.7	.2400
GCKI	0.0^{a}	0.04^{b}	0.22^{ab}	.80	17.7	.0008
HAFL	0.0^{a}	0.01^{b}	0.64^{ab}	.68	9.4	.0063
HAWO	0.03	0.14	0.01	.28	1.8	.2225
HTWA	0.0^{a}	0.0^{b}	0.74^{ab}	.78	15.9	.0011
HOWR	0.04	0.15	0.0	.25	1.5	.2780
LABU	0.15^{ab}	0.0^{a}	0^{b}	.53	5.1	.0335
MGWA	0.54^{a}	0.58^{b}	0^{ab}	.48	4.2	.0528
PISI	0.0	0.26	0.0	.29	1.8	.2204
RSTO	0.25	0.34	0.01	.32	2.1	.1759
RUHU	0.16	0.11	0.0	.30	1.9	.1997
SOSP	0.14^{a}	0.13^{b}	0.0^{ab}	.46	3.8	.0626
STJA	0.0^{a}	0.20^{ab}	0.0^{b}	.68	9.4	.0062
SWTH	0.05	0.0	0.11	.19	1.1	.3903
WCSP	0.54	0.10	0.0	.20	1.1	.3698
WIFL	0.38^{ab}	0.03^{a}	0.0^{b}	.57	6.0	.0218
WIWR	0.0^{a}	0.0^{b}	0.51^{ab}	.55	5.6	.0265

Table 5. Results of ANOVA and Multiple Range Tests across the three stand types for measures of taxonomic diversity and abundance of the total bird community, understorey birds, and overstorey birds. Stand types are: OCSS - open-canopy, structurally-simple; OCSC - open-canopy, structurally-complex; CCSC - closed-canopy, structurally-complex. Abundance is mean registrations/census/ha across the 12 stands. Data values with similar superscripts (e.g., [a]) differ significantly (P<=0.05).

BIRD COMMUNITY	TREATMENT					
	OCSS	OCSC	CCSC	R^2	F	P-VALUE
TOTAL						
ABUNDANCE	4.15	3.69	3.80	.02	0.1	.8938
RICHNESS	12.5	17.3	15.5	.32	2.1	.1780
SHANNON'S INDEX	1.88[a]	2.40[a]	2.20	.51	4.6	.0412
UNDERSTOREY						
ABUNDANCE	3.83[ab]	2.14[a]	0.80[b]	.69	10.1	.0050
RICHNESS	10.25[a]	9.00[b]	4.50[ab]	.73	11.9	.0030
SHANNON'S INDEX	1.70[a]	1.76[b]	1.07[ab]	.58	6.2	.0201
OVERSTOREY						
ABUNDANCE	.15[a]	1.38[b]	2.9[c]	.80	18.0	.0007
RICHNESS	1.25[a]	7.0[b]	10.0[c]	.89	38.0	.0001
SHANNON'S INDEX	.07[a]	1.6[b]	1.85[b]	.92	50.1	.0001

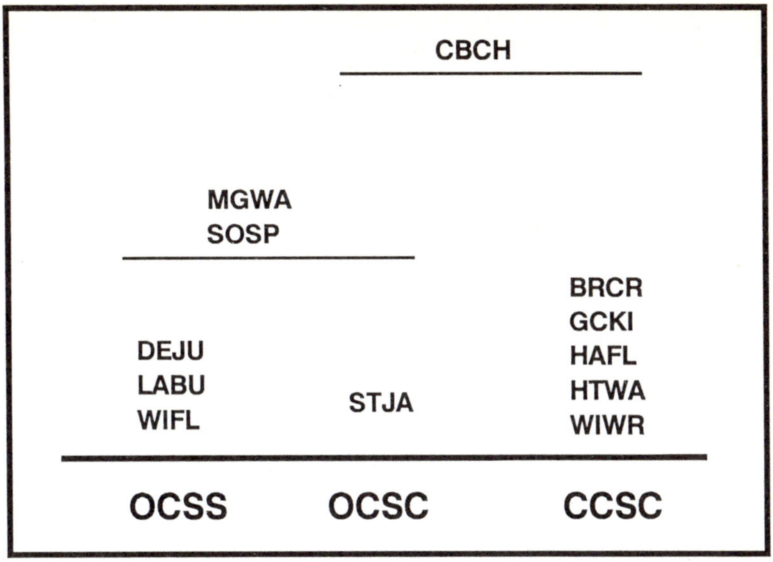

Figure 7. Significant associations among bird species and stand types. Stand types are: OCSS - open-canopy, structurally-simple; OCSC - open-canopy, structurally-complex; CCSC - closed-canopy, structurally-complex. Codes for bird species are defined in Table 3.

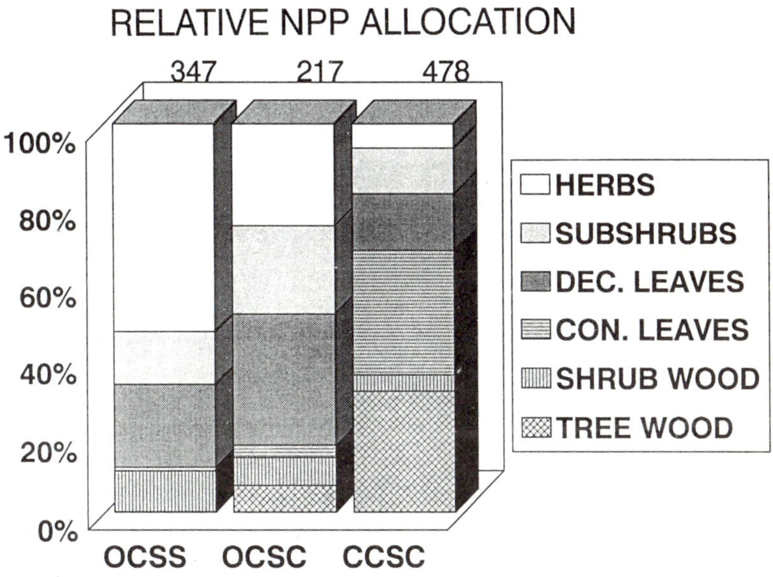

Figure 8. Relative proportion of annual NPP allocated among plant types and parts. Numbers above each bar indicate total NPP in g/m^2/year. Stand types are: OCSS - open-canopy, structurally-simple; OCSC - open-canopy, structurally-complex; CCSC - closed-canopy, structurally-complex.

Table 6. Results of linear regressions relating total, understorey, and overstorey birds to total leaf biomass. "NS" indicates that the relationships in not significant.

BIRD COMMUNITY	R^2	SIGN OF RELATION-SHIP	P-VALUE
TOTAL			
ABUNDANCE	.02	NS	.63
RICHNESS	.00	NS	.84
SHANNON'S INDEX	.00	NS	.97
UNDERSTOREY			
ABUNDANCE	.50	-	.01
RICHNESS	.77	-	.0002
SHANNON'S INDEX	.63	-	.002
OVERSTOREY			
ABUNDANCE	.51	+	.008
RICHNESS	.39	+	.02
SHANNON'S INDEX	.27	+	.08

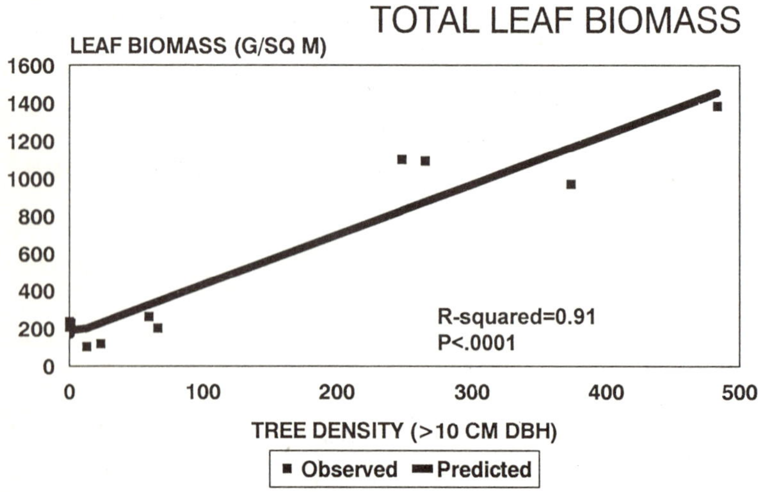

Figure 9. Relationship between annual NPP of all leaves and canopy tree density.

Energy Hypothesis

Relatively more NPP was allocated to consumable plant parts and less to wood in the open-canopy stands (Figure 8). But total NPP was greatest in closed-canopy stands, and consequently, total leaf NPP increased with canopy tree density (Figure 9). NPP of palatable leaves and herbs decreased with increasing canopy density, but these differences were not statistically significant ($n=12$, $r^2=.12$, $P<.27$; $n=12$, $r^2=.27$, $P<.08$, respectively).

Under this hypothesis, we would expect birds to respond to total biomass of consumable biomass, rather than annual NPP. However, none of the variables for the overall bird community was significantly associated with total leaf biomass (Table 6). Understorey bird abundance, richness, and diversity were negatively associated with total leaf biomass. Understorey bird abundance was positively correlated with herb biomass (Figure 10). Overstorey bird abundance, richness, and diversity increased with total leaf biomass (Table 6) and conifer leaf biomass (e.g., Figure 11).

Habitat Niche Hypothesis

Vertical habitat diversity increased with canopy tree density (Figure 12) while horizontal habitat heterogeneity was not related to tree density ($R^2=.007$, $P=.79$). None of the overall bird community variables were related to vertical habitat diversity (Table 7). Understorey birds where inversely related to vertical habitat complexity, while overstorey birds were positively related to vertical habitat complexity (Figures 13-14). No bird variables were significantly associated with horizontal habitat heterogeneity (statistical results not shown).

Distinguishing the relative strengths of bird responses to energy availability and habitat niche diversity was difficult because of strong correlations among these variables. Total leaf, conifer leaf, and deciduous leaf biomass were positively associated with vertical habitat complexity, while palatable leaf and herb biomass were negatively correlated with habitat complexity (Table 8).

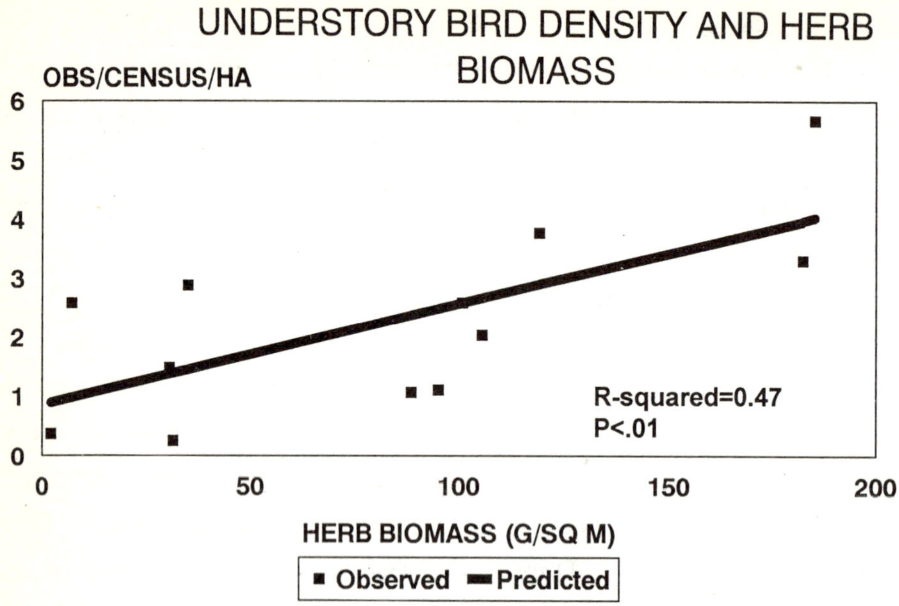

Figure 10. Relationship between the abundance of the understorey bird community and biomass of herbs.

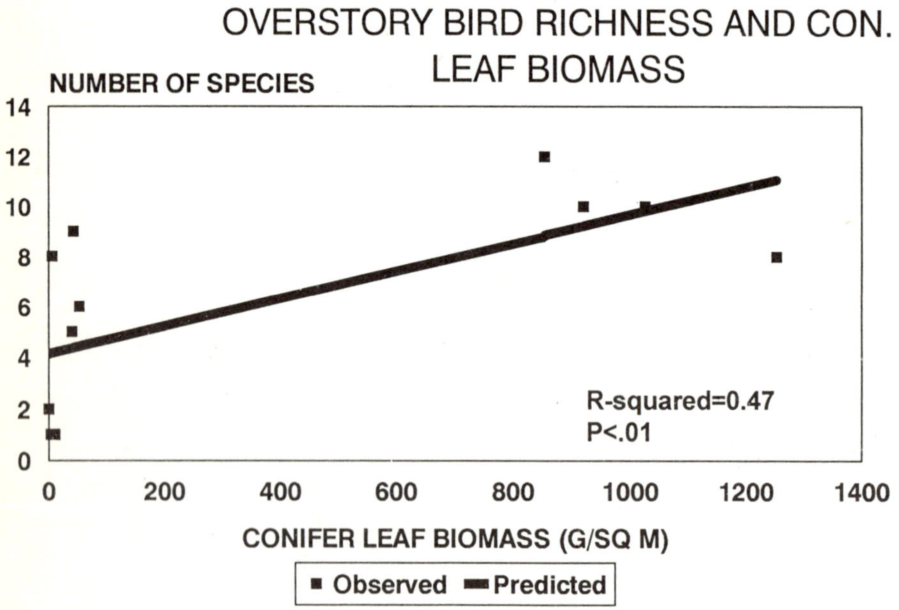

Figure 11. Relationship between overstorey bird diversity and biomass of coniferous leaves.

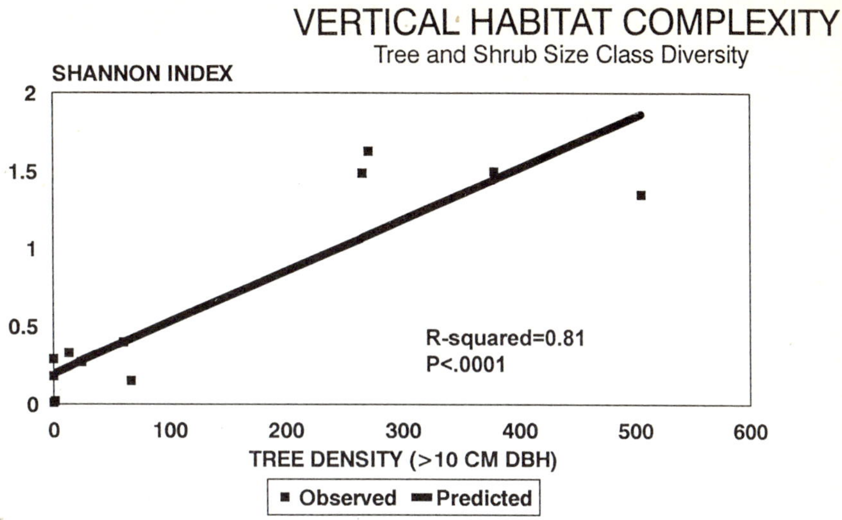

Figure 12. Relationship between vertical habitat complexity and density of canopy trees.

Table 7. Results of linear regressions relating total, understorey, and overstorey birds to vertical habitat complexity. "NS" indicates that the relationships in not significant.

BIRD COMMUNITY	R^2	SIGN OF RELATIONSHIP	P-VALUE
TOTAL			
ABUNDANCE	.07	NS	.80
RICHNESS	.03	NS	.61
SHANNON'S INDEX	.02	NS	.63
UNDERSTOREY			
ABUNDANCE	.57	-	.005
RICHNESS	.66	-	.001
SHANNON'S INDEX	.50	-	.009
OVERSTOREY			
ABUNDANCE	.71	+	.0006
RICHNESS	.60	+	.003
SHANNON'S INDEX	.40	+	.03

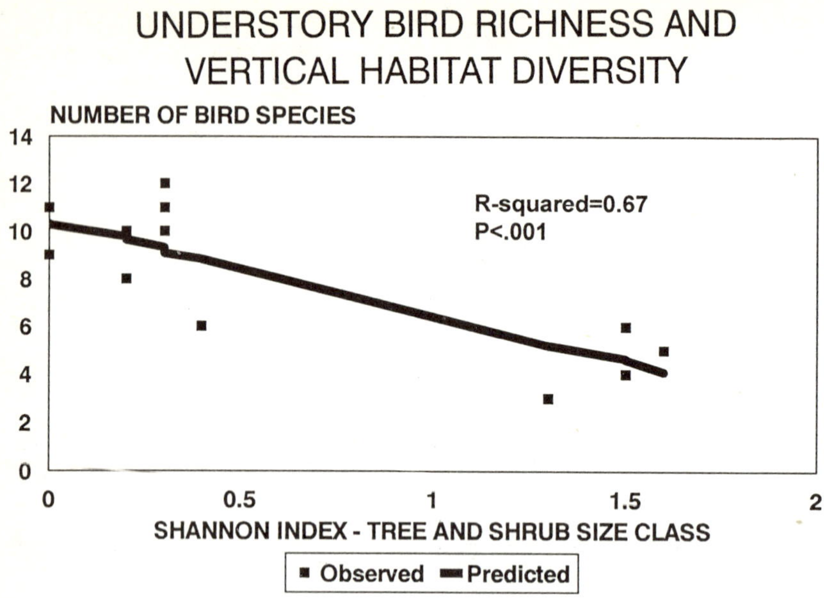

Figure 13. Relationship between understory bird richness and vertical habitat complexity.

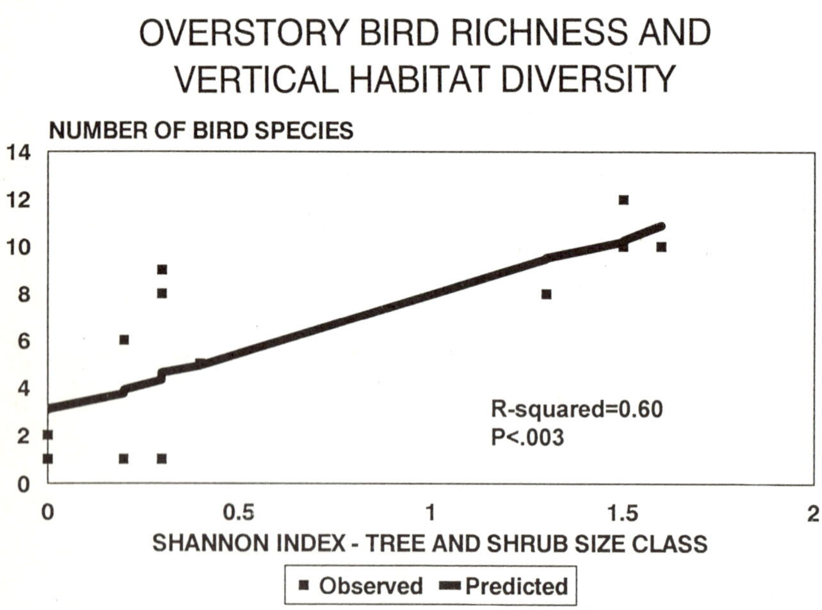

Figure 14. Relationship between overstory bird richness and vertical habitat complexity.

Table 8. Correlation coefficients and (p-values) among habitat complexity variables and plant biomass variables. Habitat complexity variables involve the diversity of tree and shrub size classes. Biomass variables are total biomass of different plant types and plant parts.

BIOMASS VARIABLES	HABITAT COMPLEXITY VARIABLES		
	RICHNESS	SHANNON DIVERSITY	SHANNON DIVERSITY SD[1]
ALL LEAVES	+0.93 .0001	+.94 .0001	-.17 .6000
CONIFEROUS LEAVES	+.94 .0001	+.96 .0001	-.14 .6500
DECIDUOUS LEAVES	+.52 .0800	+.55 .0600	+.60 .0400
PALATABLE LEAVES	-.39 .2000	-.37 .2400	-.17 .5700
HERBS	-.61 .0300	-.61 .0400	-.47 .1200

[1] SD - Standard deviation.

Nest Predation Hypothesis

Mean predation rates were 47% in OCSC stands, 24% in OCSS stands, and 6% in CCSC stands. The differences between OCSC and each of the other two treatments were significant ($P < .05$).

DISCUSSION

Bird Response to Stand Type

We found that several bird species differed in abundance among the three stand types. Those species associated with closed-canopy forest included a bark forager (Brown creeper), conifer foliage gleaners (e.g., Golden-crowned kinglet), a flycatcher (Hammond's flycatcher), and a forest understorey feeders (Winter wren). Three of the open-canopy associates avoided stands with low densities of emergent trees (Dark-eyed junco, Lazuli bunting, and Willow flycatcher). Only Steller's jay, a common nest predator, was uniquely associated with structurally-complex, open-canopy stands. These results are consistent with a study across a broader range of stand types (Hansen et al. in prep.) that found guilds of species uniquely associated with each of four stand types: OCSS, OCSC, CCSC, and closed-canopy, structurally-simple stands.

The overall bird community showed weak patterns across the stand types. Neither total bird density nor richness differed among treatments, but bird diversity was significantly higher in OCSC than in OCSS. Vega (1993) analyzing the same data set with nonparametric statistical tests, did not find significant differences in bird diversity among stand types. Our findings do not strongly support any of the three hypotheses, which predicted that diversity will be either intermediate or low at intermediate canopy densities (see Figure 2c, 3c, 4c). Stronger patterns emerged when the community was divided into understorey and overstorey species. The understorey community was significantly associated with the two open canopy stand types, while the overstorey community was associated with the closed canopy stand type. Bird diversity was relatively high in OCSC for both understorey and overstorey bird communities, and this accounts for the total community reaching highest diversity in this stand type.

<u>Evaluation of Hypotheses</u>

a) Energy and Habitat Niche Hypotheses

As predicted by the Energy Hypothesis, relatively more of annual NPP was fixed in forms available to consumers in the stands with fewer canopy trees. Closed-canopy stands, however, had substantially higher total NPP. Hence, contrary to prediction, annual NPP of available energy was greatest in these stands. These findings differ from a previous attempt to quantify NPP in the region. Turner and Long (1975) found that annual production of available energy was substantially higher in stands just prior to canopy closure than in other stands that had recently undergone canopy closure. It is likely that our open-canopy stands, all of which were created 2-5 years prior to the study, had not yet generated sufficient biomass to utilize available solar energy fully.

Vertical habitat complexity did increase with canopy tree density, as predicted by the Habitat Niche Hypothesis. This finding is consistent with many other studies in finding that habitat complexity increases during the course of succession (see review in Brown 1991).

Contrary to both the Energy and Habitat Niche hypotheses, neither bird community abundance, nor richness, nor diversity were correlated with either available NPP or vertical habitat complexity. The explanation for these findings may lie in the responses of the understorey and overstorey bird communities to these factors. Overstorey birds were strongly associated with both total leaf biomass and with vertical habitat complexity. Understorey birds were negatively associated with these variables. Moreover, understorey bird abundance was positively correlated with herb biomass. Thus, the abundances of individual bird species are closely related to either available energy or to habitat complexity, as predicted by the hypotheses. But the relationships are positive for some species and negative for other species. Consequently, no significant relationship was found when all species are combined for the community-level analysis.

Does the available energy or habitat complexity account for more of the variation in bird abundance? This question cannot be answered fully by our results because these factors were strongly correlated in our study sites. Total leaf, conifer leaf, and deciduous leaf biomass were positively associated with vertical habitat complexity, while palatable leaf and herb biomass were negatively correlated with habitat complexity. Overstorey bird abundance, richness, and diversity were somewhat more strongly associated (R^2 was .14-.28 greater) with vertical habitat complexity than with the energy variables, suggesting that this group might respond more strongly to habitat structure than to available food. The opposite may be the case for understorey birds, given that they were negatively associated with vertical habitat complexity but positively associated with herb weight.

These correlations among available energy and habitat complexity, however, suggest that birds select habitat based on both factors, in order to meet their food, cover, nesting, and other requirements. Bird species differ in food and habitat requirements, partially based on their life history attributes (Urban et al. 1992). Local patters of diversity likely reflect both the distributions of foods and habitats as well as the collective life history attributes of the species comprising the community. Both resource distributions and bird life histories may vary from place to place (Hansen and Urban 1992, Hansen et al. 1992). The important implication is that simple hypotheses about factors controlling species diversity cannot be generalized to all locations.

Whereas MacArthur and MacArthur (1961) found foliage height diversity (a measure of vertical habitat complexity) to correlate strongly with bird diversity in New England forests, we have not found this to be the case in our study sites in western Oregon. Approximately half of the birds in our area are understorey associates with food and/or habitat requirements that are best met at sites that have only a few vegetation layers, with those being near the ground. Knowledge of both resource distributions and the life history attributes of the local animal community are necessary for predicting local patterns of animal diversity (Hansen and Urban 1992).

b) Nest Predation Hypothesis

The results suggested that shrub nesting birds encounter higher nest predation rates in open-canopy stands with emergent canopy trees than in either open stands without emergent trees or in closed forests. Both the abundance of the Steller's Jay, a likely nest predator, and predation rates on artificial nests were significantly higher in the OCSC stands than in the other two stand types. Previous studies have found that predation rates on artificial nests are related to forest patch size, distance from forest edge, and edge type (Wilcove 1985, Small and Hunter 1988, Yahner and Wright 1985). Our study is the first, to our knowledge, to examine nest predation rates as a function of canopy tree density. The results suggest that emergent trees in open-canopy stands provide perches for avian nest predators and facilitate surveillance of nests of prey species, allowing the predators to achieve higher rates of nest location and predation. This effect is apparently reduced in closed forests, because trees are sufficiently dense to impair visibility for the predators.

Implications for Management

Retention of canopy trees in harvest units is an important strategy in the "ecological" forestry practices that have recently emerged in the Pacific Northwest (Swanson and Franklin 1992, Franklin 1992). The rationale for this practice is that it approximates the structural configurations that result from natural disturbances like wildfire, and that various organisms and processes are dependent on this structural pattern. Consequently, retention cuts are being widely implemented on Federal Lands in the PNW, and are advocated nationally under current USDA Forest Service policy.

Our results indicated that canopy tree density strongly influenced pathways and abundance of NPP, habitat complexity, and rates of avian nest predation. These ecological processes, in turn, were associated with patterns of density, richness, and diversity in understorey and overstorey bird communities in our study area. Total bird diversity was greatest in the open-canopy stands with some emergent trees, which might be seen as offering support for the practice of canopy tree retention in harvest units. Nest predation rates were especially high in this stand type, however. Moreover, bird species clearly responded individually to canopy tree density, with some being most strongly associated with each of the three stand types that were studied. It appears that managers can best maintain native bird species in the study area by creating an array of stand structures and ages across the landscape. The relative proportions of the stand types that will best maintain native bird species will likely vary from one geographic location to another, as a function of local resource distributions and animal life history attributes. Conserving species diversity nationally will likely require research and management strategies that are tailored to local ecosystems and animal communities.

Acknowledgements

This study was funded by the US Forest Service New Perspectives Program, the US Forest Service Long-Term Site Productivity Study, and the COPE Program (College of Forestry, Oregon State University. Cooperators include the US Forest Service Blue River Ranger District, the Cascade Center for Ecosystem Management, The National Science Foundation Research Experience for Undergraduates Program, the Eugene Bureau of Land Management Area, Rosboro Timber Co., and the Weyerhaeuser Company. We thank Jim Wernz, Pat Hounihan, Michael Nehls, and Peter Stoel for help in the field and Gody Spycher for assistance in analyzing the biomass data.

REFERENCES

Brown VK (1991) The effects of changes in habitat structure during succession in terrestrial communities. Pages 141-168 *in* Bell S, McCoy E, Mushinsky H (eds) Habitat Structure: The Physical Arrangement of Objects in Space. Chapman and Hall, New York

Cody ML (1985) An introduction to habitat selection in birds. Pages 4-46 *in*: Cody ML (ed) Habitat Selection in Birds. Academic Press, Inc., Orlando, Florida

Currie DJ (1991) Energy and large-scale patterns of animal- and plant- species richness. American Naturalist 137:27- 49.

Franklin JF (1992) Scientific basis for new perspectives in forests and streams. Pages 5-72 *in* Naiman R (ed) Watershed Management: Balancing Sustainability and Environmental Change. Springer-Verlag, New York, New York

Hansen AJ, Spies TA, Swanson FJ, Ohmann JL (1991) Lessons from natural forests: Implications for conserving biodiversity in natural forests. BioScience 41(6):382-392

Hansen AJ, Urban DL (1992) Avian response to landscape pattern: The role of species life histories. Landscape Ecology 7(3):163-180

Hansen AJ, Urban DL, Marks B (1992) Avian community dynamics: The interplay of landscape trajectories and species life histories. Pages 170-195 *in* Hansen AJ, di Castri F (eds) Landscape Boundaries: Consequences for Biotic Diversity and Ecological Flows. Springer-Verlag Ecological Studies Series, New York

Hansen AJ, Hounihan P (In Press) Canopy retention and avian diversity in the Oregon Cascades. *in* Szaro R, Johnston D (eds) Biodiversity in Managed Landscapes: Theory and Practice. Oxford University Press, London

Hansen AJ, Peterson JA, Horvath EL, Lee P (In Review) Do managed plantations support fewer vertebrates than natural forests in the Oregon Coast Range. Can. J. For. Res.

Harris LD (1984) The Fragmented Forest. University of Chicago Press, Chicago, Illinois

Huff M, Raley C (1991) Regional patterns of diurnal breeding bird communities in Oregon and Washington. Pages 177-206 *in* Ruggiero LF, Aubry KB, Carey AB, Huff MH (eds) Wildlife and vegetation of unmanaged Douglas-fir forests. USDA Forest Service General Technical Report PNW-GTR-285, Portland, Oregon

James FC (1971) Ordinations of habitat relationships among breeding birds. Wilson Bulletin 83:215-236

Lehmkuhl, JF, Ruggerio LF, Hall PA (1991) Landscape-scale patterns of forest fragmentation and wildlife richness and abundance in the Southern Washington

Cascade Range. Pages 425-442 *in* Ruggiero LF, Aubry KB, Carey AB, Huff MH (eds) Wildlife and Vegetation of Unmanaged Douglas-fir Forests. USDA Forest Service General Technical Report PNW-GTR-285, Portland, Oregon

MacArthur, RM, MacArthur JW (1961) On bird species diversity. Ecology 42:494-498

Reynolds, TT, Scott JM, Nussbaum RA (1980) A variable circular-plot method for estimating bird densities. Condor 82:309-313

Small MF, Hunter M (1988) Forest fragmentation and avian nest predation in forested landscapes. Oecologia (Berl.) 76:62-64

Spies, TA, Franklin JF, Thomas, TB (1988) Coarse woody debris in Douglas-fir forests of western Oregon and Washington. Ecology 69:1689-1702

Swanson, FJ, Franklin JF (1992) New forestry principles from ecosystem analyses of Pacific Northwest forests. Ecological Applications 2:262-274

Terborgh J (1989) Where Have all the Birds Gone? Princeton University Press, Princeton, NJ

Turner J, Long JN (1975) Accumulation of organic matter in a series of Douglas-fir stands. Can. J. For. Res. 5:681-690

Urban DL, Smith TM (1989) Microhabitat pattern and the structure of forest bird communities. American Naturalist 133(6):811-829

Urban DL, Hansen AJ, Wallin DO, Halpin PJ (1992) Life-history attributes and biodiversity: Scaling implications for global change. Pages 173-195 *in* Solbrig O (ed) Biological Diversity and Global Change. Kluwer Academic Publishers, Dordrecht, the Netherlands

Vega RM (1993) Bird communities in managed conifer stands in the Oregon Cascades: Habitat associations and nest predation. M.S. Thesis, Oregon State University, Corvallis, Oregon

Wilcove D S (1985) Nest predation in forest tracts and the decline of migratory songbirds. Ecology 66:1211-1214

Yahner RH, Wright AL (1985) Depredation on artificial ground nests: effects of edge and plot age. J. Wildl. Manage. 49:508-513

Chapter 13: Fire and Diversity in Canadian Ecosystems

Luc C. Duchesne
Canadian Forest Service
Petawawa National Forestry Institute
Chalk River, Ontario
Canada K0J 1J0

INTRODUCTION

In Canada, before settlement by Europeans, fire was a dominant feature of most landscapes, with a recurrent passage every 50 to 350 years. Short fire frequencies are typical of the boreal and Great-Lakes St-Lawrence forest regions, whereas long frequencies are found in the coastal forests of British Columbia. A particular forest site may have been visited by fire between 30 to 120 times since forest colonization 7,000 years ago. Before that fire interacted with our current vegetation since at least the Miocene or early Pliocene (30 million and 12 million years ago, respectively) (Hopkins 1967). Clearly, Canadian forest ecosystems have evolved under the selective pressure of fire, leading to fire-adapted ecosystems. The nature of these adaptations varies greatly according to each fire regime typical of biogeographic zones and local conditions.

In recent years, forest management based on holistic or ecosystem approaches has been emphasized as a measure to promote biodiversity conservation. One important premise of this approach is that managed forests should maintain their biodiversity and their health even under repeated use. Fire, because of its regulatory role on pre-settlement biodiversity, should be used as a guide for forest management. Further , it is proposed that management of Canadian forests could be conducted in a manner that emulates natural disturbance caused by fire. This stems from the widely accepted general observation that wildfire is responsible for the maintenance of biodiversity in Canadian forests at the microsite (small variations at the 1 m^2 level),

NATO ASI Series, Vol. I 20
Biodiversity, Temperate Ecosystems, and Global Change
Edited by T. J. B. Boyle and C. E. B. Boyle

ecosystem, and landscape levels. The objective of this review is to summarize some of the evidence showing the importance of fire on biodiversity in Canadian forests.

FIRE AND GENETIC DIVERSITY

The evolution of temperate ecosystems under the recurrent passage of fire has led to several adaptive mechanisms. So far, most studies on the effect of fire on the genetics of populations have dealt with commercial tree species. Some light has also been shed on the adaptive mechanisms and their genetics in animal populations have "learned" to protect themselves against forest fires.

Plant adaptations to fire, particularly commercial tree species, have been extensively studied (Rowe 1983) and many species in temperate regions show a wide array of adaptive mechanisms (Gill 1981, Keeley 1981). Serotiny of jack pine cones is such a strong adaptation to fire that it is essential for its regeneration (Cayford and McRae 1983). For example, in burns greater than 400 m in diameter the serotinous cones of jack pile offer it an advantage over non-serotinous species that must colonize from the edge (Johnson 1992). It is difficult for non-serotinous species to shed suitable seed rains beyond 200 m.

Serotiny is a genetically controlled mechanisms (Teich 1970) modulated by local fire regimes. When fires are frequent and lethal, jack pine populations accumulate serotinous individuals whereas, when the fire interval is long or fire intensity is low, populations accumulate more non-serotinous individuals (Gauthier 1991). This type of evidence is rare because few studies have addressed the regulatory effect of fire on tree populations genetics in Canada. Another evolutionary advantage of serotiny is that canopy storage of many seed crops increases the genetic diversity of the replacement stand (Lamon *et al.* 1991).

DIVERSITY AT THE MICROSITE AND STAND LEVELS

There are over 126,000 species of organisms (excluding bacteria) in Canada (Environment Canada 1991). Little work has been conducted. however, to understand the effect of fire on α-diversity, the number of taxa present in a particular ecosystem. Most studies conducted so far have been attempts to understand the relationship between fire and tree survival and growth, or the effect of fire on ungulates (Wright and Bailey 1982). In Canada commercial forest trees comprise a very small fraction of the diversity encountered in our forests. There are approximately 90 commercial forest tree species (Hosie 1980) but only a few, mostly conifer species, have been the object to extensive studies with regard to fire.

Our knowledge of the effect of fire on animals and microorganisms is even more limited. Some work has been conducted to understand the effect of fire on forage availability and quality for large ungulates (Wright and Bailey 1982) and the effect of populations of fur-bearers of economic value (Banfield 1974). Ahlgren and Ahlgren (1965, 1981) researched fire effects on soil bacteria and fungi. New approaches that emphasize other indicator groups are now emerging as shown below.

There are two basic avenues for studying the effect of fire on diversity: 1) the study of indicator groups or processes (Spellerberg 1991) and 2) the cataloguing of all species found in a particular ecosystem. Whereas a census of all taxa found in an ecosystem is interesting, it is not practical because of the high numbers involved, limited funding, and the time require. On the other hand, the use of indicator groups can pose difficulties because of the limited ecological information contained by single groups; it is often claimed that indicator groups best indicate their own state rather than the state of forest ecosystems. Unless one is studying keystone species, defined as pivotal species upon which the diversity of a large part of a community depends, one is faced with the problem of judging the value of indicator species.

Indicator groups react differently to fire disturbance

Ecological indicators are best used to investigate several groups because conclusions can vary according to the indicator used. This is illustrated by recent studies conducted at the Frontier Lake experimental site of the Petawawa National Forestry Institute in Chalk River, in Ontario, comparing the effect of prescribed fire on ectomycorrhizal fungi, litter decomposition, carabid beetle assemblages, and diversity of vegetation in a jack pine forest.

a) Mycorrihizal fungi

Ecologically, mycorrhizal fungi are an important group of soil microorganisms because they promote water and nutrient uptake, protect plants against root pathogens, and enhance plant growth (Duchesne 1993b, Harley and Smith 1983, Jeffries 1987, Le Tacon *et al.* 1987). Their ubiquity in soils makes them widely acceptable as indicators of soil diversity in such varied environments as old-growth forests and disturbed areas (Brundrett *et al.* 1993, Douds *et al.* 1993, Durall *et al.* 1993, Hendrix *et al.* 1993, Herr *et al.* 1993, O'Dell and Ammirati 1993). Investigations at the Frontier Lake experimental site show that ectomycorrhizal infectivity of a jack pine stand was increased by prescribed burning (Table 1) (Herr *et al.* 1993), whereas the vesicular-arbuscular mycorrhizal (VAM) fungi infectivity was decreased (Herr *et al.*, unpublished results). Several hypotheses pertaining to the relative resistance of mycorrhizal fungi to fire or their ecological functions in various types of ecosystems can be advanced to explain the effect of fire of mycorrhizal diversity. However, one interesting corollary is that the effect of disturbance on diversity varies according to the indicator group.

Table 1. Effect of prescribed burning on mycorrhizal infectivity, plant diversity, and litter decomposition one year after clear-cutting a jack pine forest. + shows treatment where the greatest number was found for each indicator group.

Indicator	BURNED-OVER	CONTROL
• Ectomycorrhizal infectivity	+	-
• VAM infectivity	-	+
• Vegetation		
- richness	-	+
- diversity	-	+
• Litter decomposition	-	+
• Carabid diversity	+	-

b) Vegetation as an indicator of biodiversity

Plant diversity is an obvious indicator of biodiversity following fire. In general, it is thought that high vegetation diversity indices in post-fire forests are indicators of productivity (Archibald 1993).

The passage of fire through temperate ecosystems often leads to nutrient pulses that increases plant growth for a short time in post-fire communities (Weber and Taylor 1992). Moreover, at the Frontier Lake site vegetation is important for the storage of nutrients freed by fire (Tellier, Duchesne and Ruel, unpublished results). Plant diversity, assessed using Shannon's and Simpson's indices, was greater in unburned areas compared with the burned-over plots one year after prescribed fire (Tellier, Duchesne, McAlpine and Ruel, unpublished results). These results suggest that fire decreases plant diversity at the microsite level.

c) Carabid beetles

Carabid beetles (Coleoptera: Carabidae) have been investigated with varying degrees of success as indicators of soil diversity after disturbance caused by forest fire (Holliday 1991), clear-cutting (Susted 1981, 1984, Lenski 1982, Jennings *et al.* 1986, Langor *et al.* 1991) scarification (Parry and Rodger 1986), pollutants (Stubbe and Tietze 1982, Kolbe 1988), land reclamation (Day and Carthy 1988), management effect in primeval or old growth forests (Niemelä *et al.* 1988, Terell-Nield 1990), and climate change (Elias 1991). Loreau (Chapter 5) reviews the advantages of using carabid beetles as indicators of diversity. For this discussion, it is important to stress that carabids are good integrators of a substantial amount of ecological information about the biological communities to which they belong (Day and Carthy 1988) because they are important predators in forest soils. Moreover, they constitute a large fraction of the soil arthropod biomass and are efficient natural pest control agents (Edwards *et al.* 1979, Jennings *et al.* 1986, Šustek 1981, Weseloh 1985). Worldwide, there are approximately 40,000 known species of carabids (Loreau, Chapter 5) while in Canada there are approximately 3,000 species (Lindroth 1969). Because carabids show different degrees of habitat selectivity, ranging from specialists to generalist (Niemelä 1990), carabid assemblages can be used to characterize the impact of disturbance in various habitats. Three aspects of carabid diversity are modulated by their environment: relative catches of each species within a particular forest site, absolute population size of individual species, and the presence or absence of species within assemblages.

In North America, particularly in Canada, the effects of forest disturbance on carabid beetle diversity are not sufficiently documented. Preliminary work by Levesque *et al.* (1976) showed that carabid assemblages in Québec varied in individual species population size and in species diversity among nine different forest stands of the same forest region. However, these authors did not emphasize the effect of forest disturbance in their research. Langor *et al.* (1991) demonstrated that carabid assemblages differed between mature lodgepole pine stands and younger planted stands. Carabid species were grouped in three classes based on habitat preference: species with a preference for young stands, those that preferred mature stands, and the carabids present in both groups of stands. Similar results were found in aspen (*Populus*) and spruce (*Picea*) stands (Richardson and Holliday 1982, Holliday 1991) bit, in these cases, there was disturbance caused by wildfire.

Furthermore, Holliday (1991) found that carabid assemblages of burned-over aspen and spruce sites were similar to those of undisturbed aspen sites 11 years following fire (Holliday 1991). it appears that whereas physical and biological conditions modulate carabid assemblages at the microsite level, biogeographic processes control local populations (Loreau, Chapter 5).

At the Frontier Lake experimental site, carabid assemblages unique to each were found in burned-over sites, clear-cut sites, and sites that had received neither clear-cutting nor prescribed burning. Interestingly, some species were found in greater abundance in burned-over areas, some species were found in greater amounts in clear-cut sites, others were found in greater abundance in the neighbouring undisturbed jack pine forest, yet others were equally abundant in all sites (results not shown). Figure 1 shows examples of each of these groups of species.

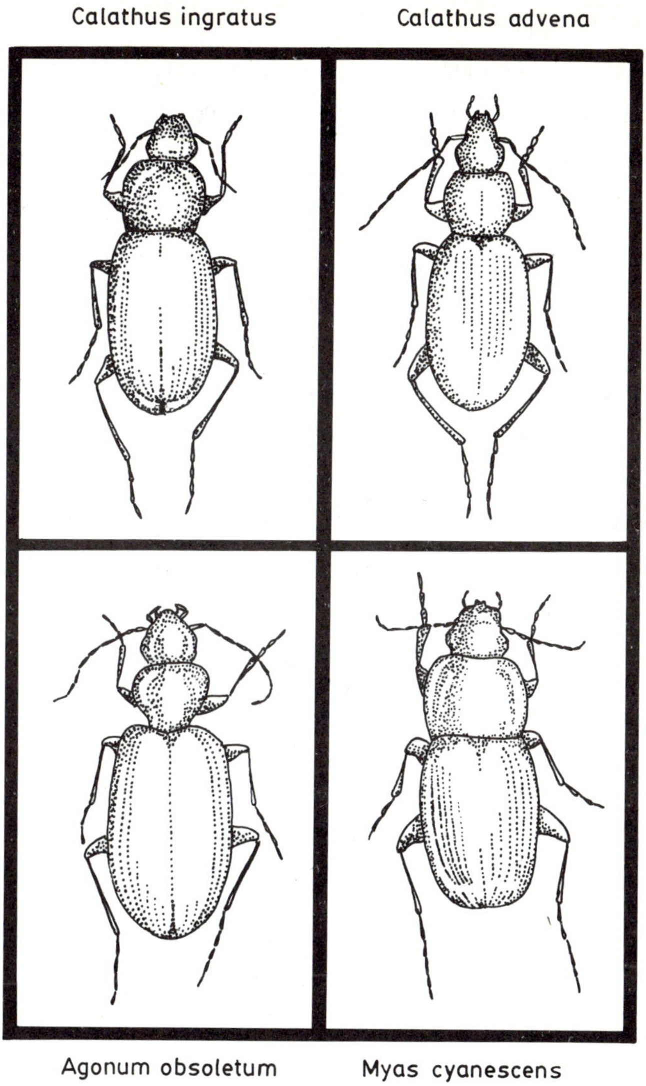

Figure 1: Species illustrating the types of response of carabid beetles to prescribed fire at the Frontier Lake experimental site: *Calathus ingratus,* found in all sites; *Calathus advena*, found mostly in clear-cut sites; *Agonum obsoletum*, found mostly in burned-over sites, *Myas cyanescens*, found mostly in undisturbed forests.

d) Litter decomposition

Litter decomposition and nutrient turnover have long been recognized as factors controlling ecosystem productivity (Harmon *et al.* 1990, Weber 1987). Because litter decomposition is affected by several chemical and physical factors, it constitutes a good candidate for biomonitoring (Swift *et al.* 1979) and was investigated at the Frontier Lake experimental site. One year after treatment, litter decomposition (measured as weight losses of pre-weighed litter bags) was greater in unburned clear-cut sits as compared with burned-over sites (unpublished results).

Clearly, results obtained from an analysis of the effect of prescribed burning on mycorrhizal fungi, vegetation, carabid beetles, and decomposition show that fire has different effects on α-diversity. From these results we can theorize that living organisms may react to fire in four different ways: (1) groups whose diversity is increased by fire, (2) groups whose diversity is decreased by fire, (3) groups whose diversity is not affected by fire, and (4) groups whose diversity reacts in unpredictable patterns (Table 2). It is important to realize that these effects may be ecosystem-specific and will vary over time. It is to be expected that, as post-fire communities evolve, their diversity levels will converge with those of pre-fire communities.

Table 2. Theoretical effect of fire on α-diversity of different taxonomic groups.

Diversity increased by fire	Diversity decreased by fire
Diversity not affected by fire	No predictable effect on diversity

DIVERSITY AT THE LANDSCAPE LEVEL

In recent years ecologists, foresters, and conservation biologists have emphasized the need to address land management at the landscape level (Dunning *et al.* 1992, Thompson and Welsh 1993, Van Wagner 1990, Westman 1990, Wien 1989). This recommendation stems from two basic principles. First, in most cases, proper management at the landscape and ecosystem levels should also help conservation at the stand and microsite levels. Second, fragmentation of habitats is identified as a major cause of loss of stand diversity (Harris and Silva-Lopez 1992). The development of natural age-class distribution is a first step towards the development of tools, based on natural diversity, that help in landscape and ecosystem management. Within an holistic or ecosystem approach to forest management, it is critical to determine a target landscape mosaic (Burton *et al.* 1992, Hof and Joyce 1992, Westman 1990).

Landscapes characterized by recognizable and repeatable mosaics are called "equilibrium" landscapes. In ecology, a territory can reach equilibrium if its surface is at least fifty times greater than the size of disturbance (Shugart and West 1991). In the boreal forest of Northern Ontario, for example, the average fire size is approximately 1,000 hectares (Ward and Tithecott 1993). Thus, landscapes must be at least 50,000 hectares to reach equilibrium. Naturally, these figures vary within and among biogeographic regions according to local and regional variations in the fire regime.

Fire-dependent equilibrium landscapes are characterized by a shifting mosaic of ecosystems where wildlife interact at the metapopulation level. Each equilibrium landscape is characterized by an age-class and community distribution (Bergeron 1991, Duchesne and Gauthier 1993, Suffling 1993, Van Wagner 1978).

In Canada, the natural age-class distribution of forests is better documented than the effect of fire on the landscape composition, although landscape models that predict post-fire mosaics are slowly emerging (for example, see Heaman 1993). Forests submitted to a constant high incidence of fire display negative exponential age distributions (Figure 2) that vary according to local and regional conditions (Duchesne and Gauthier 1993, Suffling 1992, van Wagner 1978).

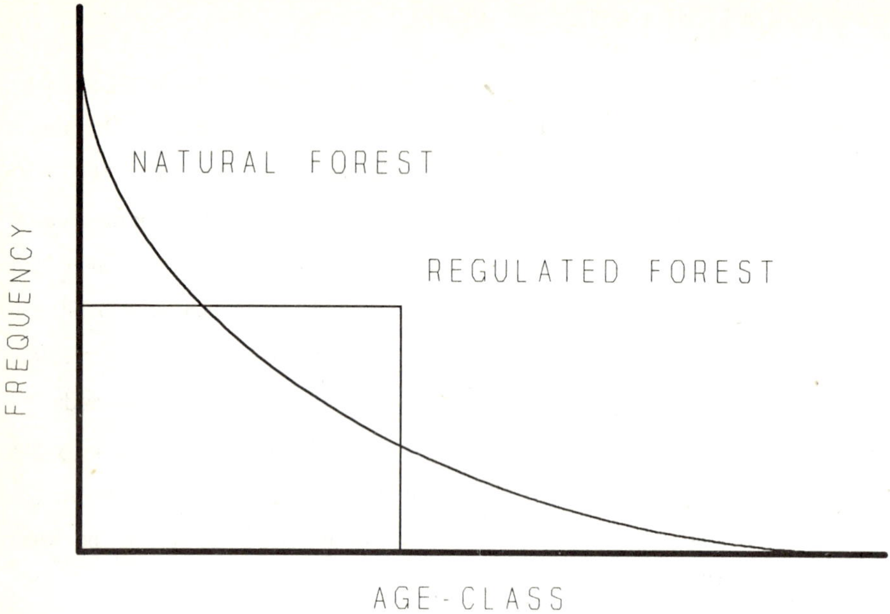

Figure 2. Comparison of the negative exponential age-class distribution with that of a regulated forest. regulated forests have an even age-class distribution whereas the negative exponential shows a large number of young stands and a small number of old stands, presumably old-growth forests.

The abundance of old-growth forests is an important ecological, political, social, and economic issue with international repercussions (Duchesne 1993a). Age-class distributions of primeval landscapes can be used to determine the natural abundance of old-growth forests in Canada's forests (Duchesne and Gauthier 1993). Such knowledge, together with a thorough understanding of the wildlife requirements, can predict the effect of forestry practices or other human activities on biodiversity. However, the factors that ultimately decide the actual quantities of old-growth forests to be conserved are determined by management objectives, themselves based on social values.

it is important to understand that natural landscapes are not necessarily optimal. A well documented example is presented by forests in Wisconsin. These are comparable to parts of the Great Lakes-St. Lawrence forest region. In pre-settlement Wisconsin, forests were made up of approximately 80% old-growth stands and 20% successional communities (Canham and Loucks 1984) harbouring white-tailed deer *(Odocoileus virgininianus)* at 2-4 animals/km^2. Today, Wisconsin forests are less that 5% old-growth and can support 2-9 animals/km^2. This increase in deer population was induced by the creation of forest edges and the incorporation of a large number of successional stands within the landscape (Alverson *et al.* 1988). Thus, current management practices devised to favour large ungulates (Caza and Neave, Chapter 1) may not always be compatible with the preservation of biodiversity based on the perspective of pre-settlement fire.

TO MANAGE DIVERSITY OR TO MANAGE FOR BIODIVERSITY?

There are two reasons why it may be profitable to promote biodiversity as a goal in forest management. First, it is possible that biodiversity and sustainability are linked (Burton *et al.* 1992, Westman 1990), at least in some forest ecosystems. Preserving biodiversity should, therefore, ensure the long term stability of managed ecosystems. Second, forest management aimed toward biodiversity may complement Canada's network of ecological reserves and parks. In the United States, for example, there is a strong case made for using managed forests to increase the effectiveness of protected areas (hansen *et al.* 1991, Wilcove 1989). This point of view may well be applicable to Canada.

However, maximizing biodiversity or managing for biodiversity should not be the general objective of forest management. At the stand levels, maximizing diversity is associated with disturbance levels leading to early successional stages. This is based on observations that species richness is greater in early successional stages than in mature forests. At the regional level, it implies that the landscape mosaic should include a large number of small patches of early successional stands. This would not be compatible with natural age-class distributions, an important component of landscape diversity.

At the landscape level, it is important to emphasize that each forested landscape features intrinsic biodiversity levels associated with unique dynamics. Each landscape type is characterized by a disturbance regime that controls biodiversity. For example, Suffling *et al.* (1988) showed that the size of forest fires and their severity varied greatly across a latitudinal gradient in the northern part of the Ontario boreal forest. The local levels of biodiversity should be used as the yardsticks for biodiversity conservation. Clearly, managing biodiversity is a better goal than managing for biodiversity.

However, it is not always desirable to emulate natural biodiversity. First, maximizing diversity at the genetic level is desirable because it provides the genetic variability for evolutionary changes (Namkoong 1991). Secondly, management of ecological reserves and parks for the protection of single species does not always require a landscape approach. it rests primarily on habitat management for the target species.

MANAGING A SHIFTING MOSAIC

Management of biodiversity from a landscape perspective in a fire-depended environment involves practices that take into account the age-class distribution of forests within a shifting mosaic. For instance, it is recognized that some while and red pine old-growth forests of the Great Lakes-St.Lawrence forest region are now degenerating into tolerant hardwood stands (Carleton and Gordon 1992, Day and

Carter 1990). White and red pine stands are mostly fire-dependent for their regeneration and, so, old-growth white and red pine stands are periodically replaced by successional communities within natural landscapes. Other stands become old-growth forests with time. This leads to a shifting mosaic of forest stands. Managing for age-class distribution with a predetermined proportion of old-growth forests is a viable alternative in this kind of fire-dependent landscape and ia an acceptable way to preserve biodiversity.

CONCLUSIONS

Tailoring forest management to the multi-scale patterns created by fire offers both limitations and advantages. Age-class management can lead to a loss of wood volume or the creation of a landscape mosaic incompatible with other forest activities. For example, maintaining a large portion of old-growth forests may affect populations of large ungulates and alter the size of harvest by hunters. natural age-class forest management may be difficult to achieve in areas of Canada where anthropogenic influences have changed the landscape structure in such a way that it would not be feasible, within the next few generations, to emulate nature. This is particularly evident in the southern and eastern parts of Canada where extensive agriculture, forestry, and urbanization practices have altered the basic landscape structure. A third reason why forests, even if left to the action of nature, would not reach an ideal age-class distribution is that space may be limited for the establishment of an equilibrium mosaic (Shugart and West 1981). landscapes smaller than 50 times the size of disturbance are intrinsically non-equilibrium landscapes with neither typical age-class distributions nor species compositions.

On a more positive note, knowing past disturbance levels caused by fire is helpful in determining he processes involved in maintaining biodiversity. After all, pre-settlement fire occurrence was responsible for the creation and maintenance of primeval forests. Given a thorough knowledge of these factors ecologists, along with biologists, will be able to predict the effect of management and global change on biodiversity.

REFERENCES

Ahlgren IF, Ahlgren CE (1965) Effects of prescribed burning on soil microorganisms in a Minnesota jack pine forest. Ecology 46: 304-310

Ahlgren IF, Ahlgren CE (1960) Ecological effects of forest fires. Bot. Rev. 26: 483-533

Alverson WS, Walter DM, Solheim SL (1988) Forests to deer: edge effects in Northern Wisconsin. Cons. Biol. 2: 348-358

Archibald D (1993) The use of FEC as a tool in prescribed burning. Prescribed burning workshop. 22-23 September, 1993. Petawawa National forestry Institute, Chalk River, Ontario

Banfield AWF (1974) The mammals of Canada. University of Toronto Press

Bargeron Y (1991) The influence of island and mainland lake shore landscapes on boreal forest fire regimes. Ecology 72: 1980-1992

Brundrett M, Abbott L, Jasper D, Ashwath N (1993) The diversity of VAM fungi in undisturbed and disturbed habitats of the Kakadu region of Australia. Proceedings of the 9th NACOM. Guelph, Ontario, 8-12 August, 1993

Burton PJ, Balisky AC, Coward LP, Cumming SG, Kneeshaw DD (1992) The value of managing for biodiversity. For. Chron. 68: 225-237

Canham CD, Loucks OL (1984) Catastrophic windthrow of the pre-settlement forests of Wisconsin. Ecology 65: 803-809

Carleton TJ, Gordon AM (1992) Understanding old-growth red and white pine dominated forests in Ontario. Ontario Ministry of Natural Resources

Cayford JH, McRae DJ (983) The role of fire in jack pine forest. *in* Wein, RS, DA McLean (eds). The role of fire in northern circumpolar ecosystems. John Wiley and Sons, New York

Day KR, Carthy J (1990) Changes in carabid beetle communities accompanying a rotation of sitka spruce. Agric. Ecosys. and Environ. 24: 407-415

Day, RJ, Carter JV (1990) Stand structure and successional development of the white pine and red pine communities in the Temagami forest. Ministry of Natural Resources

Douds DD, Galvezs L, Janke RR, Wagoner MA (1993) The effect of tillage regimes and farming system upon VAM fungi. Proceedings of the 9th NACOM. Guelph, Ontario 8-12 August, 1993

Duchesne LC (1993a) Defining Canada's Old-growth Forests - Problems and solutions. For. Chron (submitted)

Duchesne LC (1993b) Ectomycorrhizal fungi and the control of root diseases. Mycorrhizal News 5: 1-5

Duchesne LC, Gauthier S (1993) The theoretical age-class distribution of white and red pine forests in the Great-lakes St-Lawrence Forest region. For. Chron (submitted)

Duchesne LC, McAlpine RS (1993) Carabid beetles as indicators of biodiversity following prescribed fire, clear-cutting and scarification. Canadian Forest Service. Petawawa National Forestry Institute Information report (In press)

Dunning JB, Danielson BJ, Pulliam HR (1992) Ecological processes that affect populations in complex landscapes. Oikos 65: 69-175

Durall D, Jones MD, Simard S (1993) Mycorrhizal diversity in mixed and monoculture plantations of *Betula papyrifera* and *Pseudotsuga mensiesii*. Proceedings of the 9th NACOM. Guelph, Ontario 8-12 August, 1993

Edwards CA, Sunderland CD, George KS (1979) Studies of polyphagous predators of cereal aphids, J. Appl. Ecol. 16: 811-823

Elias SA (1991) Insects and climate change. BioScience 41: 552-559

Environment Canada (1991) The state of Canada's environment. Ministry of Supply and Services, Canada

Gauthier S (1991) Structure génétic et sérotinisme de populations de pin gris (*Pinus banksiana* Lamb.) soumises àdeux regimes de feux distincts. PhD dissertation. University of Montreal, Quebec

Gill M (1981) Fire adaptive traits of vascular plants. *in* Proceedings of Fire regime and ecosystems properties conf. US Department of Agriculture general technical report WO-26

Hansen AJ, Spies TA, Swanson FJ, Ohmann JL (1991) Conserving biodiversity in managed forests. Bioscience 41: 382-392

Harley JL, Smith SE (1983) Mycorrhizal symbiosis. Academic Press, New York

Harmon ME, Barker GA, Spycher G, Greene SE (1990) Leaf-litter decomposition in the *Picea/Tsuga* forests of Olympic National Park, Washington, USA. for. Ecol. Management 31: 55-66

Harris LD, Silva-Lopez G (1992) Forest fragmentation and the conservation of biological diversity. *in* Fiedler P, Jain S (eds). Conservation biology, the theory and practice of nature conservation, preservation, and management. Chapman and Hall, New York

Heaman D (1993) The role of fire in Temagami's forests. Comprehensive planning program. Ontario Ministry of Natural Resources, Temagami area

Hendrix JM, Gue B, Zhi-Qiang An (1993). Crop rotation disrupts mycorrhizal fungal communities. Proceedings of the 9th NACOM. Guelph, Ontario 8-12 August, 1993

Herr DG, Duchesne LC, Tellier R, McAlpine RS, Peterson RL (1993) Effect of prescribed burning on the ectomycorrhizal infectivity of a forest soil. Proceedings of the 9th NACOM. Guelph, Ontario 8-12 August, 1993

Hof JG, Joyce LA (1992) Spatial optimization of wildlife and timber in managed forest ecosystems. Forest Science 38: 489-508

Holliday NJ (1991) Species response of carabid beetles (Coleoptera: Carabidae) during post-fire regeneration of boreal forest. Can. Ent. 123: 1369-1389

Hopkins DM (ed) (1967) The Bering land bridge. Standford Univ. Press, Standford, CA

Hosie RC (1980) Les arbres indigéns du Canada. Ministere des Approvisionnements et Services, Canada

Jeffries P (1987) Use of mycorrhizae in agriculture. CRC Crit. Rev. Biotechnol. 5: 319-357

Jennings DT, Houseweart MW, Dunn GA (1986) Carqabid beetles (Coleoptera: Carabidae) associated with strip clearcut and dense spurce-fir forests of Maine. Coleopt. Bull. 40: 251-263

Johnson EA (1992) Fire and vegetation dynamics: studies from the North American boreal forest. Cambridge University Press, Cambridge

Keeley JE (9181) Reproductive cycles and fire regimes. *in* Proceedings of Fire regime and ecosystems properties conf. US Department of Agriculture general technical report WO-26

Kolbe W (1988) Okotoxikologische Aspekte-aufgezeigt as beispiel der coleopteren Fauna von Waldboden. Mitt. Dtsch. Ges. Allg. Angew. Ent. 6: 458-463

Lamont BB, Le Maitre DC, Cowling RM, Enright NJ (1991) Canopy seed storage in woody plants. Bot. rev. 57: 277-317

Langor DW, Niemelä J, Spence RJ (1991) Effects of forestry practices on carabid assemblages in lodgepole pine forests in western Alberta. Proc. Can. Ent. Soc. Meet. Montreal, Qué., 21-23 September, 1991

Lenski RE (1982) The impact of forest cutting on the diversity of ground beetles (Coleoptera: carabidae) in the Southern Appalachians. Ecol. Entomol. 7: 385-390

Le Tacon F, Garbaye J, Carr G (1987) The use of mycorrhizas in temperate and tropical forests. Symbiosis 3: 179-206

Levesque C, Dube J, Pilon JG (1976) Inventaire et ètude biocènotique des coleoptères carabidae de biotopes forestiers des Laurentides. Can. Nat. 103: 5889-582

Lindroth CH (1969) The ground beetles of Canada and Alaska. Part 6. Opusc Entomol. Suppl. 34

Namkoong G (1991) Biodiversity-Issues in genetics, forestry and ethics. For. Chron. 68: 438-443

Naylor B (1993) The impact of natural and prescribed fire on wildlife. *in* Prescribed burning workshop. 22-23 September, 1993. Petawawa National Forestry Institute, Chalk River, Ontario

Niemelä J (1990) Effect of changes in the habitat on carabid assemblages in a wooded meadow on the Aland Islands. Notulae Entomologicae 69: 169-174

Niemelä J, Haila Y, Halme E, Lahti T, Pajunen T, Punttila P (1988) The distribution of carabid beetles in fragments of old coniferous taiga and adjacent managed forest. Ann. Zool. Fennici 25: 107-119

O'Dell TE, Amimirati (1993) Ecology of ectomycorrhizal fungi in old-growth *Pseudotsuga heterophylla* forests of the Olympic Peninsula, W.A. Proceedings of the 9th NACOM. Guelph, Ontario 8-12 August, 1993

Parry W, Rodger D (1986) The effect of soil scarification on the ground beetle fauna of a Caledonian pine forest. Scottish For. 40: 1-9

Richardson RJ, Holliday NJ (1982) Occurrence of carabid beetles (Coleoptera: carabidae) in a boreal forest damaged by fire. can. Ent. 114: 509-514

Rowe S (1983) Concepts of fire effects on plant individuals and species. *in* Wein, RW, DA MacLean (eds). The role of fire in Northern circumpolar ecosystems. John Wiley and Son, New York

Spellerberg IF (1991) Monitoring ecological changes. Cambridge University Press, Cambridge

Shugart HH, West DC (1981) Long-term dynamics of forest ecosystems. Am. Sci. 69: 647-652

Stubbe A, Tietze F (1982) Okologische Untersuchugen an carabidengesellschaften entlang einer Trasee immissionsgeschadigter Kiefernforste in der Dubener Heide. Arch. Natur. land, Berlin 22: 27-44

Suffling R (1992) Climate change and boreal forest fires in Fennoscandia and Central Canada. Catena, suppl. 22: 111-132

Suffling R (1993) Climate change and disturbance by fire in boral and sub-alpine forests. *in* Holten JL, Paulsen G, Oechel WC (eds) Impacts of climatic change on natural ecosystems, with emphasis on boral and arctic/alpine areas. Norwegian Institute of Nature Management

Suffling R, Lihou KL, Morand Y (1988) Control of landscape diversity by catastrophic disturbance: a theory and a case of fire in a Canadian boreal forest. Environmental Management 12: 73-78

Šustek Z (1981) Influence of clearcutting on ground beetles (Coleoptera, Carabidae) in a pine forest. Cechoslovaniae 12: 243-254

Šustek Z (1984) Carabudae and Staphylinidae of two forest reservations and their reactions on surrounding human activity. Biologia (Bratislava) 39: 137-162

Swift MJ, Heal OW, Anderson JM (1979) Decomposition in terrestrial ecosystems. University of California Press, Berkeley, CA

Teich AH (1970) Cone serotiny and inbreeding in natural populations of *Pinus banksiana* and *Pinus contorta*. Can. J. Bot. 48: 1805-1809

Terell-Nield C (1990) Is it possible to age woodlands on the basis of their carabid beetle diversity. The Entomologist 3: 136-145

Thompson ID, Welsh DA (1993) Integrated resource management in boreal forest ecosystems-impediments and solutions. For. Chron. 69: 32-39

Van Wagner CE (1978) Age-class distribution and the forest fire cycle. Can. J. For. res. 8: 220-227

Van Wagner CE (1990) Six decades of forest fires in Canada. For. Chron. 66: 133-137

Ward PC, Tithecott AG (1993) The impact of fire management on the boreal landscape of Ontario. Ontario Ministry of Natural Resources. Aviation, Flood and Fore Management branch. Publication No.305

Weber MG (1987) Decomposition, litter fall, and forest floor nutrient dynamics in relation to fire in eastern Ontario jack pine ecosystems. Can. J. For. Res. 17: 1496-1506

Weber MG, Taylor SW (1992) The use of prescribed burning in the management of Canada's forested lands. For. Chron. 68: 324-334

Weseloh RM (1985) Changes in population size, dispersal behavior and reproduction of *Calosoma sycophanta* (Coleoptera: carabidae) associated with changes in gypsy month *Lymantria dispar* (Lepidoptera: Lymantriidae), abundance. Env. Ent. 14: 370-377

Westman WE (1990) Managing for biodiversity. BioScience 40: 26-33

Wiens JA (1989) Spatial scaling in ecology. Functional Ecol. 3: 385-397

Wilcove DS (1989) Protecting biodiversity in multiple-use lands: lessons from the US forest Service. Trends Ecol. Evol. 4: 385-389

Wright HA, Bailey AW (1982) Fire ecology, United States and Southern Canada. Wiley Intersciences, New York

Chapter 14: Consequences of a 150 years' history of human disturbances in some forest ecosystems

Roman J.M. Lenz
GSF - Forschungszentrum für Umwelt und Gesundheit
Neuherberg
85764 Oberschleißheim

INTRODUCTION

Human disturbances in forest ecosystems have a long history in most countries of the world. Even before the need for arable land increased, the forest served as a nutritional basis for firewood and timber, fruits, wildlife and other products. In central Europe, about two thirds of the land - predominantly the more suitable sites for agriculture, but also sites in the vicinity of industrial and urban areas - was deforested over the centuries, and in the forests which remained there was also extensive exploitation.

This history of human disturbances or impacts can be classified according to type of disturbance. The first type of disturbance was complete destruction by clear-cutting on sites where regrowth was prevented by the dry climate and shallow soils - for example, on the Dalmatian karst. The forest was thus converted into arable land and grasslands.

The second type of disturbance arose when forests were exploited by harvesting (whole) trees, raking litter, and grazing, due to an increase of population, especially during the Middle Ages. During the Middle Ages, and for centuries thereafter, utilization - predominantly grazing in forests - was the only possibility for peasant farmers to increase yields. The tragedy of commons (see Hardin 1968), subjected to exploitation and degradation was in general stopped only about two hundred years ago. In Germany, governmental restrictions on forest utilization were set up, and began to effect a reduction in the intensity of forest usage only during the second half of the

NATO ASI Series, Vol. I 20
Biodiversity, Temperate Ecosystems, and Global Change
Edited by T. J. B. Boyle and C. E. B. Boyle

nineteenth century. In some regions, however, we could litter raking, forest grazing and other activities have continued until at least the late 1950s. A hundred and fifty years ago a new phase - a third type of disturbance - started: deposition of acidifying compounds, as a consequence of fuel combustion. These depositions have rapidly increased and affected more and larger areas, partly due to the construction of taller chimneys. Consequently, these emissions of air pollutants are deposited even in mountainous and remote forested areas, in some cases resulting in the so-called "Waldsterben", or forest decline.

This chapter will focus on the consequences of the second and third types of disturbance that have occurred over the last 150 years in some regions of Germany, especially in the northeastern part of Bavaria. During this time, forest utilization was still intensive and/or results of earlier disturbances still remained - mainly, reduced humus layers and accumulated acids in the soil - affecting tree composition and nutritional status[1/]. Air pollution also became an important factor during this period.

Despite the fact that, since the early nineteenth century, forest management has been based (or has been thought to be based) on the principles of sustainability, forest exploitation (and its history), combined with the impacts of air pollution, has significantly affected productivity. In some regions we even have a kind of ecological "retrogression" (Lenz and Haber 1990), rather than succession. Hence two major processes have been interacting during this time: on the one hand, vegetation and soil organisms "improve" site conditions by enhancing storage capabilities and nutrient turnover rates, leading towards a climax ecosystem and, on the other hand, ongoing utilization plus acidic deposition tend to reduce site productivity.

In short, the actual situation, which cannot be explained without this historical and site-specific knowledge, might be described as a phase of both natural and human-accelerated succession (by reforestation). At the same time, natural disturbances associated with ecological succession have been mixed up with mostly anthropogenic disturbances causing retrogression. In this chapter, results are

---

1/ In this context I like very much the description of an ecosystems' state by its chemical identity (see Gregorius, Chapter 9)

presented of the material balance approach, which has been used to explore and quantify processes and their consequences.

THEORY AND METHODOLOGICAL FRAMEWORK

The theory of material balance was developed as a contribution to the International Biological Program (IBP) during the late 1970s and early 1980s, and are has been described and formulated explicitly by Ulrich (1989). The material balance equation - an enlarged photosynthesis and respiration equation - is used as a basis for the determination of deviations from a steady state of an ecosystem. Monitoring of these deviations and thus the quantification of fluxes and rates of the relevant process components, provides information about the total amount and the direction of changes in the material budget. Due to the fact that nearly all changes in the material budget also lead to changes in occurrence, abundance, productivity and vitality of organisms, the concept is nowadays one of the most important in ecosystem research (Figure 1).

The general material balance theory can be used to identify specific characteristics of ecosystems and the environment. For example, in colder climates turnover and mineralization rates are slower, and thus growth rates are restricted. The consequences are accumulation of dead organic materials and, at the stand level, this results eventually in die-back.

The main reason for this response might be the decrease in irradiance at ground level due to the shading associated with tree growth, thus restricting the mobilization of nutrients. Under tropical conditions irradiance will not become a limiting factor for soil organisms: turnover rates are high, and nearly all materials are stored in the living biomass above ground. Under natural conditions in the tropics, we therefore find regeneration occurring due to death of individuals, rather than as a consequence of stand level processes (Figure 2).

The consequences for sustainable utilization options are obvious: clearcutting and whole tree harvesting might be more "natural" in boreal forests than in tropical forests: the latter ones might be completely destroyed or very much retarded if soil storage and release of nutrients cannot provide enough nutrients for a regrowth of a new forest generation.

Material Balance Equation

$$aCO_2 + xM^+ + yA^- + (y - x)H^+ + zH_2O \underset{}{\overset{h \cdot \nu}{\rightleftharpoons}}$$

$$(C_aH_{2z}O_zM_xA_y)_{org.\ Sub.} + (a + ...)O_2$$

→ Type-specific turnover and reaction (e.g. buffering in soils, deposition)

→ Proton budget types, reaction types

Figure 1. Details of the material balance equation

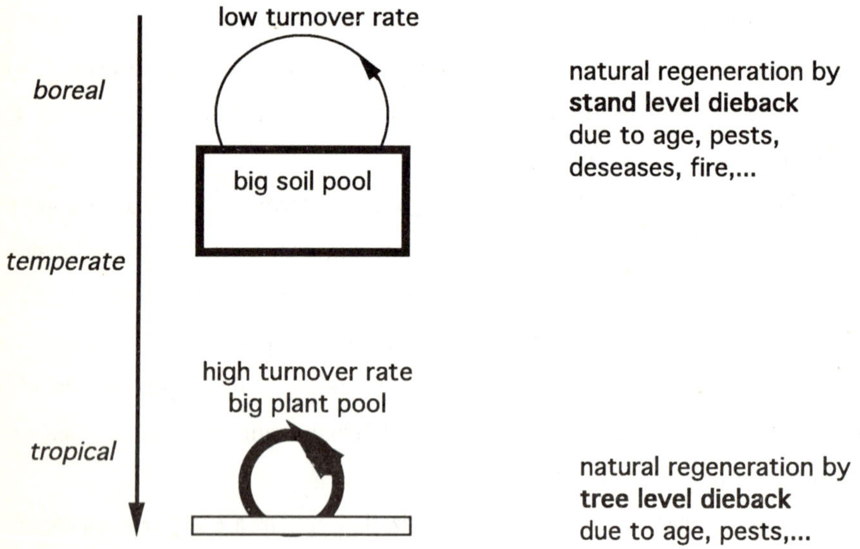

Figure 2. Characteristics of natural regeneration in relation to turnover rates in forests

Temperate forests are in general somehow intermediate: medium turnover rates and storage of materials in the soil as well as in the above ground biomass. These principles - but also the seasonality of growth and mineralization and thus the risk of decoupling of uptake and leaching from the soil - underline the importance of both components - above ground biomass and soil (including humus) - and the necessity of managing both together. Otherwise steady-state-like conditions and sustainable forest utilization cannot be achieved.

A component model, focusing on the proton budget and its internal production and consumption rates, should elucidate some processes and consequences in more detail (Figure 3). The proton budget must be taken into consideration because of the importance of electronegativity (see also Figure 1). The decoupling of production and consumption rates of electronegativity via cations and anions in an ecosystem (besides uptake and release of nutrients as such and their transport in the soil) determines rates of chemical reactions, consequently leading to a modification of availability of nutrients, storage of acidity (or alkalinity), and transport of the minerals (see literature quoted in Figure 3). On more productive sites, and at early stages of growth, forest

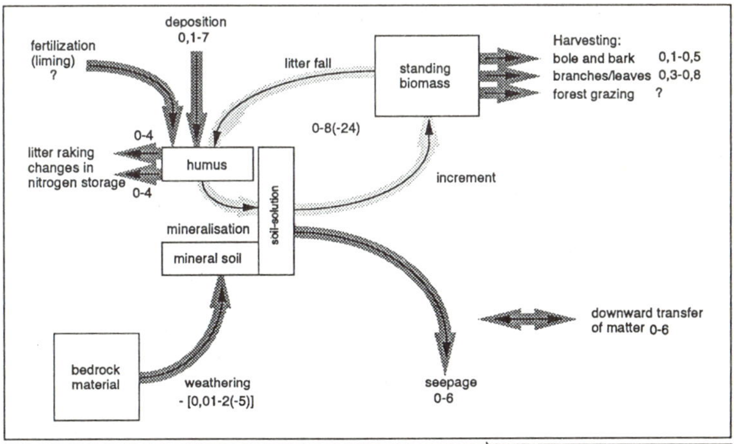

Sources:
BERDEN et al., 1987; BREDEMEIER & ULRICH, 1989; v. BREEMEN et al., 1984; FÖLSTER, 1985; GLATZEL, 1989; HAUHS, 1989; HAUHS et al., 1989; MATZNER, 1988; NILSSON & GRENNFELT, 1988; PACES, 1986; REUSS & JOHNSON, 1986; SCHULZE et al., 1989a,b; SVERDRUP & WARFVINGE, 1987a, b; ULRICH, 1987, 1989 a, b; ULRICH & MEYER, 1987

Figure 3. Internal (light arrows) and external (dark Arrows) proton production and consumption processes in forest ecosystems (with special regard to utilization): kmol/ha/year

utilization and deposition, together with the internal proton production, are the most important factors in the material balance. The only compensation for loss of alkalinity in an ecosystem is the delivery of base cations by weathering of carbonates, primary silicates and other minerals in the soil. Some measured and estimated data are summarized in Figure 3.

FOREST UTILIZATION, AIR POLLUTION, AND ACID DEPOSITION

Following the restriction of forests to the less fertile and agriculturally marginal sites, the export of organic substances and thus the net production of acidity became relatively more important, compared with better sites. Litter raking, clearcutting, coppicing, whole tree harvesting and grazing are, as mentioned above and as indicated in Fig 3, the most important impacts on the Acid Neutralizing Capacity (ANC) of soils. In the nineteenth century, the rate of reduction in the ANC were related to site and utilization conditions. Thus, the lower the productivity of the forest ecosystem, and the lower the intensity of utilization, the less harmful were the effects. Because of this, further investigation of the effects of utilization on the material balance and the proton budget do not make sense without having detailed historical data. Consequently, some examples and calculations will be given for a region of northeastern Bavaria later in the chapter. Before that, the history of acid deposition impacts on forest ecosystems will be summarized, and will also be elaborated more explicitly later. The conceptual models applied will be presented, in order to create a link between theory and application.

Air pollution and acid deposition

Although the increase in emissions of acidifying compounds in the industrialized countries seems to have stopped, the actual levels are much higher than natural levels. For the Federal Republic of Germany (within the pre-unification borders) according to Ulrich (1983), the total emission of accumulated acids since 1850 is estimated at about 300 kmol/ha (Figure 4). In remote areas like mountainous regions, where only small amounts of potentially neutralizing compounds (*e.g.* dust) are likely to be deposited, these acidic emissions might have been deposited in even higher amounts

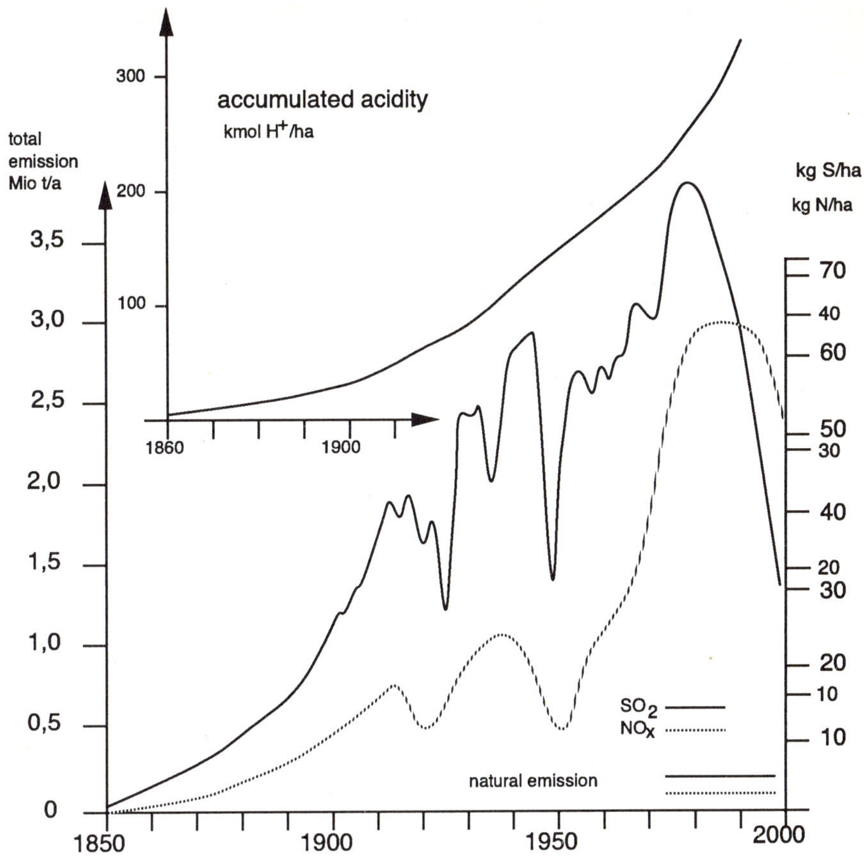

Figure 4. Annual emission of SO_2 and NO_x (in million tonnes per year, and kg per hectare per year), as well as the cumulative emission of acidity in the Federal Republic of Germany (pre-October 1990 borders) since 1850 (after Ulrich 1989b) - from 1986 onwards, estimated by UBA (1989)

due to the high filtering capacity of the forest canopy. Estimates over 150 years are fairly rough, and the actual ammounts certainly vary very much from site to site. Central Europe, together with northeastern America and east Asia, is one of the most important receptors of acidity in the world (Figure 5).

CONCEPTUAL MODELS AND ASSESSMENT APPROACHES

According to the theory, and the combination of region and site-specific factors, different conceptual models for risk assessment seem to be appropriate. A conceptual model, with various modifications will be discussed below. This model is based the principles previously described for calculating deviations from a steady state. The impacts on the material budget will also be discussed.

The Critical Loads Concept

The concept of Critical Load was described for the first time by Nilsson and Grennfelt (1986) and is defined as the load which causes no harmful effects to the most sensitive ecosystems over the long term. Nowadays great efforts are taken to apply of this concept, for example in the UN Convention on Transboundry Air Pollution (see Hettelingh *et al.* 1991). The basic and most simple model for Critical Load for acidity states: (Figure 6).

$$Ex\ (A) = A_{dep} - ANC_{w}$$

Where: CL (A) = ANC (W),

Ex(A) = Exceedance of Acidity,

A_{dep} = Acid deposition,

ANC_{w} = Acid Neutralizing Capacity due to weathering of minerals, and

CL(A) = Critical Load for Acidity

In improving our understanding of the temporal response of (forest) soils to deposition, the buffering capacity has to be taken into account. The pool of base

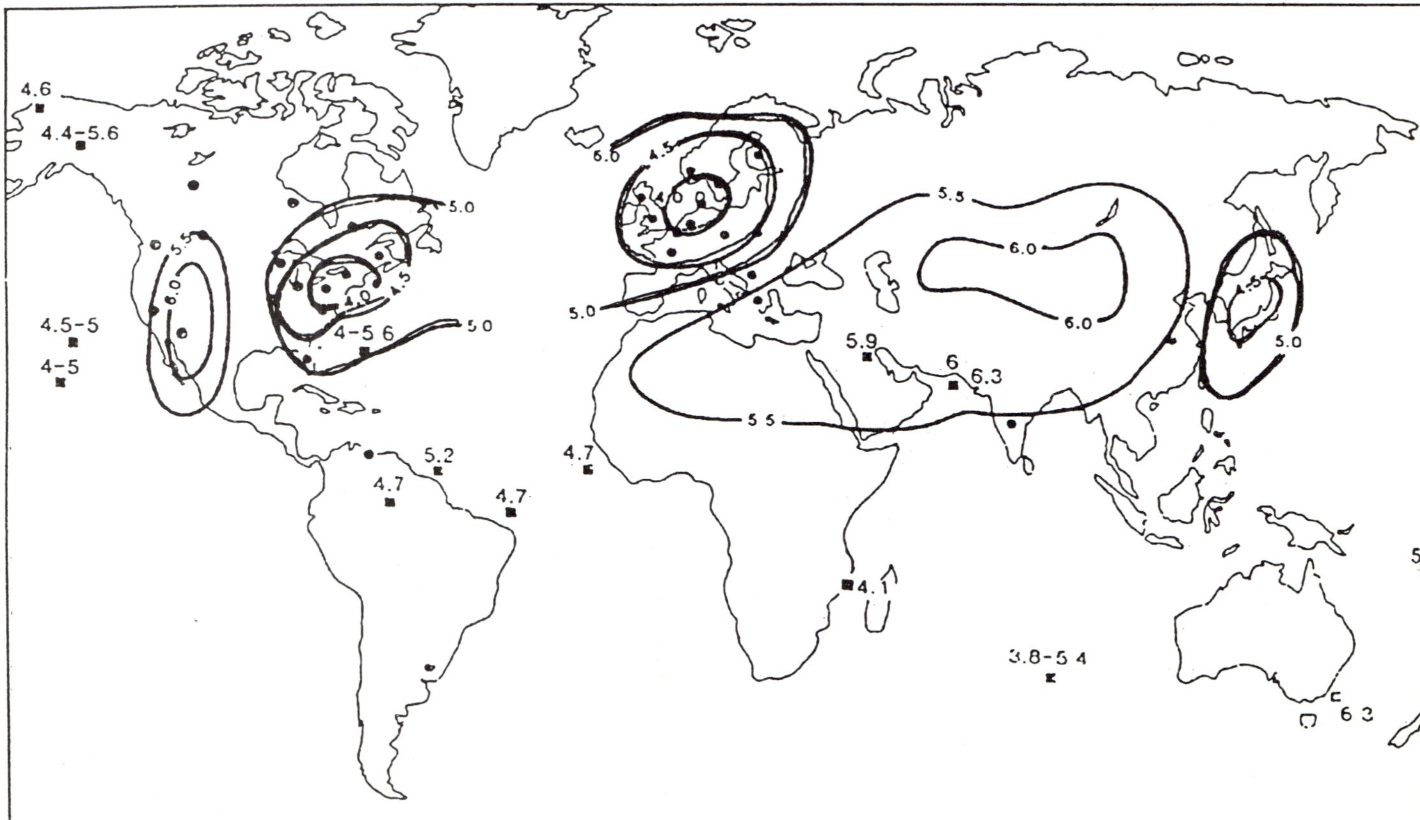

Figure 5. Global rainfall acidity, estimated by WMO (1989)

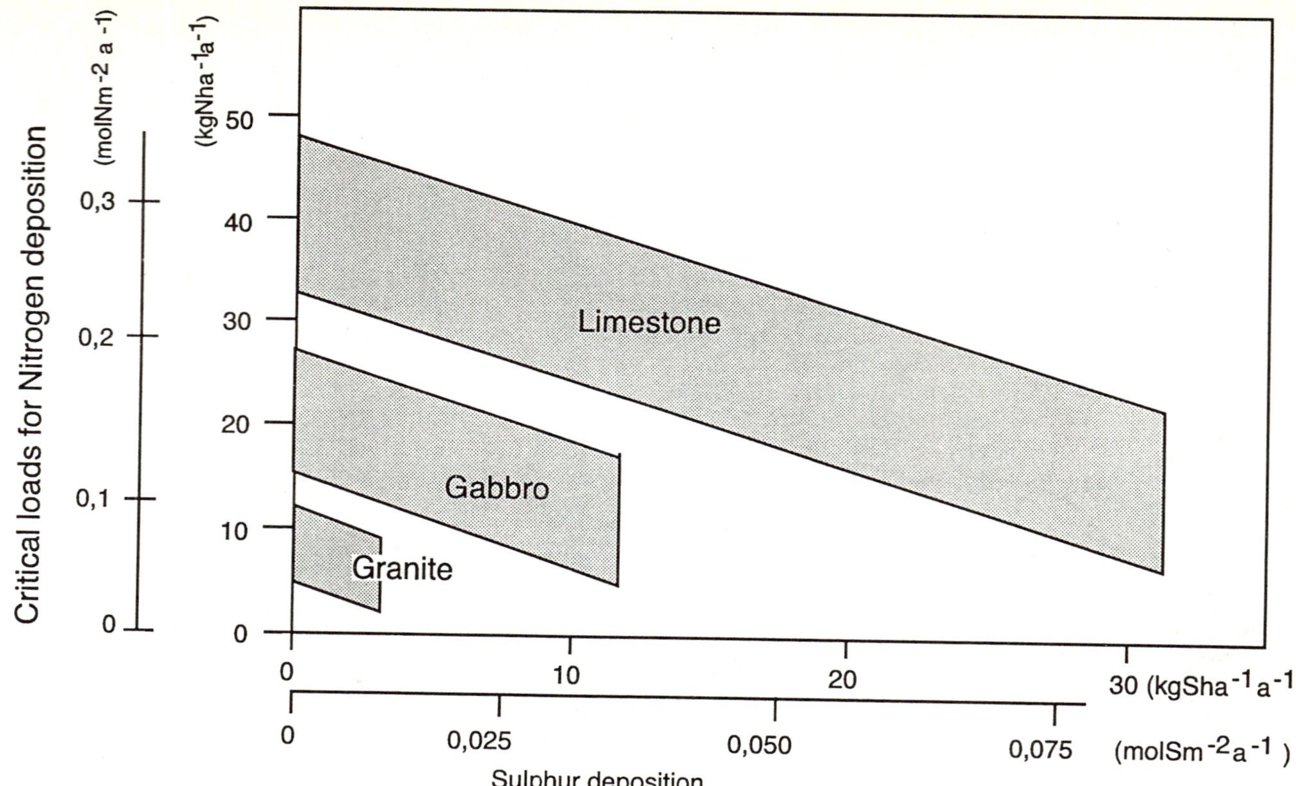

Figure 6. Critical loads for nitrogen and sulphur for conditions of base cation removal at the rate of weathering. The range represents biomass removal and net accumulation (Nilsson and Grennfelt 1988)

cations also reacts to deposition, and thus a more medium-term model can be developed, as shown in Figure 7.

As the conceptual model indicates, the better the data for the various components and processes, the more precise the temporal and spatial resolution of deviations from the steady state. Proton budget types, which are in principal representative of most important components and processes - for example the already described pools and fluxes in Figure 7 - can be also elaborated for different management regimes, climatic zones, successional states, and landscapes (see Lenz 1992, 1993). There are also many approaches in modelling and calculating biochemical reactions in the soil due to acidification, taking into account storage of sulphate as Jurbanite, and other minerals. A lot of site specific data sets are continuously improving the parameterization of such models, but they are not yet applicable to forest practices. In the meantime, there are some good models available for calculating Critical Loads and related key-factors (see UBA 1993). Hence, for purposes of risk assessment in forests some of the models might fit the data available. The conceptual models shown above are the most general ones and should allow, even with a rough database like geological maps and deposition estimates from the EMEP grid (150 x 150 km) etc., a first risk assessment of changes in the material and proton budget.

DATA AND MODEL APPLICATIONS IN TEST REGIONS

In the following, a series of maps will be shown for the test region of northeastern Bavaria. Depending on the data available the models mentioned above were applied. The spatial resolution is at forest, or even growth, districts (see maps), the temporal resolution ranges from decades to centuries.

Since the beginning of the nineteenth century, changes in forest area percentages are rather small (Map 1). A first review of the management and utilization practices over this period, described in various sources, led to the first estimate of ANC losses due to forest utilization, and it is indicated on Map 2 and in Table 1 (*cf.* Lenz and Müller 1991, Lenz and Haber 1992). The deposition was estimated (see Lenz and Haber 1990) and is shown in Map 3. The weathering of

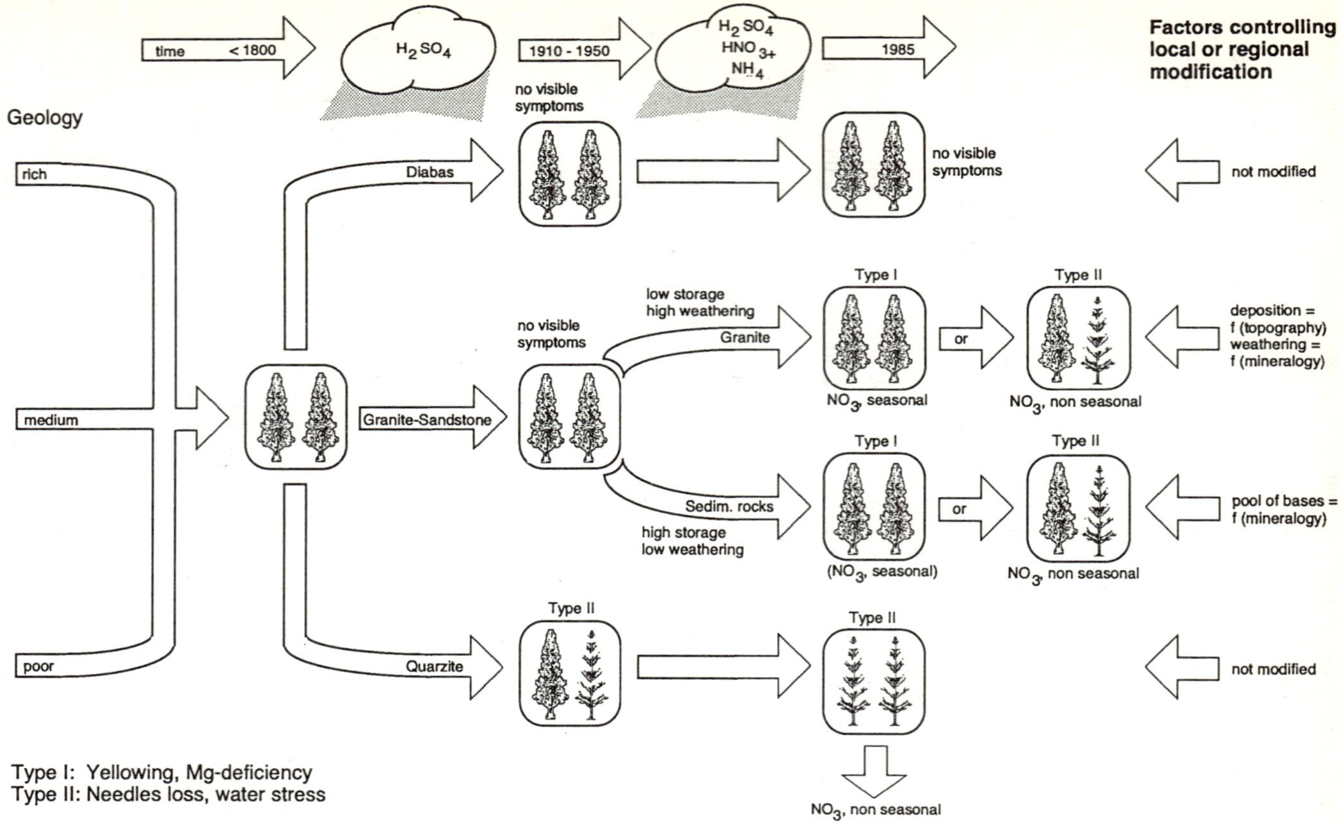

Figure 7. Decline of Norway spruce in the Hartz mountains - a hypothesis based on acid deposition and site sensitivity (Hauhs 1989)

MAP 1

Forested areas in NE–Bavaria
– changes since the early 19th century

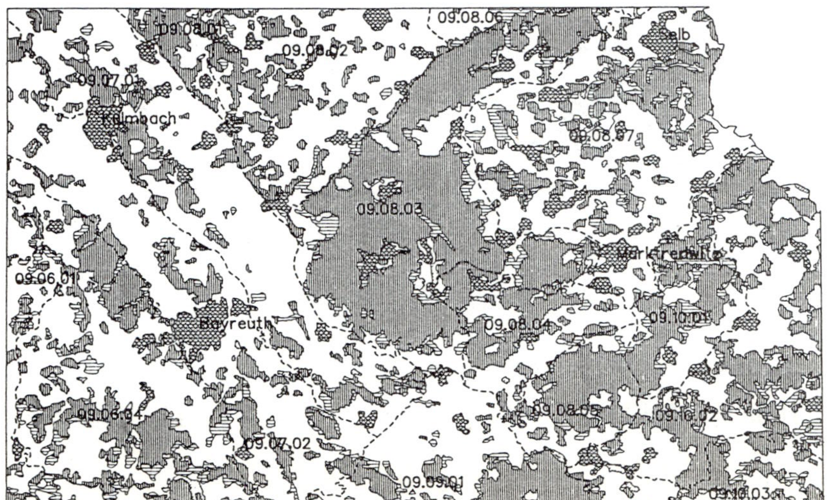

Legend

- forested areas only in the early 19th century
- afforestation
- forested areas since the early 19th century
- city, town
- forest growth district (Wuchsbezirk)

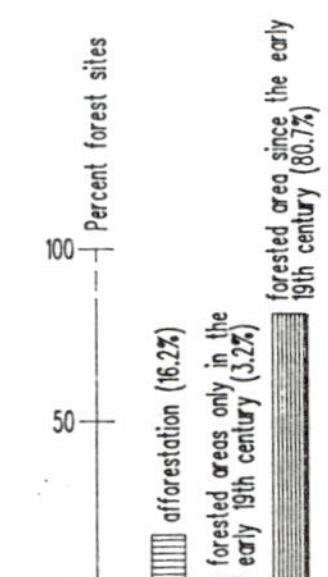

09.06	**Frankenalb und Oberpfälzer Jura**
09.06.01	Nördliche Frankenalb und Nördlicher Oberpfälzer Jura
09.06.04	Oberfränkisches Braunjuragebiet
09.07	**Oberfränkisches Triashügelland**
09.07.01	Bruchschollenland
09.07.02	Obermainhügelland
09.08	**Frankenwald, Fichtelgebirge und Steinwald**
09.08.01	Frankenwald
09.08.02	Münchberger Sattel
09.08.03	Fichtelgebirge
09.08.04	Brand-Neusorger Becken
09.08.05	Steinwald
09.08.06	Bayerisches Vogtland
09.08.07	Selb-Wunsiedler Bucht
09.09	**Oberpfälzer Becken - und Hügelland**
09.09.01	Oberpfälzer Becken - und Hügelland
09.10	**Oberpfälzer Wald**
09.10.01	Mitterteicher Basaltgebiet
09.10.02	Waldsassener Schiefergebiet und Wiesauer Senke
09.10.03	Vorderer Oberpfälzer Wald

S ~ 1:430.000

R. Lenz A. Müller S. Meigel

CHAIR OF LANDSCAPE ECOLOGY
TU MUNICH–WEIHENSTEPHAN

MAP 2

Reduction of the Acid Neutralizing Capacity (ANC) by forest utilization from 1819–1950 in NE–Bavaria

Legend

80–100 kmol/ha
101–130 kmol/ha
131–180 kmol/ha
181–240 kmol/ha

city, town
forest growth district (Wuchsbezirk)

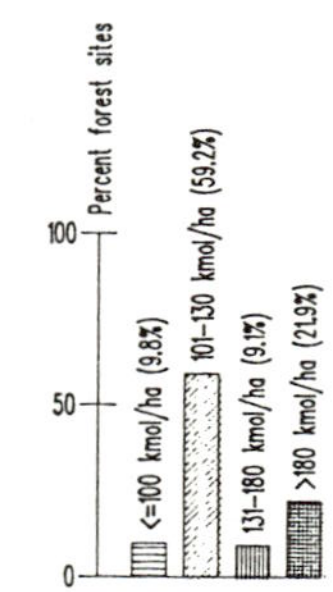

R. Lenz A. Müller M. Erhard

S ~ 1:430.000

CHAIR OF LANDSCAPE ECOLOGY
TU MUNICH–WEIHENSTEPHAN

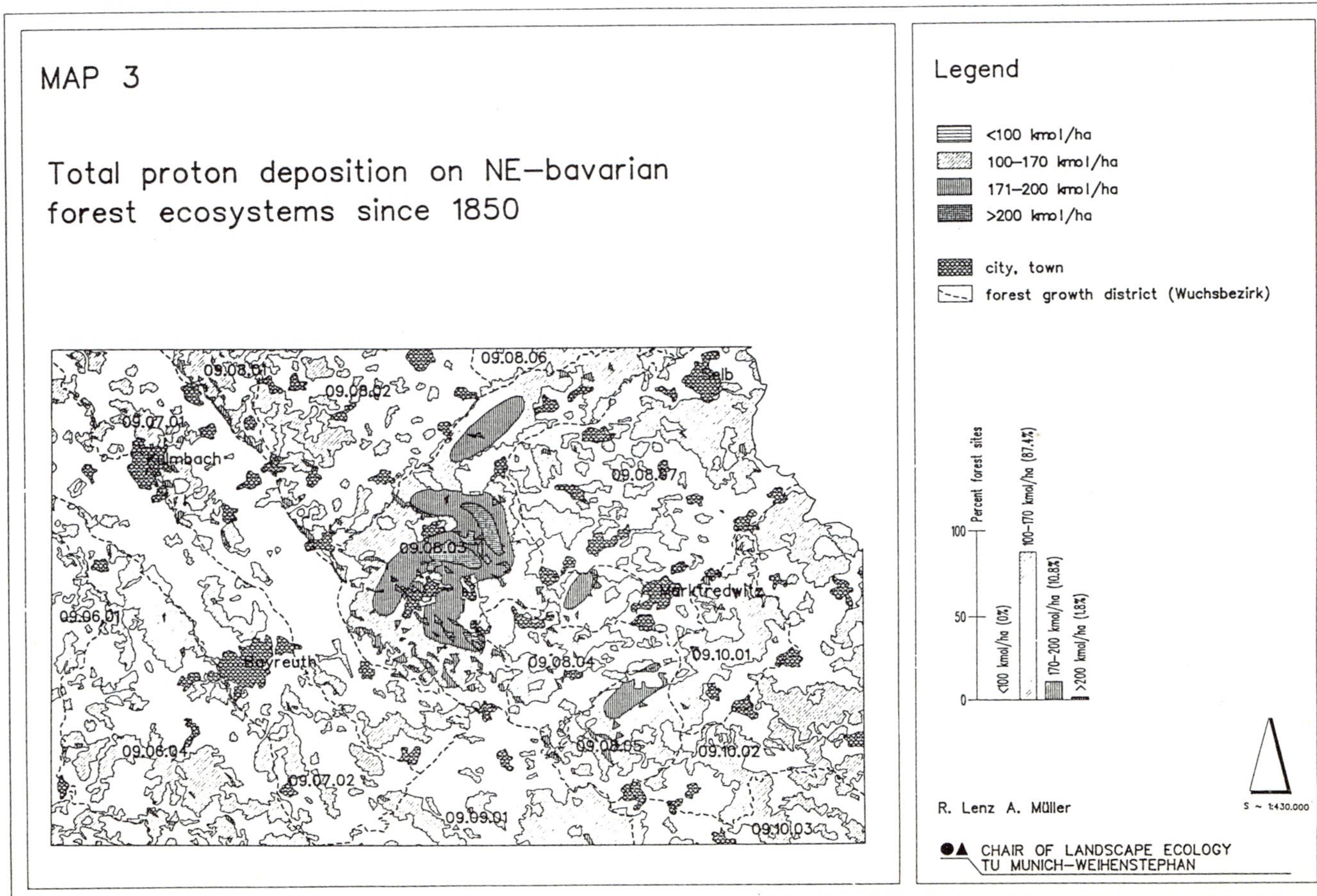
MAP 3
Total proton deposition on NE–bavarian forest ecosystems since 1850
Legend
<100 kmol/ha
100–170 kmol/ha
171–200 kmol/ha
>200 kmol/ha
city, town
forest growth district (Wuchsbezirk)
Percent forest sites
100
50
0
<100 kmol/ha (0%)
100–170 kmol/ha (87.4%)
170–200 kmol/ha (10.8%)
>200 kmol/ha (1.8%)
S ~ 1:430.000
R. Lenz A. Müller
CHAIR OF LANDSCAPE ECOLOGY
TU MUNICH–WEIHENSTEPHAN
09.08.06
09.08.02
09.07.01
09.08.07
09.08.03
Bayreuth
09.06.01
09.08.04
09.10.01
09.06.04
09.07.02
09.09.01
09.10.02
09.10.03

Table 1. Proton production rates of some forest harvesting operations, and their intensities in relation to site productivity

Yield class	bole and bark (-)kmol/ha and yr			branches, twigs (-)kmol/ha and yr			litter (-)kmol/ha and yr			forest grazing (-)kmol/ha and yr		
	low	medium	high	low	medium	high	low	medium	high	low	medium	high
six	0,1	0,2	0,3	0,17	0,33	0,47	0,08	0,15	0,38			
seven	0,11	0,21	0,35	0,24	0,41	0,58	0,09	0,18	0,44			
eight	0,11	0,23	0,37	0,24	0,49	0,69	0,10	0,20	0,50	0,05	0,13	0,38
nine	0,12	0,24	0,39	0,28	0,56	0,80	0,12	0,23	0,56			
ten	0,13	0,25	0,45	0,28	0,56	0,80	0,13	0,25	0,63			

Table 2. Changes in the ANC status in forest ecosystems of northeastern Bavaria since the early 19th century.

Growth district (1)	Weathering rate (+)kmol/ha and yr	Deposition rate (-)kmol/ha and yr	Forest utilization (-)kmol/ha and yr	ANC reduction kmol/ha and yr	Loss of lime equivalents range t/ha	t/ha
09.06.01	0,25 - 1,25	0,8	1,96	2,0 - 2,5	13 - 16,3	14
09.06.04	0,25 - 0,75	0,8 - 1,1	1,73	1,8 - 2,4	11,7 - 15,6	13
09.07.01	0,25 - 2,5	0,8 - 1,1	0,9	0 - 1,6	0 - 10,4	6
09.07.02	0,25 - 1,75	0,8	0,9	0 - 1,5	0 - 9,8	5
09.08.01	0,25 - 0,75	0,8 - 1,1	0,91	1,2 - 1,5	7,8 - 9,8	9
09.08.02	0,25 - 0,75	0,8 - 1,1	0,93	1,1 - 1,6	7,2 - 10,4	8
09.08.03	0,25 - 0,75	0,8 - 1,8	0,91	1,5 - 1,7	9,8 - 11,1	10
09.08.04	0,25 - 0,75	0,8 - 1,1	1,55	1,7 - 2,3	11,1 - 15,0	13
09.08.05	0,25 - 0,75	0,8 - 1,1	0,91	1,1 - 1,7	7,2 - 15,0	9
09.08.06	0,25 - 0,75	0,8	1,23	1,2 - 1,8	7,8 - 11,7	9
09.08.07	0,25 - 0,75	0,8 - 1,1	1,23	1,4 - 1,8	9,1 - 11,7	9
09.09.01	0,25 - 0,75	0,8	1,63	1,1 - 2,2	7,2 - 14,3	13
09.09.01/1	0,25	0,8 - 1,1	1,63	2,5	16	16
09.10.01	0,25 - 2,5	0,8 - 1,1	0,78	0 - 1,5	0 - 9,8	5
09.10.02/1	0,25 - 0,75	0,8	0,91	0,9 - 1,5	5,9 - 7,8	8
09.10.02/2	0,25 - 2,5	0,8 - 1,1	0,78	0 - 1,4	0 - 9,1	5
09.10.03	0,25 - 2,5	0,8	0,78	0	0	0
09.10.03/1	0,25 - 0,75	0,8	0,9	0,9 - 1,5	5,9 - 7,8	9

bedrock material and its mineral composition were estimated according to interpretations from Fölster (1985) (see also Lenz 1991; Map 4).

Finally, the total amount of potentially accumulated acids for each growth district was calculated (Map 5). Table 2 shows the estimates from Maps 2 to 5, and also gives some idea of the equivalent losses of lime - which should not necessarily be interpreted as an adequate measure of requirements for restoration. The loss of alkalinity also indicates which plant communities could occur or, *vice versa*, are likely to disappear when pH and base saturation are lowered. In Table 3 these changes were elaborated on the basis of an interpretation of the alkalinity needs of the potential natural ground vegetation, in comparison with the actual soil status in the same landscape unit (Lenz and Haber 1992).

Table 3. Estimates of actual (pH BODI) and desirable (pH BODS) pH values in 16 landscape units with similar PNV in northeastern Bavaria

landscape unit	PHBODI	PHBODS
8	3,5	4
9	3,5	4
12	3,5	4,5
17	5	6
20n	3,5	5
21n	5	5
24Ø	3,5	4,5
26	3,5	5
27	4	5,5
28w	5	5
32	4	5
39	4	5,5
41	4	4,5
43	3,5	3,5
9+12	3,5	4
9+8	3,5	4

MAP 5

Accumulated acids on NE–bavarian forest sites since 1819

Legend

<85 kmol/ha
85–170 kmol/ha
171–255 kmol/ha
>255 kmol/ha

city, town
forest growth district (Wuchsbezirk)

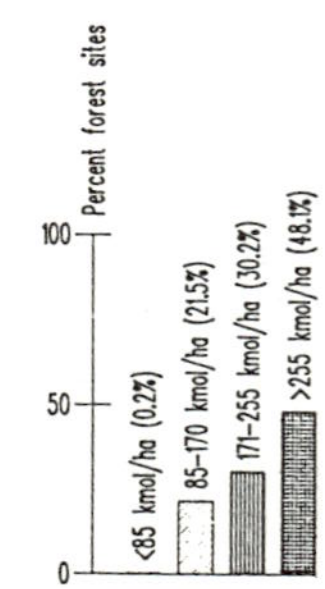

s ~ 1:430.000

R. Lenz A. Müller

CHAIR OF LANDSCAPE ECOLOGY
TU MUNICH–WEIHENSTEPHAN

MAP 4

Weathering classes on NE–bavarian forest sites

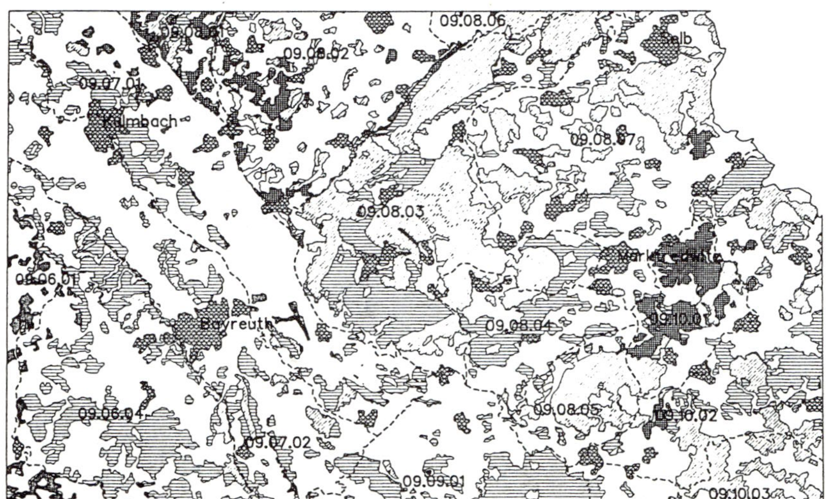

Legend

- <0.50 kmol/ha*y
- 0.50–0.99 kmol/ha*y
- 1.00–1.49 kmol/ha*y
- 1.50–1.99 kmol/ha*y
- >2.00 kmol/ha*y

- city, town
- forest growth district (Wuchsbezirk)

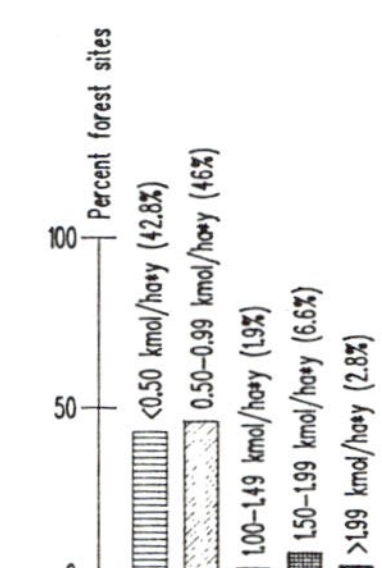

S ~ 1:430.000

R. Lenz A. Müller

CHAIR OF LANDSCAPE ECOLOGY
TU MUNICH–WEIHENSTEPHAN

The table clearly shows that there is a levelling off in the aluminium buffer range. Potentially richer soils are not as common as would be expected. This also indicates that the actual ecosystems diversity in this region, in terms of material budget, is lower than it has been previously, and also lower than it potentially might be in natural forests. Regarding biodiversity of the forests, indicated in a simplified way by the tree species composition, a comparison of the dominant species shows a similar situation (Table 4).

It has to be stated that for different groups of biota, each kind of forest utilization has it specific advantages. Some endangered species are dependent on coppice etc. Hence it is important to maintain these management practices for conservation purposes at certain sites. Probably we need a specific pattern of different forest-ecosystem types for optimal conservation, although this cannot be elaborated in a study on this scale. Nevertheless, the impacts and responses discussed here should be well understood, and efforts made such that future impacts do not degrade ecosystems further.

The model, when applied to a risk assessment of acid deposition, clearly shows an excess of Critical Loads on various sites. Summarizing these results, deposition should be reduced by 70 to 85%, which indicates also the need for a similar reduction in emissions. In general the reduction of SO_2 emissions in Germany, as indicated in Figure 4, might be expected to meet this target in the future.

CONCLUSIONS

Although the calculations are fairly rough, and some of them are based on highly variable and uncertain data, for example the deposition estimates, there is enough evidence of a high risk from disturbances. On the basis of the risk assessment, the next step - besides the necessity to improve the database and thus to refine the estimates - should be to set up targets (target loads) and develop restoration measures. Targets are political, but restoration (or better: "reclamation", see Lenz and Haber 1992) can be discussed in terms of feasibility. Only a few aspects can be mentioned here. First, the problem of time (see Figure 8) and history of ecosystem development, means that many ecosystems are unique. Secondly, the irreversibility

Table 4. Actual tree composition in percent, and potential natural dominating tree species (PNV) in forest districts of northeastern Bavaria

FoA Nr	forest district	beech	oak	fir	pine	spruce	PNV
11	Bayreuth	6	1	0	59	34	oak, beech...
12	Pegnitz	2	0	0	65	25	pine, oak...
20	Nordhalben	8	0	1	0	85	fir, beech...
31	Bad Steben	9	0	1	1	82	fir, beech...
32	Stadtsteinach	8	2	1	24	54	oak, beech...
40	Goldkronach	7	0	1	3	87	oak, beech...
51	Hersbruck	19	5	0	33	28	beech, oak...
52	Betzenstein	20	1	1	16	54	oak, beech...
60	Sulzbach- Rosenberg	15	0	1	46	35	pine, oak...
81	Kronach	7	1	2	0	82	fir, beech...
82	Rothenkirchen	6	0	0	1	86	fir, beech...
91	Fichtelberg	3	0	0	7	86	fir, spruce...
92	Hollfeld	2	1	1	41	43	beech, oak...
101	Kemnath	2	1	0	24	63	beech, fir...
102	Weißenstadt	2	0	0	5	89	fir, beech...
120	Pressath	4	1	0	59	25	pine, oak...
140	Wunsiedel	2	0	0	27	63	fir, spruce...
150	Rehau	2	0	0	22	70	fir, beech...
160	Lichtenfels	18	5	0	33	44	beech, oak...
161	Forchheim	9	11	0	51	19	beech, oak...
180	Selb	4	0	0	42	54	fir, oak...
190	Weiden	0	1	0	76	16	fir, oak...
210	Mitterteich	3	0	0	33	49	beech, fir...
230	Waldsassen	1	1	0	43	42	fir, beech...
241	Tännesberg	7	0	2	8	75	fir, beech...
242	Tirschenreuth	2	0	0	31	60	fir, beech...
243	Vohenstrauß	7	0	1	16	70	fir, beech...
250	Flossenbürg	5	0	1	12	76	fir, beech...
500	Schnaittenbach	2	1	0	60	25	pine, oak...

Sources: OFD Bayreuth, Ansbach, Regensburg (schriftl. Mitt. 1988), SEIBERT 1968, FOERST u. KREUTZER 1977, own interpretations

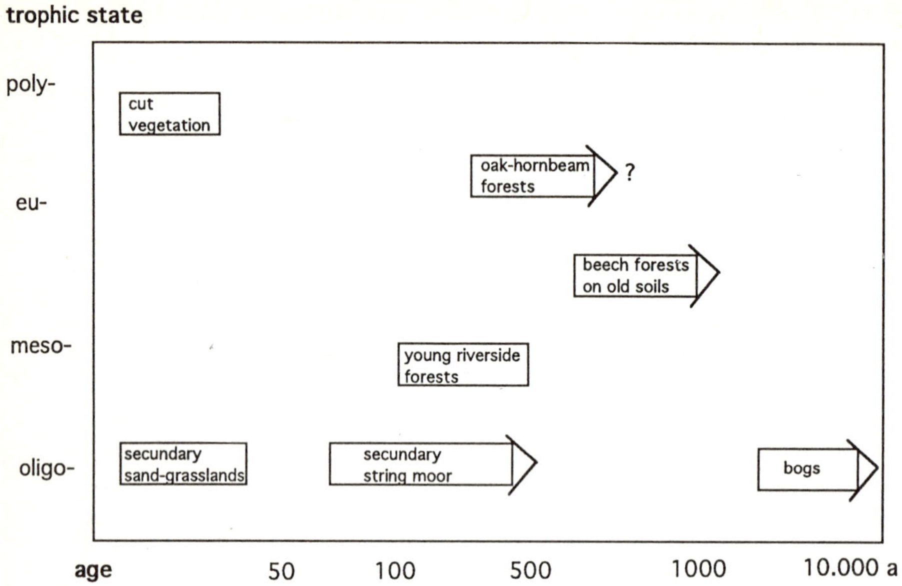

Figure 8. Trophic state and age of some terrestrial ecosystems

of changes in the soil due to the destruction of *e.g.*, clay minerals, should be taken into account. In some cases, even time and sophisticated soil amelioration by liming and adding nutrients as well as minerals, will not lead to real restoration of the whole system. Finally, biodiversity will be affected indirectly by global change of the chemical climate, forest management practices, and probably also other kinds of changes of the climate.

FINAL REMARKS ON THE TOPIC OF THE WORKSHOP

As I argued in the Introduction, current forested areas in central Europe are neither natural, nor do they cover most of the formerly forested areas. The loss and also some degradation of forests resulted in an increase in biodiversity in previous centuries, before the age of industrialization. It might be assumed that European biodiversity was unnaturally high in the eighteenth, nineteenth and early twentieth centuries: a result of a highly diverse land-use in pattern and intensity. So regional targets for biodiversity have to consider the cultural landscape, even including urban areas (see Lenz *et al.*, in prep.)!

A segregation of preservation areas from production areas cannot be the only solution for biodiversity in cultural landscapes: we would lose all the endangered species that depend mainly on extensively used land. Hence, integration is necessary, extending the concept of "single large or several small (SLOSS)" habitats to a concept of production and conservation on 100% of the area.

This is not a mere commitment to the fact that we cannot turn the wheel back towards natural ecosystems - we do not really know how diverse they have been - but a logical consequence of the fact that, in our temperate zone, biodiversity has been increased by and has adapted to human impacts. There is a lot of evidence that the biodiversity left after the glacial periods has been increased due to cultivation. There is also no doubt that the degradation of agricultural and forested landscapes, including the toxification and complete destruction of some of them, reduces biodiversity again.

Acknowledgement

I would like to thank Armin Müller, Markus Erhard, and Roswitha Weinbrunn for their valuable contributions and technical support.

REFERENCES

Berden M, Nilsson J, Rosen K, Tyler G (1987) Soil acidification - extent, causes and consequences. An evaluation of literature information and current research. Nat. Swed. Env. Protection Board, Rep. 3292, Solna, Sweden

Bredemeier M, Ulrich B (1989) Depositionsbedingte und ökosysteminterne Anteile der Säurebelastung von Waldböden. AFZ 11: 256-260

v. Breemen N, Driscoll CT, Mulder J (1984) Acidic deposition and internal proton sources in acidification of soils and waters. Nature 307: 599-604

Fölster H (1985) Proton Consumption Rates in Holocene and Present-day Weathering of Acid Forest Soils. Pages 197-209 *in* J I Drever (ed.) The Chemistry of Weathering. NATO ASI Series C: Mathematical and Physical Sciences, Vol. 149, Reidel, Dordrecht

Glatzel G (1989) Internal proton generation in forest ecosystems as influenced by historic landuse and modern forestry. Pages 335-349 *in* Ulrich B (Hrsg.) Internationaler Kongreß Waldschadensforschung: Wissenstand und Perspektiven, 2-6 Okt. 1989 in Friedrichshafen am Bodensee. Vorträge Bd. 2

Hardin G (1968) The tragedy of the Common. Science 162: 1243 - 1248

Hauhs M (1989) Decline of Norway Spruce at the Harz Mountains - A hypothesis based on acid deposition and site sensitivity. International Congress on Forest Decline Research, Friedrichshafen, 1.-6.10.1989, Poster Abstracts Vol. 2, Poster 254: 535-536

Hauhs M, Rost-Siebert K, Raben G, Paces T, Vigerust B (1989) Summary of European Data. Pages 5/1-37 *in* Malanchuk JL, Nilsson J (eds.) The Role of Nitrogen in the Acidification of Soils and Surface Waters. Nordic Council of Ministers, Miljorapport 1989:10, Stockholm

Hettelingh J-P, Downing RJ, De Smet PAM (1991) Mapping Critical Loads for Europe. CCE Technical Report No 1., RIVM Bilthoven

Lenz R (1991) Charakteristika und Belastungen von Waldökosystemen NO-Bayerns - eine landschaftsökologische Bewertung auf stoffhaushaltlicher Grundlage. Ber. Forschungszentrum Waldökosysteme. Reihe, Göttingen

Lenz R (1992) Forest Ecosystems as Proton Budget Types - A Land-scape-Ecological Evaluation, Illustrated by NE-Bavaria and Central Europe. Modeling Geo-Biosphere Processes Vol. 1: 115 - 129

Lenz R, Haber W (1990) Longterm assessment of spruce vitality in the Fichtelgebirge (West Germany) under ongoing acid deposition. Vegetatio 89: 121-135

Lenz R, Haber W (1992) Approaches for the restoration of forest ecosystems in northeastern Bavaria. Ecol. Modelling 63: 299 - 317

Lenz R, Müller A, Erhard M (1994) Veränderungen der Säureneutralisierungskapazität nordostbayerischer Waldböden - eine regionale Risikoanalyse stoffhaushaltlicher Prozesse seit Anfang des 19ten Jahrhunderts. Forstarchiv (angenommen)

Matzner E (1988) Der Stoffumsatz zweier Waldökosysteme. Ber. Forschungszentrum Waldökosysteme/Waldsterben. Reihe, Göttingen

Nilsson J, Grennfelt P (1988) Critical Loads for Sulphur and Nitrogen. Milijorapport 1988: 15. Gotab, Stockholm

Paces T (1986) Weathering and Mass Balance in Small Drainage Basins: Environmental Applications in the Bohemian Massif (Central Europe). Sci. Geol. Bull 39, 2. Straßburg: 131-150

Reuss J O, Johnson DW (1986) Acid Deposition and the Acidification of Soils and Waters. Ecol. Studies Vol. 59. Springer Verlag, New York

Sverdrup H, Warfvinge P (1987a) The kinetics of mineral weathering. Technical Report, Dept. of Chemical Engineering II, Lund Institute of Technology

Sverdrup H, Warfvinge P (1987b) The kinetics of cation production of the weathering of feldspars (Original in Swedisch). Technical Report, Dept. of Chemical Engineering II, Lund Institute of Technology

UBA (1993) Mapping Manual. Draft Report March 1993. Berlin (unpublished)

Ulrich B (1993) Interaction of forest canopies with atmospheric constituents: SO2, alkali and earth alkali cations, and chloride. Pages 33-45 *in* Ulrich B, Pankrath J (eds.) Effects of accumulations of air pollutants in forest ecosystems. Reidel, Dortrecht

Ulrich B (1987) Stability, elasticity and resilience of terrestrial ecosystems with respect to matter balance. Pages 11-49 *in* Schulze E-D, Zwölfer H (eds.)

Potentials and limitations of ecosystem analysis. Ecol. Studies Vol. 61. Springer Verlag, New York

Ulrich B (1989a) Material balance characteristics of forest ecosystems subjected to acid deposition.International Congress on Forest Decline Research, Friedrichshafen, 1.-6.10.1989, Poster Abstracts Vol. 2, Poster 339: 705-706

Ulrich B (1989b) Effects of Acid Deposition on Forest Ecosystems in Europe. Adv. Environm. Sci. (Springer) 2: 189-272

Ulrich B, Meyer H (1987) Chemischer Zustand der Waldböden Deutschlands zwischen 1920 und 1960, Ursachen und Tendenzen seiner Veränderung. Ber. Forschungszentrum Waldökosysteme. Reihe, Dordrecht

WMO (1989) Gaw and precipitation chemistry measurement activities. Fact sheet No. 5

Chapter 15: Soil arthropods for faunal indices in assessing changes in natural value resulting from human disturbances*

Roberto Pizzolotto
Universita' della Calabria
Dipartimento di Ecologia
87036 Rende (CS)
Italy

INTRODUCTION

The solution of a problem through technological innovation results in the creation of a new "space" for a new problem. This "spiral" evolution of human welfare has probably been primed (beginning from the modern age?) by the shift of every-day tools from a usage-value as survival instruments to a usage-value as profit instruments. The development of a new, man made, urban environment, and its unforeseeable and chaotic growth, led to changes in the equilibrium between natural and agricultural landscapes (Figure 1). Human activities cause a disturbance of the environment that can be either manifest immediately, or quantifiable only after long periods (storing effect). In the first case immediate monitoring equipment can be applied, but in the second case the assessment of human impact on the environment is often burdensome in terms of time and resources spent in research. Hence it follows that there is a need to find indicators of disturbance that are present in the study area and that can be measured quickly. Plants and animals answer to this need. It is likely that Clements with his "Plant indicators" (1920) was one of the first scientists who considered living organisms as test tools for the "health condition" of the environment (bioindicators). This point of view is now widely accepted, *e.g.* in Spellerberg (1991), who takes into account the importance of the integrating properties of living organisms for biological monitoring.

---

*) Research carried out with the financial help of the Italian Ministery for the University and Scientific Research (40% fund).

NATO ASI Series, Vol. I 20
Biodiversity, Temperate Ecosystems, and Global Change
Edited by T. J. B. Boyle and C. E. B. Boyle

Figure 1 - The "natural-degraded" dichotomy seems to be the most popular figure in the landscape descriptions. The equilibrium between these extremes is tied to a "something" that is usually called "naturalness". To evaluate naturalness in ecological studies (as it is likely that naturalness exists also in economy, sociology, ...) it might be helpful to take the point of view of species different from man. An index of natural value should evaluate how faunistic assemblages (guild, community, population) respond to anthropogenic disturbance, referring to an ideal condition of equilibrium (climax stage).

Restricting our attention to animals, we will consider them inside different ecosystems, as being "organized" in communities (or coenoses); that is to say groupings of species definable on faunistic (list of species) and statistical (species "weight" as number of individuals, at least) basis. In ideal conditions without disturbance, an animal community tends to develop, following a natural succession, towards the so-called climax stage. The use of bioindicators is particularly helpful in that both single species and the communities give information on both the effects of the actual environmental disturbance phenomena, and also on the trend in natural succession. Animals are reliable indicators because their life cycle, habitat choice, and feeding behaviour are the consequences of particular life strategies that have been selected through evolution. These strategies are often correlated to the stability/instability of the environment, and their commonness or rarity might give an idea about the environment quality (natural vs. degraded) and its resistance/resilience to disturbance phenomena.

Soil arthropods help to deal with only some aspects of the anthropogenic disturbance problem. The physical space in which they can be sampled is practically a two-dimensional space, very much affected by human activities. Giving a general value to the description of a soil bioindicator suggested by Den Boer (1977), it is expected that only some taxa can be used in the applied research. Good indicators of disturbances in ecological succession resulted are, *e.g.* Staphilinids (Bohac, 1988), Araneida (Ruzicka and Antus, 1989), Diplopods (Ganin, 1989), and also Acari and Collembola (Straalen *et al.*, 1988) though there are some difficulties in automatic sampling and determination. Carabid beetles are one of the most reliable taxa for applied research in the field of environmental evaluation, as they are very sensitive to several kinds of environmental change, as demonstrated by a wide spectrum of theoretical and applied studies (Thiele, 1977; Den Boer *et al.*, 1986, 1987; Stork, 1990; Desender, in print). The importance of arthropods as bioindicators is documented in the "Bioindicatores Deteriorisationis Regionis" conferences (see Bohac, 1992 for the sixth edition), and is also the subject of several studies dealing with highly degraded urban and ruderal environments (Minelli, 1974; Bornkamm et al., 1980; Santas, 1986; Kurnetsova and Krivolutsky, 1986; Klausnitzer, 1987).

It is surprising that, for conservation purposes, the natural elements from ancient science, air, water, earth and fire, are still the most elementary starting points for an interdisciplinary approach to studying the anthropogenic impact problem as a whole. This paper will deal only with some problems related to the use of a faunal index based on soil arthropods to assess the natural value of the environment.

BIOLOGICAL FEATURES - COMMUNITY FEATURES

To give a value to the environment is a problem of unit of measurement, that might be dependent on the skill and culture of the researcher. As shown by Usher (1986), the concept of value itself can be influenced by scientific and/or social factors, which can sometimes be in conflict; and criteria such as naturalness and rarity are never referred to an objective "zero value". In most cases the evaluation of environment quality is done by the application of some criteria, widely used in the field of biological conservation (Margules and Usher, 1981; Usher, 1986), that lead to a final result in which (more or less) natural (or undisturbed) conditions versus highly degraded (or with a high anthropogenic stress) conditions are distinguished. Though the concept of "natural" has a relative meaning (everything is natural depending on the perspective in which it is seen), and for some ecosystems is tied to non-equilibrium conditions (Sprugel, 1985; and consider, for example, the role of fire in Mediterranean ecosystems), attention is usually focused on the dichotomy "natural-degraded" in expressions such as: degree of naturalness, natural value, environment quality, degree of conservation, and so on. This suggests that, for the purposes of biological conservation, there is a need to quantify, at the operative level, an environmental feature that in this paper we call "naturalness". The idea that man takes part in this naturalness seems to be largely accepted (Westhoff, 1971; Usher, 1986; Lapka *et al.*, 1992), since his existence seems not to be an "accident" in the history of the Earth (not in man's opinion, at least!).

The need to define an evaluation scale sometimes leads to the application of a "naturalness score" based on what the researcher considers as "evidence" from field studies. It is the researcher who, in this way, fixes an ordinal scale of naturalness, the reliability of which depends on the "evidence", and consequently it is not possible to measure the distance between two degrees of naturalness. The choice of a scoring criterion depends on the data set on which the evaluation is based (Bedward *et al.*, 1991), inasmuch as naturalness is the "manifestation" of a complex of factors, rather than an entity with physical dimensions measurable in an absolute way. The problem to cope with is to start from "evidence" (measurable entities in field research) with a scientific value (*i.e.*, verifiable, with possibility of repetition), and to come to an estimate of naturalness (a pure number) that gives a synthetic representation of that "evidence".

In biological conservation, it seems reasonable to consider naturalness (of an environment, of a community, etc.) as a characteristic reflecting some meaningful biological features (of that environment, community, etc.) and quantifiable by weighting those features by what we call the natural value (of the environment, community, etc.). The natural value is, in practice, the criterion on which decisions are base, *e.g.* to the question: which among these areas should be conserved, saved, restructured?

It seems too simple to use only animals, and among them only soil arthropods, to tackle evaluation of the conservation problem (see also Soulé 1987), but we are aware that an animal is not just a piece of monitoring equipment. An animal (a living organism) shows a range of biological features, or life strategies, that enable it to evolve, that is to say to survive in an environment subject to fluctuations. These biological features may have an ultimate value, if we refer to the geological time that speciation requires, or a proximate value, if we consider the specimens as we observe them at the present time (actual perspective); hence a species plays, at the same time, the role of historical and ecological indicator (Brandmayr and Pizzolotto, 1988, 1990; Pizzolotto and Brandmayr, 1990). The weight of these ultimate and/or proximate values for each biological feature may be high, depending on the single species taken into account. For these reasons "to trust" in a single species may be too simple, while it seems more useful to take into account the "experience" of different species that, as the result of a non-sympatric speciation (in the sense of Pielou, 1979), have evolved different life strategies to solve the same ecological problem (think, for example, of sympatry after different glacial survival processes). This suggests that, for the purposes of environment evaluation, it should be better to focus attention at the level of the community *i.e.* an assemblage of species definable on a faunistic-statistic basis. Considering that many organisms with their relationships form an ecosystem, we might think of the community as a useful tool to reduce the degree of complexity of the ecological space, going from the dimensions of an entire ecosystem to that of a community.

The evaluation method described in the following is based on some meaningful biological features (ecological and chorological features) of carabid species; they are: wing size (dispersal power), endemic species, species at the fringe of distribution, species diversity, spermophagous species (feeding also on seeds). These features are well-correlated with the degree of evolution and naturalness of carabid communities, and they are easy to quantify for evaluation of faunal resources. A brief discussion of the features of the method is given in Pizzolotto (1993a). The meaning of dispersal power in soil arthropods has been the subject of many papers; syntheses are in den Boer (1980, 1990); Brandmayr (1983a, 1991), Wagner and Liebherr (1992). For the purposes of naturalness evaluation, dispersal power can be correlated to the "youth" of a community, to the structural degradation of the environment and to the stability of a stage in a natural succession (see Figure 2). Evidence of correlation with anthropogenic disturbance is also given by Mossakowski *et al.* (1990). Endemism is a clear consequence of the "game" that animals play with natural selection. The presence of endemic species can be considered as a fundamental (constituent) feature of a community; while the presence of species at the fringe of distribution can be interpreted as the consequence of both particular geomorphological and ecological factors influencing the ecosystem, and of the biology (proximate and ultimate value of its life strategies) of the species. The more endemics and "fringe" species there are, the higher the natural value of a community. Species with an opportunistic diet (like some spermophagous carabids) are able to take advantage of conditions in which the prey-predator trophic web is altered by natural and/or anthropogenic factors; and in some cases they can be used as indicators of that alteration. Species diversity is one of the most frequently used criteria in biological conservation (Usher, 1986).

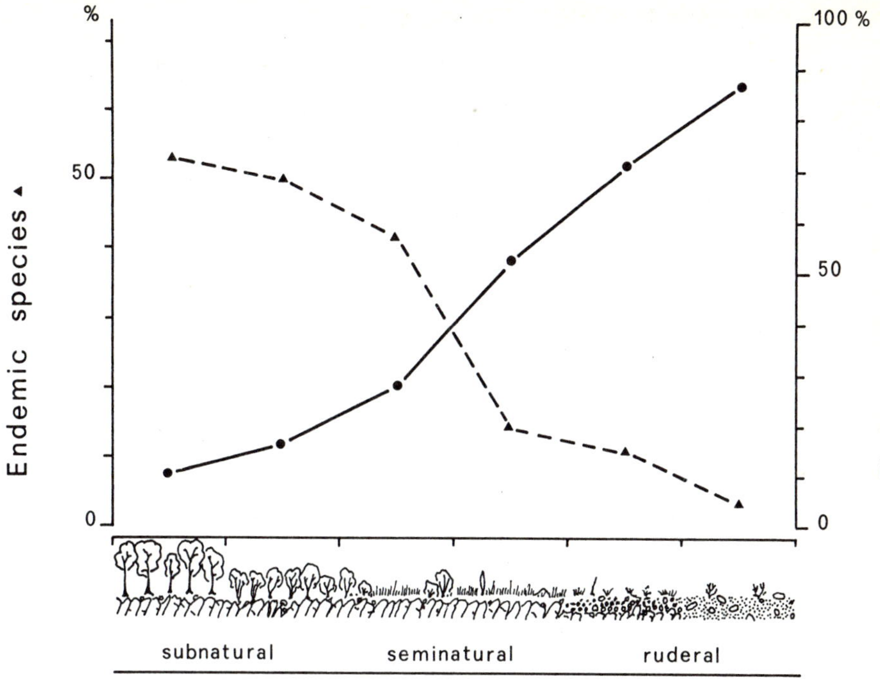

Figure 2 - Percentage of flying species (full line) and endemites (dashed line) in a secondary succession on the Karst near Trieste (from Brandmayr, 1983a)

SOIL ARTHROPODS FOR AN INDEX OF NATURAL VALUE

A procedure to assess the natural value of the environment is outlined in Figure 3. In Italy this procedure has been applied using some community features of carabid beetles, but we think it can be easily extended to other taxa of the soil fauna (*e.g.*, earth-worms, Myriapods, Arachnids, Staphylinids, Curculionids; and also Orthopters and Lepidopters).

A brief history of the development of the method.

This method was developed originally from Brandmayr (1975), who pointed out, on the basis of field research results, the importance of carabids for environmental rehabilitation in agricultural and forest biotopes. Brandmayr and Colombetta (1981) proposed a quantitative evaluation criterion (not depending on generic "expertise judgment") based on two community features: number of endemic species, and number of species able to fly (*i.e.*, with a high dispersal power). The same criterion has been followed to draw a map of the naturalness of the Karst environment near Trieste (northern Italy) (Toselli *et al.*, 1982), and to evaluate the anthropogenic disturbance on ecological succession at a landscape level (Brandmayr, 1983b). The dispersal power of a pteridimorphic carabid beetle (*Notiophilus biguttatus*), evaluated (as the number of full-winged specimens) in beech forest populations of central and southern Italy, was correlated with the degree of anthropogenic disturbance and to the age (geological time) of the species (Brandmayr and Pizzolotto, 1988). For an environmental impact assessment (EIA) on the building of a river dam in Calabria (southern Italy), the main evaluation steps of the method (first published in a technical report in Brandmayr *et al.*, 1988; then reprinted in Cagnin et al., 1991) include the assessment of species at the fringe of distribution and species diversity, taking into account endemism and dispersal power. The same method has been adjusted for bird communities of an Alpine valley (Mingozzi and Brandmayr, 1991). Recently this method has been applied to build up a geographical information system (GIS) and for an EIA in the Karst near Trieste (Pizzolotto, 1993a). A version improved by multivariate analysis has been proposed by Pizzolotto (1993b).

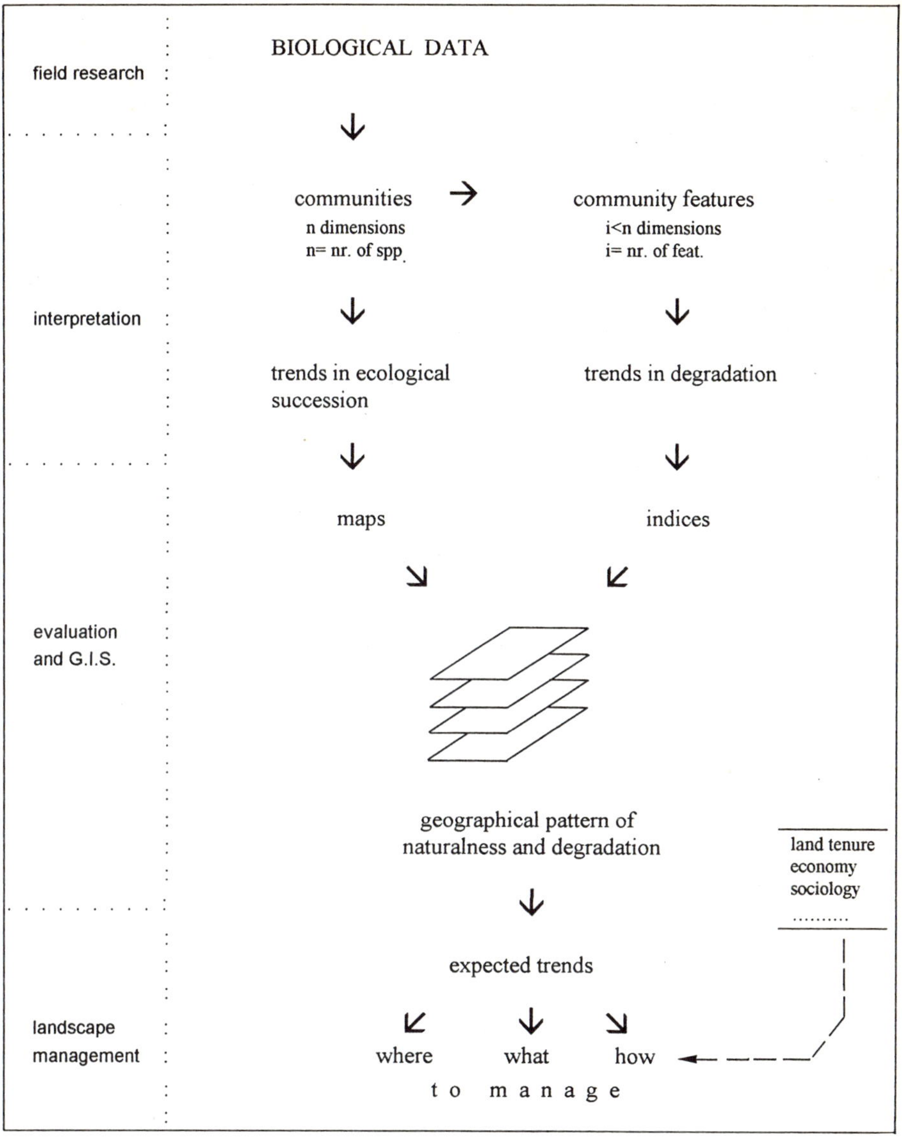

Figure 3 - The proposed evaluation method is structured in four steps, as follows: i) collection of biological data from field research; ii) data analysis and interpretation in terms of natural succession and ecological series; iii) computation of an index of the natural value; iv) reconstruction of a geographical pattern of naturalness and degradation, as a tool to hypothesize future scenarios, and to plan landscape management.

The steps for the evaluation.

The evaluation method (Figure 4) is based on biological data coming from field research; that is to say, a list of species with different distribution over some sample sites (the study area). A reconstruction (real or hypothesized) of the main bioclimatic gradients characterizing the study area might be useful to select the sites to be sampled; *i.e.* along a transect (cross-section) following those gradients (Brandmayr *et al.*, 1991; and see also Figure 5). Data are synthesized in a zoo-sociological table, as for example in Figure 4, where the number of individuals belonging to seven species is given for three sample sites. Then the "situation" for each species having some biological feature is recorded (in Figure 4 endemic species have been marked with "end"); and this leads to table in which the species are described by features. To turn from biological features to the community features level, means to turn from presence/absence (qualitative datum) of a feature for a given species to the weight (quantitative datum) of that feature in a community. In Figure 4, a weighting simulation, in which both the number of species and individuals showing a feature have been taken into account, is given for sample site A. The relative importance (percentage values) of the feature "endemism" has been calculated in three ways: a) as the ratio between the number of endemic species and the total number of species in the site (*i.e.* what is the origin of the fauna of this biotope, or environment?); b) as the ratio between the number of endemics in the site and the number of endemics in the investigated area (*i.e.*, what is the role played by this biotope, or environment, in the endemism phenomenon?); c) as the ratio between the number of individuals belonging to endemic species and the total number of individuals in the site (*i.e.*, are the endemic species dominant or rare at the present time?). The three percentage values have been assigned to classes on a five-graded scale, getting three weights for the feature "endemism" in site A. A similar procedure is followed for the other features; then the same is done for each sample site. The weighting procedure results in a new table (Pizzolotto, 1993b), the table of natural value (TNV), in which the sample sites are described by the features (Figure 4). At this level we can speak in terms of community features (or biological features of a

Zoosociological Table

	sample sites A	B	C	
sp. 1	5	0	0	end
sp. 2	1	3	1	-
sp. 3	4	1	0	end
sp. 4	2	0	1	-
sp. 5	2	5	4	-
sp. 6	4	1	5	end
sp. 7	0	0	2	end
N. of spp.	6	4	5	
N. of ind.	18	10	13	

WEIGHT OF ENDEMICS IN SITE A

a) number of endemic spp. in A = 3 / n. of spp. in A = 6
b) n. of end. spp. in A = 3 / n. of end. spp. in the investigated area = 4
c) n. of individuals of end. spp. in A = 13 / n. of ind. in A = 18

a) 3 / 6 * 100 = 50	----->	0 - 20 → 1 21 - 40 → 2 41 - 60 → 3 61 - 80 → 4 81 - 100 → 5	-----> 3
b) 3 / 4 * 100 = 75			4
c) 13 / 18 * 100 = 72			4

Table of the Natural Value

	sample sites A	B	C
feat. 1a	3	3	2
feat. 1b	4	3	3
feat. 1c	4	1	3
feat. 2a	3	1	2
feat. 2b	1	3	4
feat. 3	5	3	1
total	20	14	15
mean	3.3	2.3	2.5

--->

Multivariate Analysis

Index of Natural Value (INV)

Geograph. Infor. System (GIS)

Environmental Management

Figure 4 - The main steps to calculate the index of natural value (INV) are: i) to compile a zoo-sociological table based on field research data, in which the presence/absence of some biological features is recorded for each species (end = endemic species); ii) to evaluate the weight of the biological features at a community level (community features); iii) to compile the table of the natural value (TNV). On the basis of the TNV, the INV can be computed in more than one way, as explained in the text.

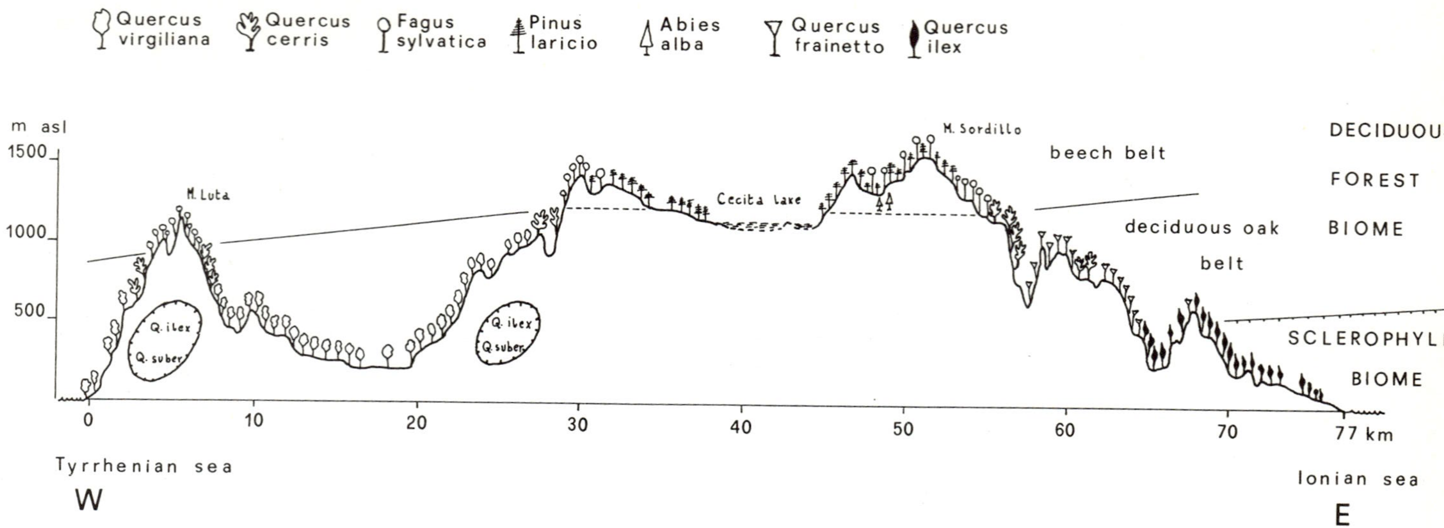

Figure 5 -To select the sites where to sample, it might be helpful to have a reconstruction of the main bioclimatic gradients characterizing the study area. This is an example of the mapped biomes on a cross-section of the northern part of Calabria (Italy). Similar gradients (potential biomes) can be reconstructed on the basis of standard climatic parameters (from Brandmayr *et al.*, 1991).

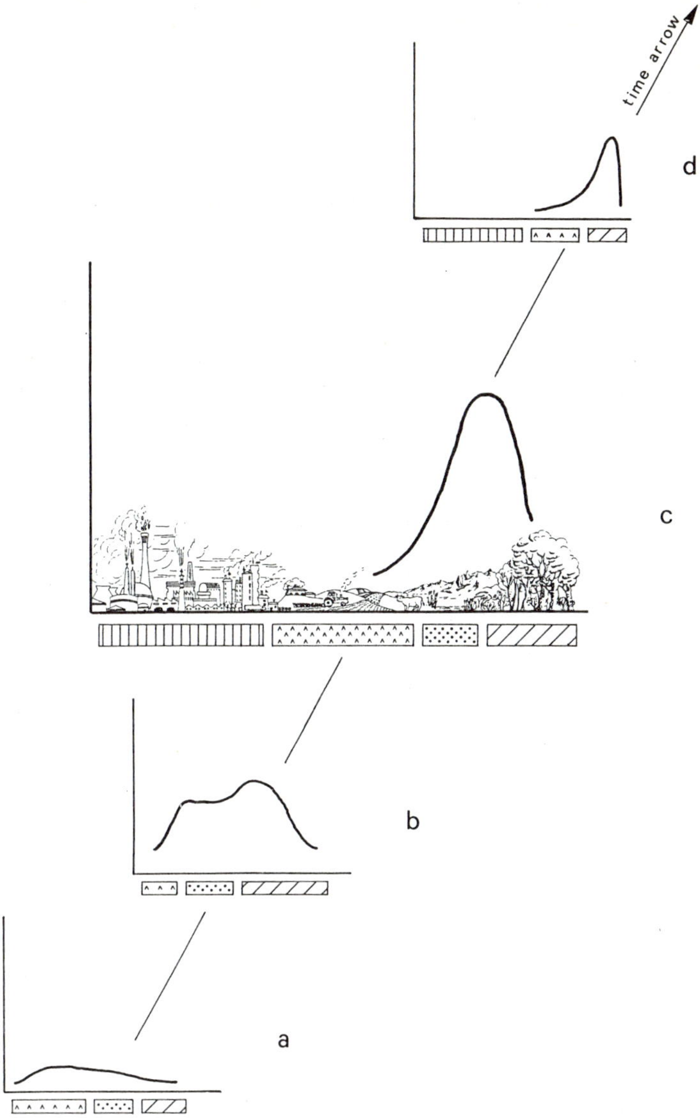

Fig. 6 - The presence and abundance of a species in an ecosystem is the result of the "evolutionary history" of that species. The appearance or the spreading of an ecosystem (a - b) becomes a new "ecological space" to which that species adapts itself , showing a particular preference for it. On the basis of its life strategies, a species will answer to the environmental changes affecting that ecosystem in historical times. The result will be an abundance change in historical times, and the shifting of its habitat choice in evolutionary times (a - d). In the frame of a community, a species is a resource that, on the basis of its own features, determines the natural value of the community.

community) in the sense that the weight of the features is a synthesis of the "behaviour" of a group of species.

The importance of using both the number of species and individuals comes from the fact that these values give information about the "ecological situation" from both the historical and the present time point of view. Considering each species sample as a representative state of the multidimensional "space of the communities" we can say that the observed feature is a resource of "that" community, in that (sample) site, at that moment of natural succession; but at the same time it is also the "result" of the story of each species forming the community (Figure 6). As there have been some very important factors determining (in geological time) the presence of a species in an environment, so it is likely that degradation factors actually (in historical times) affect the existence of a species in a territory, modifying its geographical distribution, or hampering colonization by endemic species. By means of features reflecting this disturbance, it is possible to evaluate both the historical (species number) and actual (individuals number) aspects of the "ecological situation" in the study area.

The TNV may be considered a reduced (in terms of dimensions) description of the "ecological space", in which the variables are related only to the environmental evaluation (in Figure 4 variables change from seven to six, but in real cases a much higher reduction takes place). To evaluate if the TNV is structured in a similar way to the zoo-sociologic table, reflecting similar relationships among communities, Pizzolotto (1993b) clustered the sample sites on the basis of the two tables (Figure 7) and found a significant correlation between the two classifications. The TNV enables us to compute an index of natural value (INV) that gives an estimate of the community importance for the "ecological situation" in the study area. A simple way is to compute the arithmetic mean of the columns in the TNV, and use these means as an INV (Pizzolotto, 1993a; but see also Mingozzi and Brandmayr, 1991). Another way (Pizzolotto, 1993b) is to partition the space of the natural value by means of an ordination method (Figure 8), then to outline the trend(s) in degradation, and to assign the INV in accordance with the outlined trend(s) (*e.g.*, by subdividing the first axis of a Correspondence Analysis into score classes). If the TNV has been based

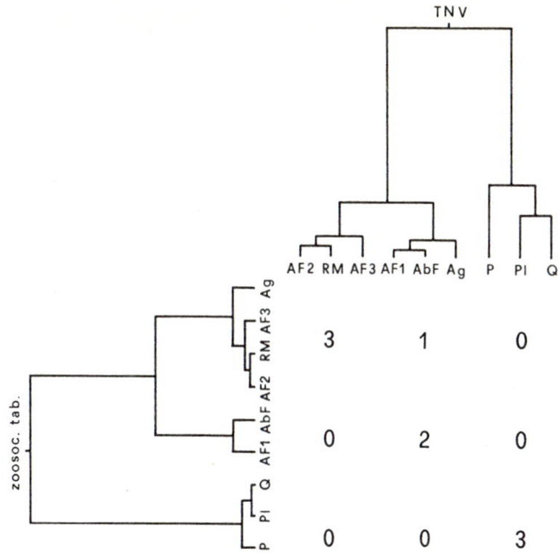

Figure 7 - Cross table resulting from comparing the sample sites classifications based on a zoo-sociological table (Table 1) and a TNV (Table 2). A significant correlation between the tables has been found. The c^2 of the cross-table is significant with a probability of $p=0.9$ ($c^2=13.5$, for 4 dof). Six vegetation types have been sampled: a dry pasture (P); a black-pine forest (Pl); an oak wood (Q); three thermophilous beech forest stands (AF), and a moist beech stand (RM); a beech-fir forest stand (AbF); a wet-meadow alder grove (Ag) (from Pizzolotto, 1993b).

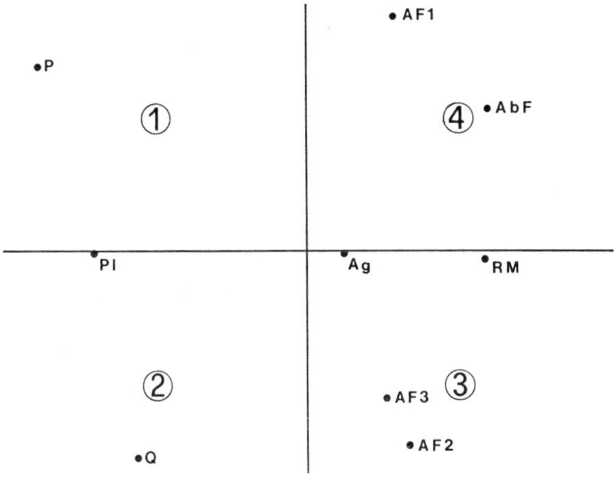

Figure 8 - PCA ordination of the TNV of sample sites of Figure 7. The PCA arranges the sample sites along the first axis according to a gradient of decreasing degradation from left (P site) to right (AbF site); but only 87% of the total dispersion is deducted by the first and second axes together. PCA is interpreted taking into account the angular distance among the sample sites rather than to refer the degradation gradient to a linear pattern. Worthy of notice is that axes I and II partition the sample sites in a similar way as in Figure 7 (Table 2). The INV has been assigned to the quadrants of the PCA space counterclockwise starting from value 1 assigned to quadrant II. Site P received value 1; sites Pl and Q value 2; sites AF2, AF3, RM and Ag value 3; sites AbF and AF1 value 4 (from Pizzolotto, 1993b).

on many features, it might be profitable to apply Concentration Analysis (Feoli and Orloci, 1979) to detect the main groups of features responsible for the degradation trend(s).

DISCUSSION

In this paper only some possibilities for the computation of an INV have been proposed, according to what we call the community features method. We think that it is important first to focus attention on meaningful biological features, if a reliable INV is needed; then the ways of computation will give an appropriate index according to the geographical scale, or the studied taxon, or the type of disturbance (structural, chemical, agro-ecosystems, etc.).

The INV is a helpful tool if related to a thematic map. To draw an exact reproduction of the faunal landscape is often a problem, because a lot of sites should be sampled to cover the study area. However it is possible to produce maps in which only the main lines of faunation are drawn. Vegetation maps are a powerful aid for such purposes when correlations between animal coenoses and vegetation have been found (*e.g.*, as demonstrated in many works for carabids, Staphilinids, Curculionids, Lepidopters). This enables us to consider the extent of a community as that corresponding to the vegetation type with which it is associated. A good use of these maps is to organize them in a geographical information system (GIS), so as to combine different landscape "descriptors", and to get a geographical pattern of naturalness and degradation, from which to start to hypothesize future scenarios (Figure 3). The procedure of superimposed thematic maps has been applied by Pizzolotto (1993b) for an EIA on the building of a river dam in Calabria. A similar procedure as in Figure 3 has been applied by Pizzolotto (1993a) for the compilation of a raster based GIS (Figure 9), with the purpose of assessing the impact related to the building of research laboratories in the Karst, and suggesting some environmental improvements in the surrounding areas so as to counterbalance the loss of natural value caused by the buildings.

Figure 9 - Carabid coenoses characterizing the main aspects of the karstic landscape near the city of Trieste. This is an example of how to get a map of the faunation on the ground of a vegetation map by means of a geographical information system.

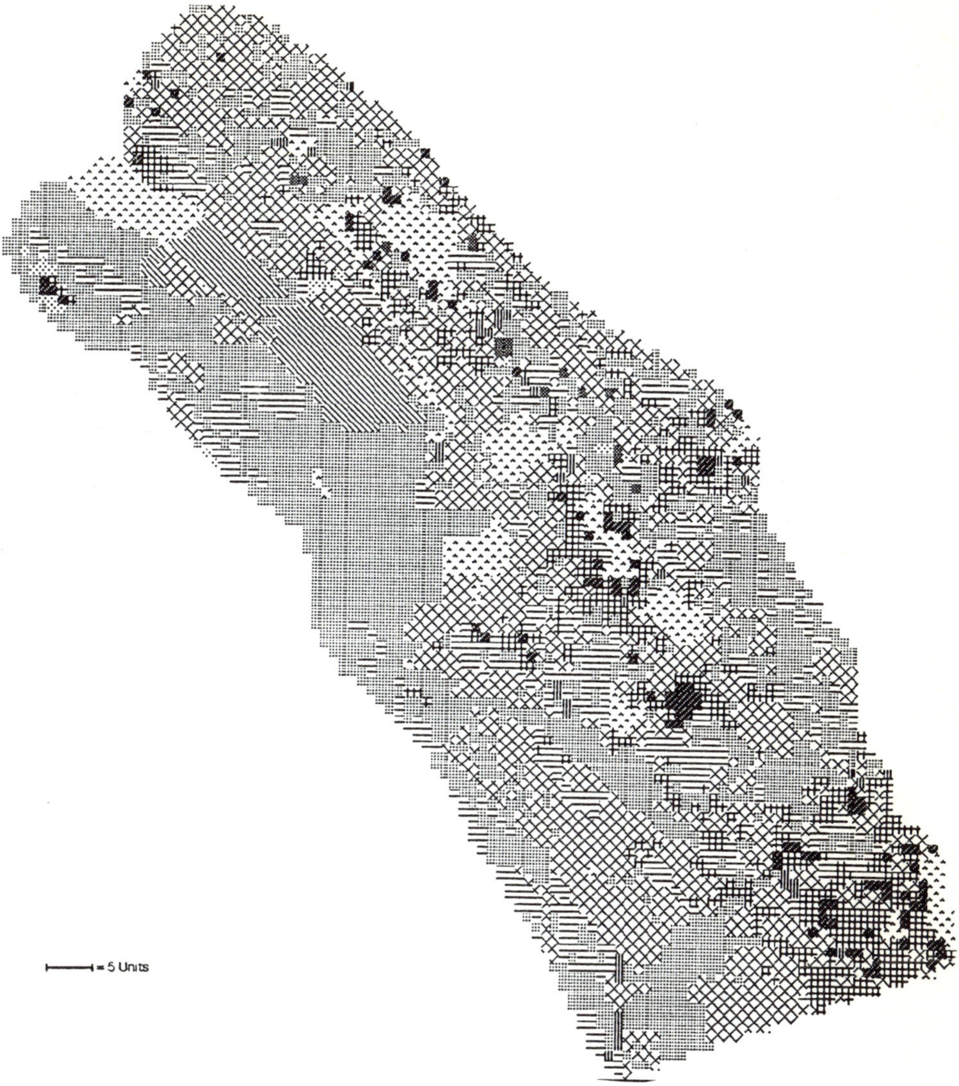

	cells	habitat	leading spp
	1728	Oak and Pine woods	C. caelatus, M. chalybaeus
	21	mesophilous forests	Abax ovalis
	16	"doline" aband. fields	Pt. strenuus, Poec. coerulescens
	556	bushed grasslands	-
	1630	grasslands, pastures	Poec. koyi, Cal. glabricollis
	505	mowed grasslands	Harpalus flavicornis
	114	fields	Poec. cupreus, Harp. rufipes
	42	ruderal environments	-
	191	tank area	Harpalus smaragdinus
	341	urban areas	-

The application of the INV in cases of land rehabilitation can be supported by the evaluation of the "potential coenoses distance" (Brandmayr, 1983b); if reliable data on species habitat choice are available. The species structure of the potential ecosystem is suggested, and by means of a distance index (*e.g.*, Sörensen index) the affinity among real (sampled) and potential coenoses is calculated.

Reviews on methods and criteria used in biological conservation are given by Margules and Usher (1981), Pearsall *et al.* (1986), and Ploeg (1986). Many other methods have been proposed based on the combination of the many evaluation criteria. In some of them ecological categories of species are defined, and the evaluation is based on them (*e.g.*, Bohac, 1988; Rushton *et al.*, 1993). Biometric parameters may also be taken into account (Šustek, 1987). A common procedure is to focus attention on the commonness and rarity of attributes of the area to be evaluated (Bedward *et al.*, 1991; Sankovskii, 1992). The use of biological features as described in this paper is frequent, but the information is not always synthesized in a single value or index.

The method that we use has been developed trying to focus attention on parameters to be checked in a relatively short time, and independently of the skill of the researcher. The collection of much field data, in many environments, for (possibly) many species, is needed to assign a group of species to ecological categories. The definition of ecological categories is more subject to researcher interpretation than the choice of biological features. The use of ecological categories well known from the literature, is a good method when working with single indicator species, and at a local geographical scale, as species with wide distribution might show different habitat choice with, for example, latitude.

The INV can be applied on the basis only of presence or absence of the species, a fact that facilitates data collection and the evaluation of large areas, where only (reliable) species inventories exist. The parameters used to assess naturalness, as described above, have a clear ecological and evolutionary meaning, being related both to the life strategies of the single species and to the stability (resilience/resistance) of the community. Their importance reflects the alternation of r or K strategy species in natural succession (Brandmayr, 1983a, b). In this way it is possible to try to "observe" degradation phenomena from the point of view of species different from man, and to evaluate what is the response of the faunistic assemblages (guild, community, population) to anthropogenic disturbance.

The method suggested in this paper is based on descriptive models that are a helpful simplification of ecological landscape complexity. The many environments of the study area are ordinated in an ecological series of bioclimatic belts (Figure 5), dynamic aspects are explained in terms of natural succession (*e.g.*, the ecological sere of Figure 2), and the highest natural value is applied to the climax stage (an ideal condition of equilibrium between climate and biological coenoses). Research in the field of biological conservation is aimed mainly at the application of the results to so-called environmental management, hence the use of models is really helpful when these can be updated. Worthy of attention is the shift from evaluation to map drawing and interpretation of the geographical patterns of naturalness and degradation. Actual ecosystems are not yet under threat of complete disappearance, but their geographical range has changed (sometimes in a radical way) and it is expected to continue to change. Catastrophic events like a meteorite impact or the disappearance of Atlantis are not necessary for this to happen; but events that cause "catastrophes" like the global increase (expected in the next hundred years) of several degrees in mean temperature are sufficient.

ACKNOWLEDGEMENTS

I am particularly indebted to my sister Patrizia who kindly drew Figure 1, and to my wife Valentina who gently agreed to spend part of our honeymoon waiting for the completion of this manuscript.

REFERENCES

Bedward M, Pressey RL, Nicholls AO (1991) Scores and Scores Classes for Evaluation Criteria: A Comparison Based on the Cost of Reserving all Natural Features. Biological Conservation, 56: 281-294

Boer PJ den (1977) Dispersal Power and Survival. Carabids in a cultivated countryside. Miscellaneous Papers, 14. Landbouwhogeschool, Wageningen, the Netherlands

Boer PJ den (1990) The Survival Value of Dispersal in Terrestrial Arthropods. Biological Conservation, 54: 175-192

Boer PJ den, Loevei GL, Stork NE, Sunderland KD (eds.) (1987) Proceedings of the 6th European Carabidologist's Meeting. Acta Phytopathologica et Entomologica Hungarica, 22. Akademiai Kiado, Budapest

Boer PJ den, Luff ML, Mossakowski D, Weber F (eds.) (1986) Carabids Beetles, Their Adaptations and Dynamics. Gustav Fischer, Stuttgart, New York

Boer PJ den, van Huizen THP, den Boer W, Daanje B, Aukema B, den Bieman CFM (1980) Wing Polymorphism and Dimorphism as Stages in an Evolutionary Process (Coleoptera, Carabidae). Entomologia Generalis, 6: 107-134

Bohac J (1988) Die Ausnutzung von Raubkäfergemeinschaften (Coleoptera, Staphylinidae) für die Indikation der Umweltqualität. Pages 160-164 *in* Bohac J, Ruzicka V (eds.). Bioindicatores Deteriorisationis Regionis. Proceedings of the Vth International Conference. Ceske Budejovice

Bohac J (ed.) (1992) Bioindicatores Deteriorisationis Regionis. Proceedings of the VIth international conference. Ceske Budejovice

Bornkamm R, Lee JA, Seaward MRD (eds.) (1980) Urban Ecology. Blackwell Scientific Publications, Oxford

Brandmayr P (1975) Un gruppo di invertebrati del suolo, i Coleotteri Carabidi, in relazione al grado di trasformazione dei biotopi agrari e forestali del basso Friuli: sua importanza per la ricostruzione ambientale. Informatore Botanico Italiano, 7: 237-243

Brandmayr P (1983a) The main axes of the coenoclinal continuum from macroptery to brachyptery in carabid communities of the temperate zone. Pages 147-169 *in* Brandmayr P, den Boer PJ, Weber F (eds), The synthesis of field study and laboratory experiment in carabids, Report of the fourth meeting of European Carabidologists

Brandmayr P (1983b) Entomocenosi come indicatori delle modificazioni antropiche del paesaggio e pianificazione del territorio: esempi basati sullo studio di popolamenti a Coleotteri Carabidi. Atti 12° Congr. naz. ital. entomol., Roma 1980: 263-283

Brandmayr P (1991) The reduction of metathoracic alae and of dispersal power of carabid beetles along the evolutionary pathway into the mountains. Pages 363-378 *in* Lanzavecchia G, Valvassori R (eds.). Form and function in zoology. Selected Symposia and Monographs U.Z.I., 5

Brandmayr P, Colombetta G (1981) Criteri possibili per una valutazione quantitativa del carattere primario e della capacita' di ricostituzione spontanea di popolamenti indicatori a Coleotteri Geoadefagi del Carso Triestino. Atti 1° Conv. Ecol. Territ. Carsici: 183-188

Brandmayr P, Pizzolotto R (1988) Indicatori "storici" ed ecologici nella coleotterofauna terricola delle foreste dell'Appennino. Atti XV Congr. naz. ital. Entom.: 589-608

Brandmayr P, Pizzolotto R (1990) Ground beetle coenoses in the landscape of the Nebrodi Mountains, Sicily (Coleoptera Carabidae). Il Naturalista Siciliano, s.IV, XIV (suppl.): 51-64

Brandmayr P, Codogno M, Pizzolotto R (1991) Bases for the ecological mapping of natural resources in Calabria: biomes and their subunits on the cross-section Catena Costiera - Sila Grande - Sila Greca (in Ital.). S.IT.E. Atti, 12: 389-393

Brandmayr P, Cagnin M, Pizzolotto R, Mingozzi A, Tripepi S (1988) Acquedotto del Menta. Studio per la Valutazione di Impatto Ambientale. Allegato d2) Le risorse biologiche e il paesaggio. ELC-electroconsult, Milano

Cagnin M, R Venanzoni, P Brandmayr, R Canullo, F Dessì-Fulgheri, F Pedrotti (1991). The botanical, zoological and landscape resources in the study of environmental impact (EIA) "Menta Dam" (Aspromonte-Italy) (in ital.). S.IT.E. Atti, 12: 739-748

Clements FE (1920) Plant Indicators. The relation of plant communities to process and practice. Carnagie Institution of Washington

Desender K (ed.). (in print). Carabid Beetles - ecology and evolution. Kluwer Academic Publishers, Dordrecht

Feoli E, Orloci L (1979) Analysis of concentration and detection of underlying factors in structured tables. Vegetatio, 40: 49-54

Ganin GN (1989) Role of diplopodes in disintegration and transformation of the forest litter in the south of Khabarovsk territory. Zool. Zh., 58, 1: 145-149

Klausnitzer B (1987) Oekologie der Grosstadtfauna. Gustav Fischer, Stuttgart, New York

Kurnetsova, Krivolutsky (1986) Invertebrates as a bioindicator of the environment in Moscow. Pages 109-113 *in* Paukert J, Ruzicka V̇, Bohac J (eds.). Bioindicatores Deteriorisationis Regionis. Proceedings of the IVth international conference. Ceske Budejovice

Lapka M, Gottlieb M, Cudlinova E (1992) Social indication of man niche. Pages 60-71 *in* Bohac J (ed.). Bioindicatores Deteriorisationis Regionis. Proceedings of the VIth international conference. Ceske Budejovice

Margules C, Usher MB, (1981) Criteria used in assessing wildlife conservation potential: a review. Biological Conservation, 21: 79-109

Minelli A (1974) Studio preliminare della fauna di Treviso con riflessioni sulla fauna degli ambienti urbani. Atti Ist. Veneto Scienze, Lettere ed Arti. Cl. Scienze Matematiche e Naturali, CXXXII: 115-156

Mingozzi A, Brandmayr P (1991) L'evaluation cartographique des ressources faunistiques: un example applique aux ornithocenoses d'une vallee alpine. Rev. Ecol. Alp., Grenoble: 1-21

Mossakowski D, Fraembs H, Baro A (1990) Carabid beetles as indicators of habitat destruction caused by military tanks. Pages 237-243 *in* Stork NE (ed.). The Role of Ground Beetles in ecological and environmental studies. Intercept, Andover,Hampshire

Paoletti MG, Stinner BR, Lorenzoni GG (eds.) (1989) Agricultural Ecology and Environment. Elsevier, Amsterdam, Oxford, New York, Tokyo

Pearsall SH, Durham D, Eagar DC (1986) Evaluation methods in the United States. Pages 111-133 *in* Usher MB (ed.). Wildlife Conservation Evaluation. Chapman and Hall

Pielou EC (1979) Biogeography. Wiley & Sons, Chichester

Pizzolotto R (1993a) Ground beetles as a tool for environment management: a geographical information system based on Carabids and vegetation for the Karst near Trieste (Italy). *in* Desender K (ed.). Carabid Beetles - ecology and evolution (ed. K Desender). Kluwer Academic Publishers, Dordrecht (in print)

Pizzolotto R (1993b) Carabid Beetle (Coleoptera, Carabidae) Coenoses for Evaluation of Faunal Resources and Impact Assessment in the Aspromonte National Park of Calabria (Italy). COENOSES: (in print)

Pizzolotto R, Brandmayr P (1990) The Carabid Groupings of the Nebrodi Mountains in Sicily: Ecological and Historical Indicators. Pages 201-207 *in* Stork NE (ed.). The Role of Ground Beetles in Ecological and Environmental Studies. Intercept, London

Ploeg SWF Van der (1986) Wildlife conservation evaluation in the Netherlands: a controversial issue in a small country. Pages 161-180 *in* Usher MB (ed.). Wildlife Conservation Evaluation. Chapman and Hall, New York

Rushton SP, Eyre MD, Luff ML (1993) Modelling the consequences of land use change on the distribution of ground beetles. *in* Desender K (ed.). Carabid Beetles - ecology and evolution. Kluwer Academic Publishers, Dordrecht (in print)

Ruzicka V, Antus M (1989) 25-year changes in steppe spider communities in the surroundings of Lochkov at Prague. Pages 154-159 *in* Bohac J, Ruzicka V eds.). Bioindicatores Deteriorisationis Regionis. Proceedings of the Vth international conference. Liblice, Prague

Sankovskii A (1992) Toward evaluation of natural objects. Environmental Management, 16: 283-287

Santas P (1986) Soil communities along a gradient of urbanization. Rev. Ecol. Biol. Sol., 23 (4): 367-380

Soulé ME (1987) Introduction to Viable Population for Conservation. Pages 1-10 *in* Soulé ME (ed.). Cambridge University Press

Spellerberg IF (1991) Monitoring ecological change. Cambridge University Press

Sprugel DG (1985) Natural disturbance and ecosystem energetics. *in* Pickett STA, White PS (eds.). The Ecology of Natural Disturbance and Patch Dynamics. Academic Press, London

Stork NE (ed.). (1990) The role of Ground Beetles in Ecological and Environmental Studies. Intercept, Andover-Hampshire

Straalen NM van, Kraak MHS, Denneman CAJ (1988) Soil microarthropods as indicators of soil acidification and forest decline in the Veluwe area, The Netherlands. Pedobiologia, 32: 47-55

Šustek Z (1987)Changes in body size structure of carabid communities (Coleoptera, Carabidae) along an urbanisation gradient. Biologia, 42: 145-156

Thiele HU (1977) Carabid Beetles in their environments. Springer Verlag, Berlin, Heidelberg, New York

Toselli E, Ballarin L, Brandmayr P, Gombach ML, Juretig L, Perco F, Semeraro R (1982) La carta geologica della zona prevista per l'AREA di ricerca scientifica e tecnologica (a.R.S.T.) nella provincia di Trieste: mappa indicizzata di valutazione iniziale dello stato dell'ambiente naturale. Atti XVIII Convegno Naz. Assoc. Ital. Cartografia: 33-48

Usher MB (1986) Wildlife conservation evaluation: attributes, criteria and values. *in* Wildlife Conservation Evaluation Pages 3-44. Chapman and Hall, London, New York

Wagner DL, Liebherr JK (1992) Flightlesness in Insects. Trends in Ecology, Evolution, 7: 216-220

Westhoff V (1971) The dynamic structure of plant communities in relation to the objectives of conservation. Pages 3-14 *in* Duffey E, Watt AS (eds.). The Scientific Management of Animal and Plant Communities for Conservation. Blackwell Scientific Publications, Oxford

Chapter 16: Changes in the Species Diversity of Ground Beetles (Coleoptera, Carabidae) Following Disturbances

A. Georges
CNRS UA 0696
Laboratoire d'Evolution des Systemes Naturels et Modifies
Campus de Beaulieu
35042 Rennes cedex
France

INTRODUCTION

Disturbances affect community structure at various hierarchical levels and spatial scales, which usually complicates the interpretation of species occurrence and the research of causal relationship to environmental variables. Nevertheless species composition and abundance analyses assume primary importance for conservation purposes and trends in species composition in fragmented landscapes suggest that a more comprehensive view is required for maintenance of diversity (Noss, 1983).

Different sources of disturbance may be important among different habitats. In France, it is only during recent decades that draining of the marshes came to our attention as a disturbance. Traditionally, wetlands have been considered lands which can be brought to good use only if they are drained and reclaimed for more intensive grazing or arable farming, but recent changes in agricultural politics led to the removal of some lands from production. Thus there is a need to assess changes in land-uses and to quantify conservation criteria.

Our question was: what are the effects of these different types of disturbances that operate at different scales and are interdependent on species communities, using carabid beetles (Coleoptera: Carabidae) as an example? Carabid beetles appear especially suitable to provide information for forecasting, and assisting in the prospective management of wetlands. Ecological data are available for many species

NATO ASI Series, Vol. I 20
Biodiversity, Temperate Ecosystems, and Global Change
Edited by T. J. B. Boyle and C. E. B. Boyle

and they can be used to test a wide range of ecological problems (Thiele, 1977, Den Boer, 1986).

Rushton *et al.* (1989), Luff and Rushton (1989) and Eyre *et al.* (1990) used classification and ordination methods to assess similarities in carabid communities to define factors affecting the distribution of species assemblages. Refseth (1980) considered this group well fitted to yield the most valuable results for habitat classification. Luff *et al.*(1992), after studies in managed grasslands, estimated that ground beetles were a viable group to be used quantitatively as indicators of environmental quality and change.

The aim of our studies was to analyse as wide a variety of land-use as possible using ground beetle species assemblages to assess differences associated with engineering works and agricultural practices in wetlands. To achieve an available picture of long term faunal changes in wetlands, old quantitative data sets would be needed for comparison. For invertebrates, such data are generally lacking. Consequently, we have to base our conclusions on comparing the composition of present assemblages in areas that are at different stages of undergoing change.

In this paper, we summarize the results of a quantitative survey of carabid beetles throughout Marais poitevin reclaimed salt marshes in western France (Figure 1). This area appeared to be a typical area for evolution of reclaimed salt marshes. The diversity observed is the spatial manifestation of time succession and engineering works creating new biotopes. The first dykes were built during the Middle Ages, and the exclusion of the sea initiated the formation of two groups of ecosystems:

a) The eastern part of Marais poitevin is composed of low lands regularly flooded by the rivers running from the surrounding watershed. In this area, drainage is by ditches bordered with hedges composed of pruned ashes and poplars.

b) The western part of Marais poitevin is composed of polders and open fields. This area is protected from equinoctial tides by dykes along the coast and from the winter rising of the rivers by embankments.

These two features have their own patterns of development; notably, following

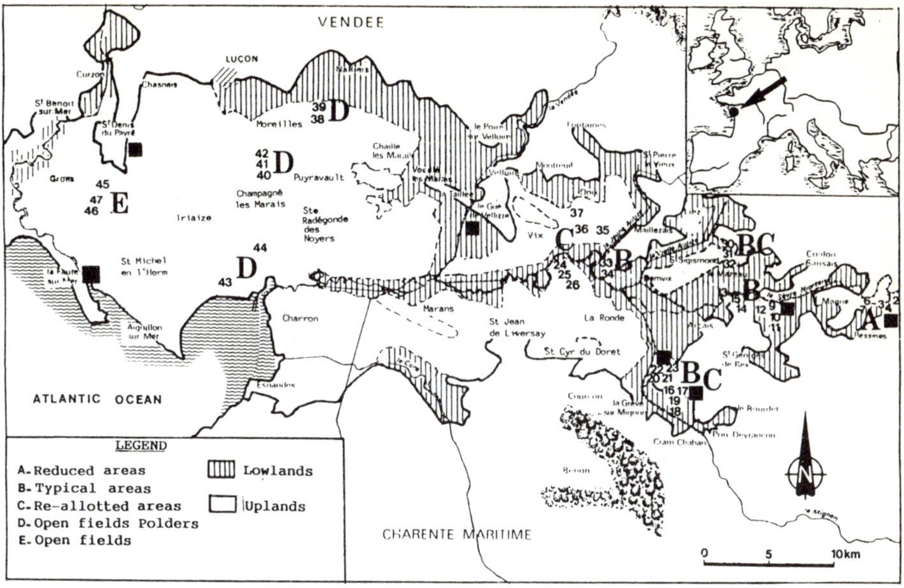

Figure 1. Map of the sample sites

engineering works over the centuries, and also following consolidation, re-allotment and other agricultural practices during the last decades.

It is helpful to analyse disturbances and species diversity at three levels of information (even if they are closely interdependent):

1. changes in communities along environmental gradients and different land forms;
2. changes in communities within small areas (a parcel of few hectares) and uniform land-uses; and
3. effects on communities of the edges and interconnections among the patches in the landscape.

These three levels correspond with the three basic scales of diversity generally recognized (Wolda, 1983, Noss, 1983).

The effects of environmental conditions on carabid communities at these three levels are well known:

a) at the alpha scale (within habitat) the factors influencing carabid species communities are those which have direct effect on microclimate most notably soil substrate (Eyre *et al.,* 1990) and vegetation (Jennings *et al.,* 1986);

b) at the gamma scale (regional), Cachan *et al.* (1979) suggested that along gradients from climax forest to *Picea* forest and from cultivated lands to urban spaces, the species richness of a region increases with cultivated lands and decreases with urbanization; and, finally

c) at the beta scale, studies of the effect of hedgerow networks on carabid distribution emphasized the importance of interconnections and edges effect among the patches in the landscape (Pollard, 1968, Devaux, 1976, Wallin, 1985).

In the present study the concept of community must be interpreted as an assemblage of species which occurs in a number of sites with a particular pattern of ecological factors and no further properties are implied. We sampled in representative sites according to the main landscape patterns occurring in the area. Sample sites were naturally arranged on gradients with both complexity and wetness decreasing from the eastern part to the western part of the studied area and the sampling design facilitated the comparison between rather undisturbed old wetlands and different man-made new areas.

From the eastern part to the western part of the area, five features were previously identified:

A (sites 1 to 7) - In the extreme east, reduced areas and the lowest lands in Marais poitevin that are regularly flooded. Shady places on river banks and along ditches.

B (sites 9 to 23 and 30 to 34) - In typical landscape patterns with a geometrical network of drainage ditches associated with hedges composed of pruned ashes and poplars. Peat lands settled before the Middle Ages. Uplands adjacent to the plain have been generally re-allotted during recent decades.

C (sites 24 to 26 and 35 to 39) - In a central position, clayey lowlands, drained during the sixteenth and seventeenth centuries. Woodlike areas in a destructured hedgerow network.

D (sites 38 to 44) - Open fields and polders on clayey substrate. Dykes were gradually built from the Middle Ages to the present day.

E (sites 45 to 47) - In the western part of Marais poitevin, this area has been settled since the first embankments were constructed during the Middle Ages. Open fields and sparse trees. Arable lands are on clayey, well drained soil.

Investigations were carried out during seven years (from 1981 to 1987) and several sample designs were used. The samples were obtained using pitfall traps, this method included some limits but was convenient for extensive use (Greenslade, 1964, Luff, 1975).

We first present gradient sampling sites and hierarchical classifications. We will then look at discontinuous faunistic variations. And the discussion will be devoted to a schematic model across the landscape that supports these interdependent disturbance processes.

Primary investigations (Figure 2) indicate that carabid data represented by Shannon's diversity index are well correlated with the *a priori* classification of the sites on a wetness gradient (soil water content). Nevertheless, the simultaneous

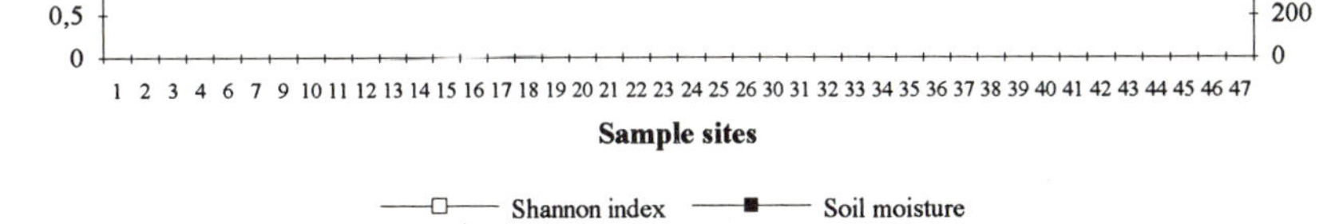

Figure 2. Species diversity in each sample site compared with soil moisture along the transverse cross section.

analysis of species richness indicates that such an index is not suitable to compare localities or to survey temporal modifications in a community; it over-estimates the diversity in small catches and under estimates it in large samples (Southwood, 1978, Butterfield and Coulson 1983, Rushton et *al.* 1990), and we did not employ it further in the study. For simplicity we used the terms richness (number of species within a defined area), relative abundance (per species in the total catches) and occurrence (percentage of sample units where a species is present).

CHANGES IN CARABID COMMUNITY ALONG ENVIRONMENTAL GRADIENTS AND DIFFERENT LAND FORMS

A total of 47 sites differing in soil substrate, vegetation type, hedgerow and drainage network were selected to cover as wide a range as possible of different landscape patterns present in Marais poitevin (Figure 1). The responses of various carabid species to differing wetland management regimes were investigated in sites with well known environmental conditions. The description of the sample sites included two series of parameters:

1. Physical and chemical characteristics of the soil on each site were determined from cores of 30 cm depth taken near the cells of pitfall traps (residual soil moisture, porosity, fine sand coarse sand, lime, clay, organic matter, were determined for each core).
2. Landscape elements: agricultural uses (pasture, meadows, cultivated lands and woodlots), hedgerow network, drain density and their pattern were recorded both for sample fields and surrounding ones in order to estimate landscape heterogeneity.

Carabid beetles were sampled by pitfall trapping on each site. Grids of four traps were localized in the middle of the field. Species nomenclature follows that of Bonadona (1971).

A correspondence analysis (Benzecri, 1973) was performed on a matrix using 54 species and 41 species (only data from March 1984 to March 1985 were used for this analysis) followed by a hierarchical classification (Roux, 1985) on samples using the leading factors (in this cluster analysis the distance used was the Euclidian distance and the aggregation criterion was the mean of weighted distances). To improve the quality of interpretation, we selected species in relation to their global and local contribution to diversity (Lam Hoai *et al.* 1987), and the pooled data arranged by classification were used for comparing species assemblages in the habitat groups interpreted by the cluster analysis.

The interpretation is easy on sample factor planes of the correspondence analysis (Figure 3): F1 (eingenvalue 0.683) separates wetlands from rather dry lands, and F2 (eingenvalue 0.533) separates typical structures from disturbed areas. Site groups from cluster analysis, as in the correspondence analysis, appeared to be related to moisture conditions and to features of landscape spatial structures.

Figure 4 shows the dendrogram formed by clustering the sites, the number of species in each group and the main characteristics of the group. Based on factor analysis of the samples and the grouping of the samples, five groups appear quite interpretable ecologically; with the most interpretable classification being achieved at level 0.84. We separated sites into groups representing particular ecological patterns, on a well defined wetness gradient with:

1. Lowland sites separated into three groups:

    a) peat, clayey and mud substrate sites on river banks or along drain ditches; reduced woodlike habitats adjacent to the plain or open fields;

    b) peat substrate sites in a typical network of drainage ditches and hedges; and

    c) peat and clayey substrate sites, generally in re-allotted areas with a wide and discontinuous hedgerow network.

2 - Upland sites separated into two groups:

    a) clayey substrate sites, more or less wet, on open fields with sparse trees; and

    b) clayey substrate sites rather well drained on polders.

Overall, soil water content seems to be related to the composition of carabid assemblages over the sites.

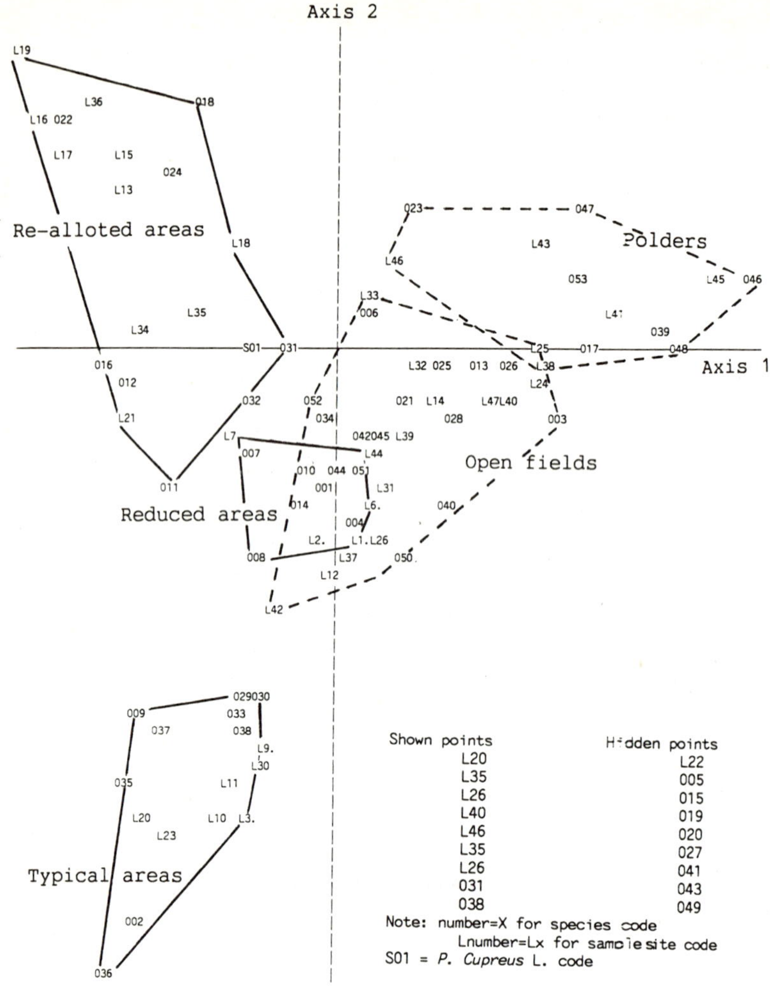

Figure 3. Projection of the sample sites and species in relation to the first axis (axis 1 by axis 2) of the correspondence analysis

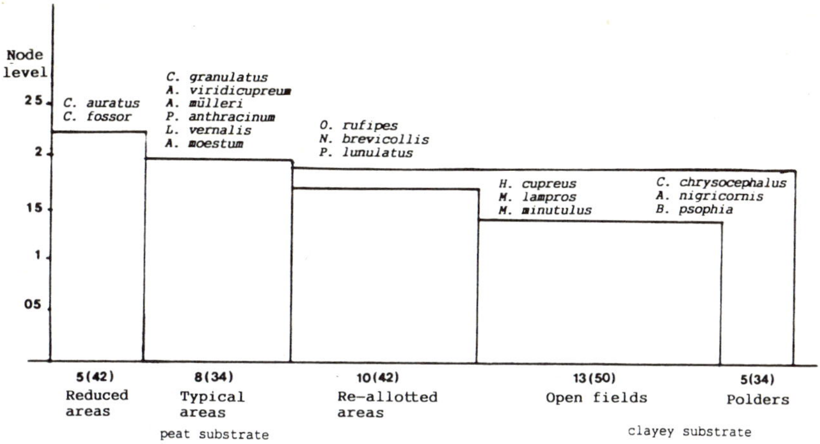

Figure 4. Hierarchical classification of the sample sites according to carabid beetles data showing the number of sites in each division and the main environmental characteristics. The number of species in each group is shown in parentheses.

These results show a well defined gradient of wetness from low- lands to uplands in the eastern part of the area and from old polders to more recent ones in the western part of the area. The method emphasized the importance of the wetness gradient and provided information on the possible role of the disturbance (changes in the hedgerow and drainage networks, underground water control) affecting species diversity.

Distribution of carabid species in the habitat groups

Table 1 shows the pooled carabid data of the five habitat groups arranged by the cluster analysis. Species richness is highest in the samples from open fields, followed by reduced areas and re-allotted areas and, in last position, samples from typical areas and rather recent polders.

Globally there are no consistent trends in the species caught over the different sites plots. We attempted to obtain groups of indicator species by investigating invertebrate overlap along the main gradient by selecting species in the original data from the three years of sampling. Thus we selected species which show both a relative abundance of more than 0.40% and which occur in all sample sites in a proportion greater than 40%. In this way, we removed rare species, or species which may have occurred by chance in the catches, and species which occurred locally and not in the majority of the samples representing the habitat (by the arbitrary choice of the percentage, this method has limitations but is adequate for comparisons between communities).

Thirty-one species were selected (Table 1). The abundance patterns of these species in each group of sites arranged by the hierarchical classification are presented in Figure 5. The selected species are not uniformly dispersed over the habitats but show some habitat preferences:

Group 1 - rather hygrophilous, preferring wet substrate in woodlike habitats. Samples from slightly disturbed areas with typical hedgerow network composed of pruned ashes and poplars: *Carabus granulatus* (L.), *Dyschirius globosus* (Herbst), *Agonum mulleri* (Herbst), *Lagarus vernalis* (Panzer), *Platysma anthracinum* (Ill), *Agonum viridicupreum* (Groeze), *Agonum moestum* (Dufts).

Group 2 - eurytopic species, generally occurring both in shady places and in cultivated land, suggesting that the relative abundance of these species may vary with management and changes in wetness. Samples from upper lands in re-allotted areas: *Nebria brevicollis* (F.), *Ophonus rufipes* (De Geer), *Harpalus distinguendus* (Dufts.), *Harpalus affinis* (Schrank), *Anysodactylus binotatus* (Dej.), *Platysma vulgare* (L.), *Philochthus lunulatus* (Steph.).

Group 3 - species living in all kinds of open, moist, more or less vegetated ground. Samples from the lower lands on shady places along ditch or river banks: *Clivina fossor* (L.), *Carabus auratus* (L.), *Metallina lampros* (Herbst.), *Amara aenea* (De Geer).

Group 4 - slightly hygrophilous or rather xerophilous species occurring in grassland. Samples from old polders and open meadows: *Adelosia macrum* (March.), *Harpalus cupreus* (Dej.), *Microlestes minutulus* (Schm.-Goeb.), *Parophonus mendax* (Rossi), *Phyla obtusum* (Serv.), *Zezea erythrochnema* (Zimm.).

Group 5 - rather xerophilous species, occurring in open country. Samples from the rather more recent polders: *Anchomenus dorsalis* (Pont.), *Brachinus psophia* (Serv L.), *Chlaenius chrysocephalus* (Rossi), *Diachromus germanus* (L.), *Agostenus nigricornis* (F.), *Ophonus rotundicollis* (Frm. et Lab.).

Table 1. Total number of each ground beetle species caught in the site groups interpreted from cluster analysis.

		EASTERN PART OF THE AREA Peat substrate			WESTERN PART OF THE AREA Clayey substrate	
		Typical areas	Re-allotted areas	Reduced areas	Open fields	Polders
	Selected species					
2 CAGR	*Carabus granulatus L.*	258	12	3	12	0
35 AGVI	*Agonum viridicupreum Goeze*	76	21	1	5	0
37 AGMU	*Agonum mülleri Herbst*	92	23	14	20	2
9 DYGL	*Dyschirius globosus Herbst*	29	16	1	2	1
29 LAVE	*Lagarus vernalis Panzer*	64	14	5	20	12
30 ARST	*Argutor strenuus Panzer*	22	9	8	9	3
33 PLAN	*Platysma anthracinum Ill*	312	70	6	115	57
38 AGMO	*Agonum moestum Dufts.*	99	29	2	44	7
16 ANBI	*Anisodactylus binotatus Dej.*	20	79	0	3	0
23 HAAF	*Harpalus affinis Schrank*	0	32	2	10	38
24 HADG	*Harpalus distinguendus Dufts.*	1	98	3	21	12
5 NEBR	*Nebria brevicollis F.*	26	159	88	41	34
22 OPRU	*Ophonus rufipes De Geer*	91	2053	17	60	122
11 PHLU	*Philochthus lunulatus Steph.*	2	33	29	12	0
32 PLVU	*Platysma vulgare L.*	4	17	3	17	8
45 AMAE	*Amara aenea De Geer*	4	14	11	66	0
1 CAAU	*Carabus auratus L.*	1	2	15	9	0
8 CLFO	*Clivina fossor L.*	22	25	47	29	0
14 MELA	*Metallina lampros Herbst.*	32	46	26	108	17
31 ADMA	*Adelosia macrum March.*	0	13	1	14	10
25 HACU	*Harpalus cupreus Dej.*	3	25	13	90	55
52 MIMI	*Microlestes minutulus Schm.-Goeb*	3	16	21	31	8
19 PAME	*Parophus mendax Rossi*	2	3	8	35	22
13 PHOB	*Phyla obtusum Serv.*	7	20	18	28	51
40 ZEER	*Zezea erythrochnema Zimm.*	0	0	8	6	3
39 ANDO	*Anchomenus dorsalis Pont.*	1	7	1	35	91
53 BRPS	*Brachinus psophia Serv. L.*	1	92	18	238	1037
48 CHNI	*Agostenus nigricornis F.*	30	8	0	24	224
17 DIGE	*Diachromus germanus L*	6	8	2	26	41
20 OPRO	*Ophonus rotundicollis frm et Lab*	4	31	1	28	93
46 TRCH	*Chlaenius chysocephalus Rossi*	0	8	0	117	907
	Other species					
3 CAPU	*Carabus purpurascens F.*	0	0	6	1	5
4 LEFU	*Leïstus fulvibarbis Dej.*	2	1	1	2	1
6 NOBI	*Notiophilus biguttatus F.*	0	8	11	2	10
7 LOPI	*Loricera pilicornis F.*	0	1	0	1	0
10 TROB	*Trechus obtusus Erich.*	0	2	14	5	1
12 BEQU	*Bembidion quadrimaculatum L.*	1	9	6	0	0
15 ASFL	*Asaphidion flavipes L.*	0	0	0	1	0
18 PAMA	*Parophonus maculicornisDufts.*	0	3	0	0	1
21 OPPU	*Ophonus puncticeps Frm. et Lab.*	0	0	6	18	6
26 HADM	*Harpalus dimidiatus Rossi*	0	0	2	13	5
27 HAAT	*Harpalus attenuatus Steph.*	1	6	2	2	0
28 ACME	*Acupalpus meridianus L.*	0	0	0	1	0
34 STMA	*Steropus madidus F.*	0	0	1	1	0
36 AGMA	*Agonum marginatum L.*	1	0	0	0	0
41 AMNI	*Amara nitida Sturm.*	0	0	0	1	0
42 AMNO	*Amara anthobia Villa*	0	1	0	11	1
43 AMEU	*Amara eurynota Panzer*	0	2	2	2	0
44 AMFA	*Amara familiaris Dufts.*	0	3	15	9	0
47 AGOL	*Agostenus olivieri Crotch*	0	0	0	0	1
49 OOHE	*Oodes helopoïdes F.*	3	1	2	1	0
50 PACM	*Panagaeus crux major L.*	1	0	0	3	0
51 BABI	*Badister bipustulatus F.*	0	0	1	3	0
54 POCU	*Poecilus cupreus L.*	1421	5259	492	492	1335
	Number of species	34	42	42	50	34
	Number of sites	8	10	5	13	5

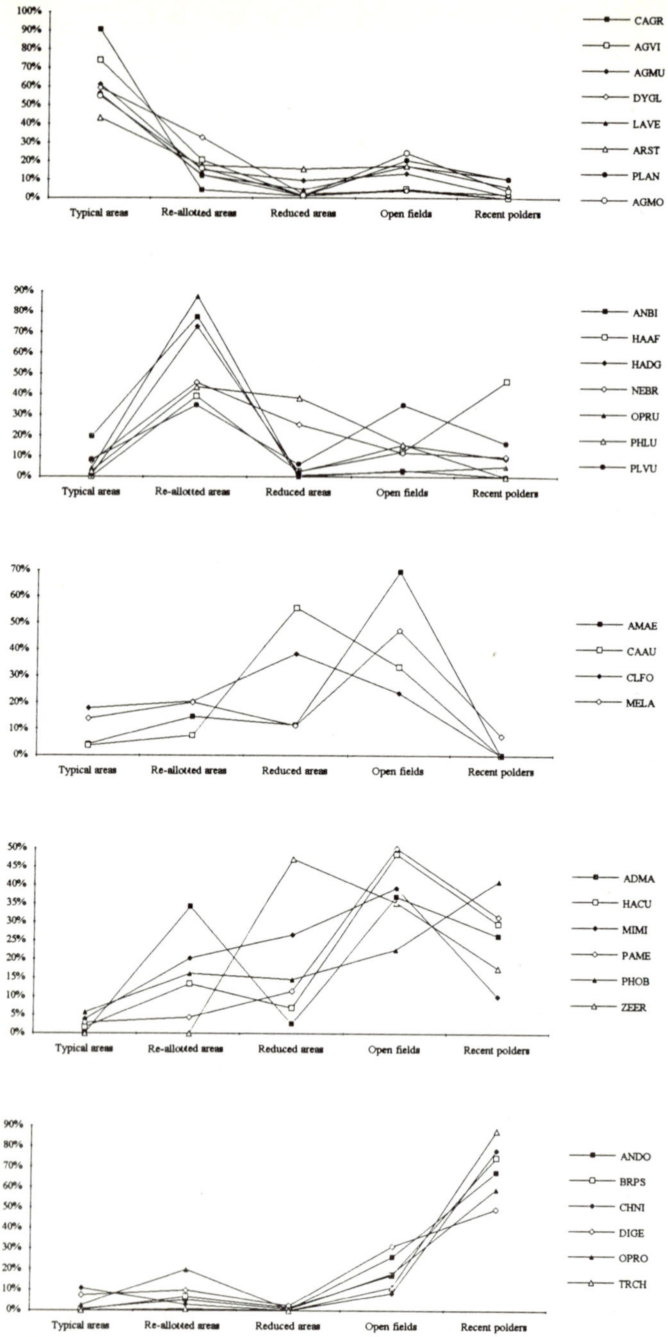

Figure 5. The abundance patterns of the 31 selected species in the site groups interpreted from the cluster analysis. Species codes are related in Table 1.

The selected species represented species with well known preferences The proportion of species preferring damp conditions is higher in group 1. Species of genus Carabus are confined essentially to peat soil, *Carabus granulatus* occurred only in group one, and *Carabus auratus* was recorded only in group 3.

The heterogeneous nature of the re-allotted areas on the one hand and of the open fields on the other hand is evident from the wide distribution of the species (group 2 and group 4 respectively) and the evasive species composition. In group 2, the occurrence of ubiquitous species like *Nebria brevicollis* and *Ophonus rufipes* was probably due to the proximity of cultivated upland to some sites of this group. Some dominant species (*Platysma anthracinum, Agonum moestum*) in the typical structures (group 1) were also recorded in great proportion in open fields (group 4), which suggested changes in soil substrate conditions of some sites.

Whilst possible fluctuations of local populations at individual sampling may be masked by pooling of the data, we studied more precisely some sample sites along a transect crossing three embankment lines settled in 1874, 1770, and 1650 from littoral to old back wetlands. Rather xerophilous species occurred in these sites (Figure 5), but we also noticed the presence of *Zezea erythrochnema* which indicated a high level of salinity, which increases after the compaction of the ground by cattle, and the occurrence of *Platysma anthracinum* and *Agonum moestum* in the lowland old pastures where animal stocking was reduced. The occurrence of species preferring damp conditions highlighted the difficulties of drainage as the distance to the sea increases.

Species composition of carabid communities shows some affinities between sample sites located in old polders and sample sites located in old eastern wetlands; nevertheless, even if polders of longer standing show some features of ageing, they do not equal the maturity level of the older back wetlands (notably no members of the genus *Carabus* were present).

In short, the evidence based on correspondence analysis is further corroborated by classifications which show that all environmental factors measured were of importance in species distribution. At the regional scale, both species composition and abundance assume primary importance in a sustainable development. Overall, the comparison indicates that the highest number of species is mainly due to interrelation between sites and possibilities of dispersion.

CHANGES IN CARABID COMMUNTIY WITHIN SMALL AREAS (A PARCEL OF A FEW HECTARES AND UNIFORM LAND-USE)

In addition, in order to investigate the extent to which local factors influence carabid distribution, a new sampling design was carried out in fields selected in the previous sampling features during the same period. Traps were placed at 12m intervals on a line from one edge to the opposite one.

Effect of the underground water drainage on the field.

A comparative study was carried out on two fields identically farmed (5ha of maize on peat soil) but in site A there was no control of water level and in site B underground water was drained by pumping. Table 2 compares the relative abundance of carabid species between the two sample sites.

It was obvious that many species in the freely drained site were no longer present in the managed site. In the unmanaged site dominant species are essentially hygrophilous. In the managed site dominant species are essentially rather xerophilous - in this last site we found *Brachinus psophia,* a low number of *Ophonus rufipes* and no *Carabus granulatus*.

Effect of the field structure.

Pitfall traps were located on a transverse cross section from the upper part to the lower part of the field, representing bare areas and woodlike habitats, respectively. Figure 6 shows a well defined gradient from the upper part to the lower part of the transect. Species richness was higher in the bottom of the field and the proportion of species with a lowland distribution increased from the top to the bottom of the transect.

Table 2. The occurrence of species in unmanaged peat substrate site (A) and in peat substrate site with control of water level by pumping (B).

Species	**Sample site A**	**Sample site B**
Poecilus cupreus	27,5	43,67
Platysma vulgare	21,92	10,14
Ophonus rufipes	10,77	1,41
Metallina lampros	0,58	2,53
Lagarus vernalis	1,15	0,84
Trechus obtusus	0,38	0,84
Bembidion quadrimaculatum	0,38	0,84
Agostenus nigricornis	0,38	0,84
Notiophilus biguttatus	0,38	1,41
Anchomenus dorsalis	0,38	1,97
Phyla obtusum	0,38	2,25
Platysma anthracinum	16,35	-
Carabus granulatus	11,34	-
Loricera pilicornis	3,27	-
Nebria brevicollis	1,15	-
Agonum moestum	0,96	-
Agonum mülleri	0,7	-
Agonum marginatum	0,38	-
Agonum viridicupreum	0,38	-
Clivina fossor	0,38	-
Dyschirius globosus	0,38	-
Celia praetermissa	0,38	-
Agostenus vestitus	0,38	-
Philochthus lunulatus	0,38	-
Microlestes minutulus	-	12,11
Brachinus psophia	-	5,35
Parophonus maculicornis	-	4,79
Harpalus dimidiatus	-	3,09
Carabus purpurascens	-	1,69
Adelosia macrum	-	1,69
Stomis pumicatus	-	1,69
Amara ovata	-	1,41
Haraplus affinis	-	0,84
Haraplus rubripes	-	0,84
Chlaenius chrysocephalus	-	0,84
Apototomus rufus	-	0,84
Drypta dentata	-	0,84
Zuphium chevrolati	-	0,84
Badister bipustulatus	-	0,84
Number of species	24	26

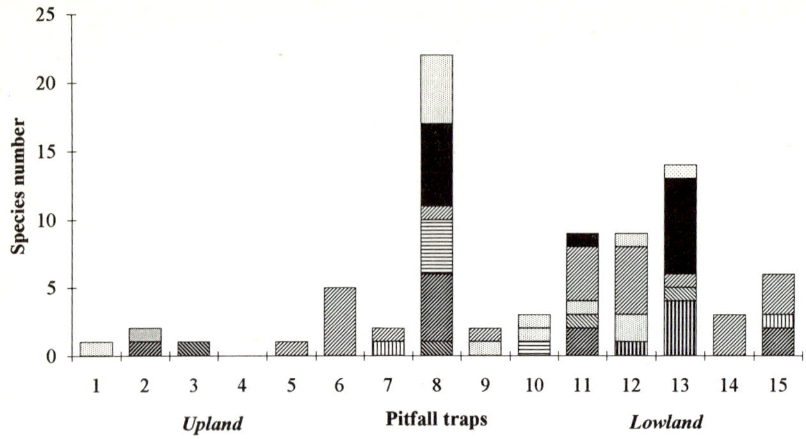

Figure 6. Effects of the field structure in carabid beetle distribution

EFFECT OF INTERCONNECTIONS AMONG THE PATCHES IN THE LANDSCAPE ON CARABID COMMUNITY

Hedgerow and ditches network influence.

At the regional scale it is obvious that no single site structure accounted for the variability of the species composition. The importance of edge structures (hedges) has been widely studied in terms of their effect on carabid distribution (Pollard, 1968, Devaux, 1976, Wallin, 1985). Their influence on the microclimate within the field has been recognized (windbreak and humidity gradient), but beyond that, the influence of the extent of the areas and the features of the surrounding sites are not related.

Since few individuals were caught in one site in each sample period, direct comparisons were carried out between the species composition of the original sites (characterized by edges and drainage quality) (Figure 7). In site A, without management, including a dense hedgerow network, the number of species is quite homogeneous between the middle of the field and close to the edges. In site B, with management and a wide hedgerow network, the number of catches is higher close to the edges (on the banks of ditches and along the hedges). In site C, located in open fields, the number of species is higher near the ditches and there is also an increase

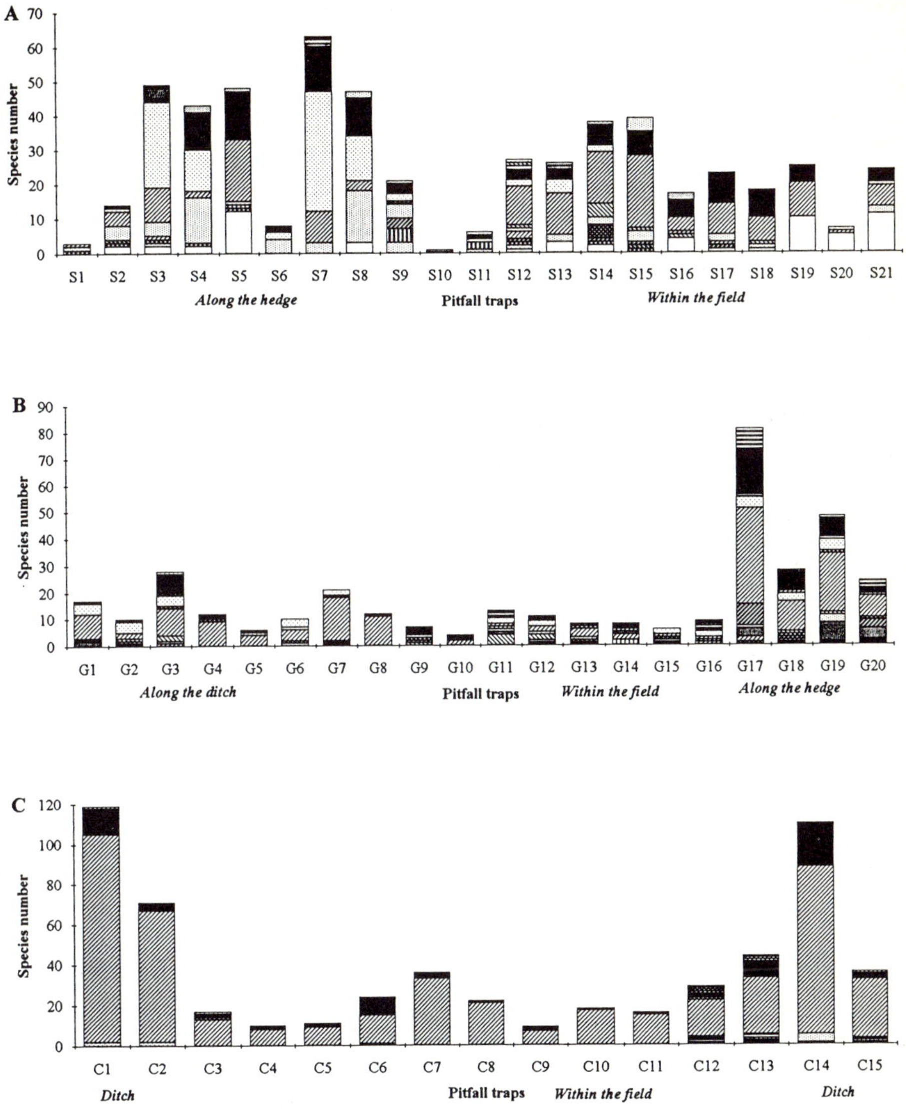

Figure 7. Effect of the underground water drainage
A. Free drained peat site substrate included in a dense hedgerow network
B. Peat substrate site with underground water control included in a wide hedgerow network
C. Clayey substrate site in open fields

in the catches in the middle of the field. Ditches and hedges modified soil water conditions and influenced Carabid beetles distribution.

Edge structure does not always have the same effect on Carabid distribution and, at beta scale, carabid diversity is also influenced by disturbances occurring at other levels of organisation.

DISCUSSION

The results presented here are in agreement with those found by Desender *et al.* (1984) in Belgium and by Luff *et al.* (1989) in England along humidity gradients. At the regional level, the coexistence of communities that belong to moist areas (marsh and riparian environments), to littoral areas, and to cultivated lands (chalky plain) explains the ecological richness of the studied area. If we use species richness as a criterion, lands of a longer standing are no more valuable habitats for carabid beetles than re-allotted areas and new man-made sites. The results confirm that disturbances (consolidation, drainage works) result in a change in ground beetle communities, with large relatively slow dispersing species being replaced by smaller, highly active colonising species (Eyre *et al.* 1990). Thus there will be a trend toward an ecological impoverishment of the lowlands through a reduction of habitats and a penetration of euryecious species.

A schematized distribution of carabid beetles in Marais poitevin illustrates these tendencies (Figure 8).

Two main features appeared during the study. First, in the eastern part of the area, a well defined assemblage characterized rather undisturbed areas driven by autogeneous processes; beyond that, different assemblages were noticed in lowlands in relation to soil quality, and in uplands in relation to consolidation and re-allotment impacts. Secondly, a well defined assemblage characterized globally the western part of the area but trends in species changes were not obvious because disturbances are patchy in time and space. Dated human settlements allowed a reconstitution of salt marsh processes and we can define a gradient in relation to the distance to the seashore and the increase of soil water content.

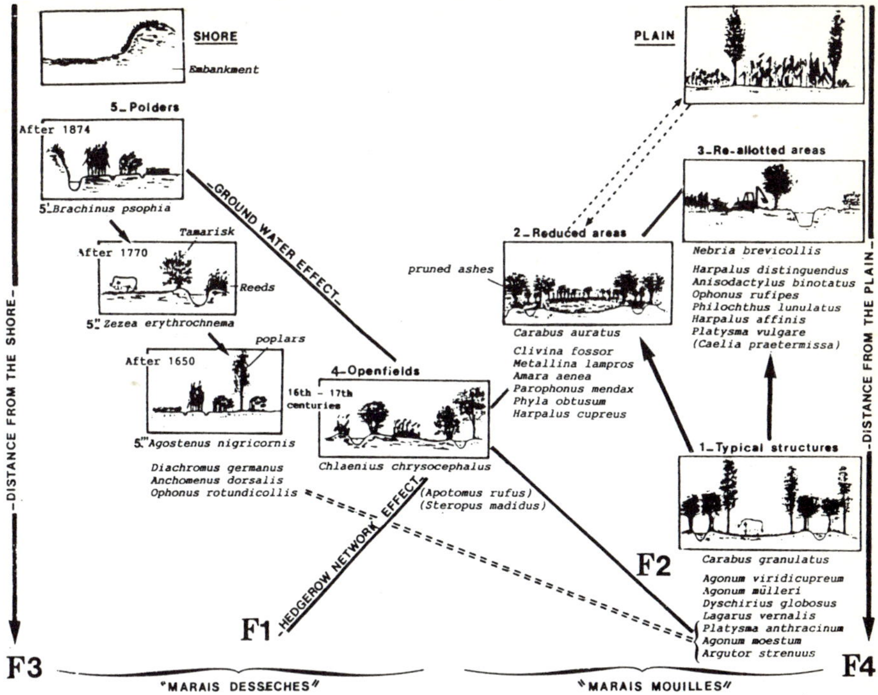

Figure 8. Schematic distribution of carabid species in Marais poitevin.

CONCLUSION

The classification both demonstrated the variation within and between habitats and emphasised the importance of several levels of investigation for conservation purposes. Habitats can be classified according to species assemblages at a regional level, but information about species diversity at the alpha level seems necessary for comparison.

The examples of carabid distribution induce us to support the maximization of habitat complexity on the regional scale since the synchronic approach also highlighted patterns that included reduced areas and species with high extinction probability (*e.g.* genus *Carabus*).

Synchronic analysis allows the comparison of the structure and functioning of ecosystems presently occurring in the studied area. Nevertheless, it is not easy to place communities chronologically in a spatiosequence. Human activities initiated new sequences and the coexistence communities in several parts of the study area does not necessarily imply that they belong to the same sequence. Thus, to clarify the determinism of the spatial organisation of the communities it may be necessary to complete such a study by diachronic analysis.

These results emphasized the value of studies on community dynamics (Pavé, Barbault, 1992). Such an approach highlights mechanisms in local processes for maintenance or change in biodiversity; it could take into account both the hierarchical levels in biodiversity and its functional understanding, and it allows the linking of biological problems for conservation purposes to global changes. Thus, inventories and suitable observation stations are important to evaluate the effects of disturbances on biodiversity.

Acknowledgments

I am indebted to the following persons and institutions: the Ministry of Environment for financial support and the staff of Parc Naturel Regional of Marais poitevin.

REFERENCES

Benzecri JP (1976) L'analyse des données. 2 L'analyse des correspondances. Paris, Dunod

den Boer PJ (1986) Carabids as objects of study. Pages 539-551 *in* Carabid beetles, their adaptations and dynamics. Gustav Fisher, Stuttgart, New-York

Bonadona P (1971) Catalogue des Coléoptères carabiques de France. Nouvelle revue d'Entomologie, Laboratoire de Zoologie, Univ. P. Sabatier, Toulouse, 177pp

Butterfield J, Coulson JC (1983) The carabid communities on peat and upland grasslands in northern England. Holartic Ecology 6: 163-174

Cachan P, Clement A (1979) Recherche d'une formulation générale de paramètres distributifs des peuplements plurispécifiques dans les milieux homogènes. Les peuplements carabidologiques des prairies du plateau lorrain. Bulletin de l'E.N.S.A.I.A., Nancy, 21: 95-113

Desender K, Pollet M, Segers R (1984) Carabid beetle distribution along humidity gradients in rivulet-associated grassland (Coleoptera, Carabidae). Biol.Jb. Dodonaea 52: 64-75

Deveaux D (1976) Répartition et diversité des peuplements en crabiques, en zone bocagère et arrasée de l'ouest. C.R. Table ronde C.N.R.S. "Ecosystèmes bocagers", Rennes: 377-384

Eyre MD, Luff ML, Rushton SP, Topping CJ (1989) Ground beetles and weevils (Carabidae and Curculionoidea) as indicators of grassland management practices. J. Appl. Ent. 107: 508-517

Eyre MD, Luff ML, Rushton SP (1990) The ground beetle (Coleoptera, Carabidae)

fauna of intensively managed agricultural grasslands in nothern England and southern scotland. Pedobiologia 34: 11-18

Greenslade PJM (1964) Pitfall trapping as a method for studying populations of Carabidae (Coleoptera). Journal of Animal Ecology 33: 301-310.

Jennings DT, Houseweart MW, Dunn GA (1986) Carabid beetles (Coleoptera: Carabidae) associated with strip clearcut and dense spruce-fire forests of Maine. Coleopt. Bul., 40: 251-263

Lam Hoai T, Lasserre G, Amanieu M (1987) Utilisation de la diversité pour décider du choix des espèces à retenir dans un lot faunistique. Acta Oecologica: 8(1): 17-28

Luff ML (1975) Some features influencing pitfall traps. Oecologia (Berlin) 19,121-130

Luff ML, Rushton SP (1989) The ground beetle and spider fauna of managed and unimproved upland pasture. Agriculture, Ecosystems and Environment, 25: 192-205

Luff ML, Eyre MD, Rushton SP (1992) Classification and prediction of grassland habitats using ground beetles (Coleoptera carabidae). Journal of Environmental Management 35: 310-315

Noss RF (1983) A regional landscape approach to maintain diversity. BioScience 33(11): 700-706

Pave A, Barbault R (1992) Ecosystems et environnement. Lettre du Programme Environnement du CNRS, 7: 7-21

Pollard E (1968) A comparison between the *Carabidae* of a hedge and field site and those of a woodland glad. Journal of Applied Ecology, 5: 649-657

Refseth D (1980) Evcological analyses of Carabid communities -potential use in biological classification for nature conservation. Biological conservation 17: 131-141

Roux M (1985) Algorithme de classification Masson (ed), 151pp

Rushton SP, Luff ML (1988) The use of multivariate ordination techniques to asses the effects of chlorpyrifos on ground beetle and spider communities in grassland. BPBC mono 40 Environmental effects of pestiides: 175-181.

Rushton SP, Luff ML, Eyre MD (1989) Effect of pasture improvement and management on the ground beetle and spider communities of upland grasslands. Journal of Applied Ecology, 26: 489-503

Rushton SP, Eyre MD, Luff ML (1990) The efects of scrub management of the ground beetls of oolitic limestone grassland at Castor Hanglands National Nature Reserve, Cambridgeshire, UK. Biological conservation 51: 97-111

Southwood TRE (1978) Ecological method with particular reference to the study of insect population, 2nd ed. Chapman and Hall, London

Thiele HU (1977) Carabid beetles in their environments. A study on habitat selection by adaptations in physiology and behaviour. Springer Verlag Berlin. Heidelberg, New-York

Wallin H (1985) Spatial and temporal distribution of some abundant carabid beetles (Coleotera carabiae in cereal fields and adjacent habitats). Pedobiologia, 28: 19-34.

Wolda H (1983) Diversity, diversity indices and tropical cock-roaches. Oecologia (Berlin) 58: 290-298

Chapter 17: Biodiversity of the Wadden Sea (Denmark, Germany, The Netherlands): recent changes and future projections

W.J. Wolff[a,b], N. Dankers[c], K.S. Dijkema[c], P.J.H. Reijnders[c] & C.J. Smit[c] [1]
DLO Institute for Forestry and Nature Research
P.O. Box 23
6700 AA Wageningen
The Netherlands

INTRODUCTION

The Wadden Sea is a shallow estuarine area situated along the North Sea coasts of Denmark, Germany and The Netherlands. Average tidal ranges vary between 1.4 and 3.4 m. The Wadden Sea occupies about 8,000 km^2 and about half of this area consists of bare tidal flats. Salt marshes occur only above mean high water mark and cover about 300 km^2.

The Wadden Sea is considered to be one of the major nature areas of western Europe. Its tidal landscapes where natural processes such as erosion and accretion are visibly active, seem relatively unaffected by humans. The extensive tidal flats contain extremely large numbers of benthic invertebrates and are characterized by high biomasses. Also the subtidal areas are rich in individuals and biomass. On this invertebrate biomass fish and birds feed in very large numbers. Many North Sea fish and crustacean species use the Wadden Sea as a nursery. The bird population includes both breeding birds, such as gulls, terns,

---

[a] DLO Institute for Forestry and Nature Research, P.O. Box 23, 6700 AA Wageningen, The Netherlands
[b] Dept. of Nature Conservation, Agricultural University, Wageningen, The Netherlands
[c] DLO Institute for Forestry and Nature Research, P.O. Box 167, 1790 AD Den Burg, Texel, The Netherlands

NATO ASI Series, Vol. I 20
Biodiversity, Temperate Ecosystems, and Global Change
Edited by T. J. B. Boyle and C. E. B. Boyle

and several species of shorebirds, and non-breeding migratory species. The latter species mainly breed in the Arctic and visit the Wadden Sea as a stopover site during migration or as a wintering site. Non-breeding migratory species include geese, ducks, many species of shorebirds, and several other species. It is estimated that altogether about 7 million shorebirds visit the Wadden Sea annually. Wolff (1983) gives an extensive description of the ecology of the Wadden Sea.

This paper attempts to describe changes in the biodiversity of the Wadden Sea during the period that man was able to interfere with its natural processes, i.e. the last thousand years, and to predict how biodiversity might develop from now under expected conditions of global change and local human impact. As will appear below, the Wadden Sea is very much under human influence, so next to global change of environmental conditions, developments in the local human society determine the fate of this coastal sea to a very large extent. Because of our ignorance about most of the development of our society over the next century, predictions about the development of the biodiversity of the Wadden Sea over a century make little sense. Even predictions for a period of 10-20 years are difficult.

PAST DEVELOPMENT OF BIODIVERSITY

The past development of biodiversity in the Wadden Sea has been summarized in two review papers by Wolff (1992a, in press). His conclusions are repeated below.

Natural situation

In a completely natural situation a Wadden Sea landscape would show the following sequence of belts of different landscape types when going from the North Sea to the higher inland areas:

- beaches and dunes on barrier islands
- salt marshes on barrier islands
- tidal flats and channels
- salt marshes

- brackish marshes
- freshwater marshes and swamps
- peat moors
- high-lying soils with forests and other dry vegetation types.

In the past, with rising sea level, these belts of different types of landscapes have moved inland and they would still do so when man had not interfered. At the same time sediment, mainly derived from barrier islands and the shallow sea floor in front of them, has been imported into this moving Wadden Sea. The outcome of these two processes has been dependent on the amount of sediment available, the tidal range and on the slope of the coast. Sometimes areas of open water and tidal flats filled in, sometimes they enlarged.

Embankments and reclamation

Until about 1000 years ago man hardly interfered with the geomorphological development of the Wadden Sea. Although the coastal area was fairly densely populated, people lived on naturally elevated areas and on artificial mounds and did not seriously interfere with the tidal movements. At that time, however, the invention of dike-building introduced a major change (Wolff, 1992b). In the beginning only rather small areas in the highest parts of the coastal region were were surrounded by low dikes. Soon however these areas were connected by other dikes and at the end of the 12th century the larger part of the vegetated landscape was separated from the tidal Wadden Sea by a contiguous system of dikes. Outside these dikes large salt marsh areas still occurred. Gradually, however, these and newly accreting salt marshes were reclaimed and turned into farmland (Dijkema, 1987). Thus the connection between the Wadden Sea and the adjacent belt of brackish and fresh wetlands was broken.

This separation of the Wadden Sea from its coastal marshes and peat moors may well have been the most important intervention in its history. However, since it was long ago, we are inclined to neglect this major change of the Wadden Sea ecosystem and we tend to consider the present-day Wadden Sea as the normal situation.

The result of all embankments and reclamations is a very marked and firm separation between the tidal Wadden Sea on the one hand and the non-tidal land and freshwater areas on the other hand. In the remaining, estuarine part of the Wadden Sea these geomorphological changes probably did not cause major changes of biodiversity, but at the landward side of the dikes a landscape of fresh and brackish wetlands was changed into agricultural landscapes at the cost of great losses of biodiversity. The dikes and dams constructed in between Wadden Sea and mainland also constitute a major barrier for transport of materials and for migratory aquatic organisms.

The engineering works of the past have had an influence on the tides and the tidal currents. Usually the changes were only slight, but in the case of the closure of the Zuiderzee in The Netherlands the tidal amplitude in the remaining part of the Wadden Sea increased by about 50%. This change in the tidal dynamics may have been an additional factor in the disappearance of seagrasses from the Dutch Wadden Sea (Van den Hoek *et al.*, 1979). They also caused erosion of the few salt marshes which still occurred at that time.

Due to the construction of dikes and dams the gradual transition between the sea on the one hand and the fresh water on the other hand became much more abrupt. Gradual transitions of salinity with a variety of brackish water types disappeared and with them the ecological conditions of these different types of waters vanished. This is especially true for the types of brackish water with only slow and/or slight temporal changes of salinity, such as the former Zuiderzee, a brackish lagoon of about 3700 km^2 in The Netherlands. Thus several characteristic brackish-water species such as the Zuiderzee crab *Rhithropanopeus harrisii* tridentatus and the nudibranch Corambe batava have virtually disappeared from the Wadden Sea area (Dankers *et al.*, 1981).

Introduction of species

Table 1 presents an incomplete list of species known to have been introduced from abroad into the Wadden Sea area by human intervention. Most species were inadvertently introduced, some like the cord grass Spartina townsendii and the Japanese oyster *Crassostrea gigas* deliberately for, respectively, stimulation of accretion and oyster farming. In most cases we have no indication that the original Wadden Sea ecosystem was upset because of the new species. Only in the case of *Spartina* rather far-reaching changes in the vegetation of salt marshes were observed (Beeftink, 1977).

Most species introduced from other continents arrived here in unknown ways. In some cases, e.g. those of the Chinese mitten crab *Eriocheir sinensis* and the New-Zealand barnacle *Elminius modestus*, it is reasonably well established that transport in ballast water or on the hull of ships has been the cause and the same way of transport may be suspected for many other species (Carlton & Geller, 1993). Another way of introduction is transport on imported shellfish. This is, of course, true for the shellfish species themselves, such as the Japanese oyster *Crassostrea gigas*, but also for the species attached to the shells, such as the Japanese alga *Sargassum muticum*.

Other new species were able to colonize the Wadden Sea because their habitat was introduced there. The best example is the construction of artificial rocky shores in the form of breakwaters, moles, and stone-covered dike slopes. Before hardly any hard substrates occurred in the area. Many rocky-shore species such as the kelp *Laminaria saccharina* have colonized these shores (Den Hartog, 1959).

Summarizing it may be concluded that the Wadden Sea nowadays lodges quite a number of species of algae and invertebrates which did not occur there before. Most of these species have in common that they are relatively small and short-living and that they have short reproduction cycles. They did not change the ecosystems very much, and they certainly increased local biodiversity. On the

Table 1. Some species introduced into the Wadden Sea area due to human intervention. After Wolff (1973), Den Hartog & Van der Velde (1987), Essink (1985) and Essink & Kleef (1988).

Algae	
Asterionella japonica	Pacific
Codium fragile	Pacific
Sargassum muticum	Pacific
Vascular plants	
Spartina townsendii (*S. anglica*)	America - England
Coelenterata	
Cordylophora caspia	Caspian Sea
Garveia franciscana	America
Mollusca	
Corambe batava	America
Crepidula fornicata	America
Crassostrea gigas (*C. angulata*)	Pacific
Teredo navalis	?
Ensis directus	America
Petricola phola	America
Mya arenaria	America
Polychaeta	
Marenzelleria viridis	America
Crustacea	
Balanus improvisus?	?
Elminius modestus	Pacific
Eriocheir sinensis	Pacific
Rithropanopeus harrisii	America
Tunicata	
Molgula manhattensis	America
Styela clava	America

other hand, because of their usually already widespread distribution, their addition hardly enhances biodiversity on the larger scale.

Extinction of species due to human activities

On the other hand, human activities have had such a strong impact on the Wadden Sea ecosystem, that several species have become extinct. In many cases the cause has been identified or at least can be suspected.

The fossil and archaeological records show that Grey whales (*Eschrichtius gibbosus*) once occurred in the North Sea area. The Grey whale is now extinct in the Atlantic Ocean and its adjacent seas (Mitchell 1974, Mitchell and Mead 1977). De Smet (1976) presents evidence for mediaeval whaling in the North Sea and the English Channel and believes that both the Grey whale and Atlantic right whale (*Eubalaena glacialis*) may have been involved. He assumes that these slow species had been exterminated in the North Sea and English Channel area already in the Middle Ages. De Smet therefore suggests that the Grey whale once used the Wadden Sea as a breeding ground.

Subfossil remains of Grey seals (*Halichoerus grypus*) are rather common in prehistoric settlements and it is likely that our ancestors hunted this species. Because its pups stay on dry land for several weeks, these will have been an easy prey, and hence it may be assumed that in the Wadden Sea area it became an early victim of overexploitation (Reijnders, 1978; Reijnders *et al.*, 1992). This provides a likely explanation for the absence of Grey seals from the Wadden Sea from the late Middle Ages until quite recently.

How quickly overexploitation of marine mammals can reduce their populations, shows in the example of the Harbour seal (*Phoca vitulina*) in the Netherlands in the fifties. In that period virtually all juveniles were killed because of their furs, resulting in a population decrease of about 60-70% in about 10 years (Van Haaften, 1974; Reijnders, 1976).

Slowly reproducing large birds are also very vulnerable to exploitation or deliberate killing. This may explain why several species which elsewhere in the Northern Hemisphere have a large breeding distribution do not occur in the

Wadden Sea area, although the area seems to be suitable. Examples are Little Egret (*Egretta garzetta*), White-tailed eagle (*Haliaetus albicilla*), Osprey (*Pandion haliaetus*), Black Kite (*Milvus migrans*), and Caspian tern (*Hydroprogne caspia*). According to Pliny (Voous, 1960) Dalmatian Pelicans (*Pelecanus crispus*) also have occurred in the Wadden Sea area.

Also smaller birds have become victim of exploitation, mainly because their eggs were collected by the coastal population. Although a few colonies of gulls and terns (Eijerland-Texel, Rottumeroog, Sylt?) were exploited on a sustainable basis (Swennen, 1982), many other colonies will have disappeared as have populations of species breeding as single pairs. This exploitation of eggs explains why at the beginning of this century Eider duck (*Somateria mollissima*), Lesser black-backed gull (*Larus fuscus*), Herring gull (*L. argentatus*), Common gull (*L. canus*), and Avocet (*Recurvirostra avosetta*) were quite rare along the shores of the Wadden Sea (Leege, 1905).

Fish species may also have become rare because of exploitation, especially the larger, slowly reproducing species. This probably applies to Thornback ray (*Raja clavata*), Sting ray (*Dasyatis pastinacea*), and Dogfish (*Scyliorhinus caniculus*). In this case it is probably not, or at least not in the first place, the result of Wadden Sea fisheries, but that of intensive bottom-trawl fisheries in the entire North Sea, in which these species mainly occur as by-catch. The effects of river fisheries on several anadromous species probably have been similar. It may be suspected that up to the Second World War these fisheries had a strong impact on populations of species such as Sturgeon (*Accipenser sturio*), Salmon (*Salmo salar*), Allis shad (*Alosa alosa*), and Houting (*Coregonus oxyrhynchus*). However, nowadays the river fisheries have largely disappeared and their possible effects are overruled by the effects of river regulation and population.

A clear example of overexploitation in the Wadden Sea itself is the disappearance of the Wadden Sea oyster (*Ostrea edulis*) beds. Once flourishing oyster fisheries existed in the Dutch Wadden Sea and in Nordfriesland in Germany. These have completely disappeared because of overfishing. Thus the classic ecosystem study of Moebius (1877) can never be repeated because its subject has been extinguished. Recently intertidal mussel beds have disappeared from the Dutch Wadden Sea due to overexploitation. Reise (1982) and Reise *et al.* (1989) have documented this change in the composition of the benthic fauna in the Nordfriesland Wadden Sea in Germany. Long-living species, such as several species of molluscs, gave way to opportunistic species, mainly polychaetes. This pattern is not restricted to Nordfriesland, however. It is also recognizable in Dutch coastal waters (Horst, 1883-84; Wolff, 1973; Dekker, 1989). For the deeper parts of the Wadden Sea, Reise attributes this change mainly to fisheries and especially those forms with heavy bottom gear. It is not only a change in species, but also a change of ecosystems. In the old days oyster beds, Sabellaria reefs, sublittoral seagrass beds and "meadows" of hydroids were easily recognizable systems with a characteristic flora and fauna. These features now hardly or longer exist in the Wadden Sea.

Extinction of species due to habitat loss

Some species have disappeared because their habitats disappeared. One example is the Zuiderzee race of herring (*Clupea harengus*), which lost its spawning grounds because of the closure of the Zuiderzee. Within a few years it became virtually extinct (Havinga, 1954). After the herring disappeared the bottlenose dolphin (*Tursiops truncatus*) also failed to appear in spring (Verwey & Wolff, 1981). The disappearance of invertebrate brackish-water species in the same area due to the same cause has already been mentioned above. At the time of the closure of the Zuiderzee, the seagrass beds in the remaining part of the Dutch Wadden Sea disappeared and never returned. Their disappearance probably was due to the so-called "wasting disease", a wide-spread natural phenomenon at that time; the failure to return is probably due to the closure of the Zuiderzee which changed hydrographical conditions in the remaining area. With the disappearance of their seagrass habitat several plant and animal species also disappeared from the entire Wadden Sea: among others the sea stickleback (*Spinachia spinachia*), the deep-snouted pipefish (*Syngnathus typhle*), and the gastropods *Rissoa membranacea* and *Lacuna vincta*.

Effects of pollution

Pollution has obtained much attention, but so far its effects have been less far-reaching than those of some other human interventions in the Wadden Sea. No species is known with certainty to have disappeared from the Wadden Sea because of pollution.

Chlorinated hydrocarbons, however, have had very strong effects on the populations of many Wadden Sea birds, especially in the Netherlands but also in Germany (Koeman, 1971; Becker & Erdelen, 1987). Species such as Eider duck, Spoonbill (*Platalea leucorodia*), and Sandwich tern (*Sterna sandvicensis*) decreased very sharply in the sixties, and only started to increase again when the cause of pollution had been removed. However, after 25 years the tern populations have not yet returned to their original levels.

Since the sixties the Dutch population of Harbour seals showed a continuous decline due to the effects of polychlorinated biphenyls (PCB's) (Reijnder, 1986) until the decrease was balanced in the eighties by immigration of seals from Germany. At it lowest level, less than 10% of the original Dutch seal population was left. Whether the Habour porpoise (*Phocoena phocoena*), another marine mammal, has become extinct in the Wadden Sea because of PCB pollution, as has been suggested several times, is doubtful (Reijnders, 1992).

Introduction of species for cultivation

Cultivation in the Wadden Sea at present is restricted to blue mussels (*Mytilus edulis*) and Japanese oysters. Especially with mussel cultivation, it is clear that it changes the environment and thus promotes possibilities for a number of species. Apart from the mussels, Eel (*Anguilla anguilla*) and Starfish (*Asterias rubens*) can be considered as examples. It has also been suggested that Eider ducks have increased because of mussel cultivation.

Effects of nature conservation and environmental protection

Wadden Sea conservation dates back to 1909 when the island of Jordsand, at that time part of Germany, was protected. Under Danish rule the protection was restored in 1922. Protection of the islands of Griend in the Netherlands and of Mellum in Germany followed in 1917 and 1924, respectively. Thus originally conservation in the Wadden Sea aimed at the protection of the most important breeding colonies of coastal birds, but after the Second World War the salt marshes with lower numbers of birds also gradually became protected. Protection of the Wadden Sea itself started in Germany with the reserve "Wattenmeer ostlich Sylt" in 1973. However, this and other large reserves established in the German Wadden Sea in later years lacked a management organisation. The Netherlands protected most of the the Dutch Wadden Sea by application of the Nature Conservation Act in 1981 and at the same time a management organisation was set up. Three national parks in Germany followed in respectively 1985, 1986 and 1990. Denmark protected its Wadden Sea in 1979 as a game reserve and in 1982 as a nature reserve.

Protection of breeding colonies has proved very successful. Breeding populations of almost all species of coastal birds have shown an exponential increase since the beginning of this century (Swennen, 1982).

The recent protection of the Wadden Sea itself probably has prevented some negative developments, but it has not yet produced clear positive ecological changes. Perhaps the return of the Grey seal as a breeding species may be viewed as such (Reijnders *et al.*, 1992). Three major reasons for the lack of positive changes are the short period since Wadden Sea conservation started, the exceptions allowed to several conservation measures, and the fact that conservation problems have shifted from local impacts such as overexploitation, which can be counteracted by the creation of reserves, to widespread negative impacts on the environment such as pollution by chlorinated hydrocarbons. Such impacts are not restricted by reserves but instead need a strict environmental policy.

Indeed, environmental measures have already resulted in positive ecological changes. After the widespread mortality of seabirds in the western part of the Wadden Sea due to discharge of pesticides at the Rhine estuary, measures taken to decrease these discharges have resulted in a strong increase of the populations concerned (Smit & Wolff, 1981; Swennen, 1982).

Measures to reduce the discharge of organically enriched waste-water have resulted in lasting recolonization by benthic invertebrates of tidal flat areas depopulated earlier because of frequent absence of oxygen (Essink, 1984).

Conclusions on past changes of biodiversity

The following general conclusions may be drawn on the past changes of biodiversity in the Wadden Sea:

- in a seemingly natural area humans appear to have had a very strong impact on biodiversity; already in the Middle Ages half of the ecosystem had been destroyed;
- many large, long-living and slowly reproducing species (K-selected species) have disappeared from the Wadden Sea, whereas many small, short-living and rapidly reproducing species (r-selected species) have colonized the area or increased in numbers; and
- nature conservation and environmental protection so far have only had limited results with regard to biodiversity. Positive responses have mainly been shown by more opportunistic species (r-species).

FUTURE PROJECTIONS OF BIODIVERSITY

Possible scenarios for development of the Wadden Sea

The abiotic environment of the Wadden Sea is expected to change as a consequence of climate change. The ecological consequences of the expected changes for the Wadden Sea and other coastal seas and estuaries have been investigated at a workshop in Texel, the Netherlands, in 1988 (Beukema *et al.*, 1990), assuming that the worldwide scenario developed and used by the Intergovernmental Panel on Climate Change can also be applied to the local situation of the Wadden Sea. However, until now, Global Circulation Models are not sufficiently detailed to make meaningful predictions for small areas. Bearing this in mind we have, nevertheless, used the current worldwide IPCC scenario to predict the main changes in the Wadden Sea.

- Present tectonic subsidence of the Wadden Sea area and worldwide "normal" rise in sea level are expected to be reinforced by sea level rise due to climate change. Locally and temporally this trend is reinforced again by subsidence of the tidal flats and slat marshes due to extraction of natural gas. For the next 10-20 years the rise of sea level is not expected to move outside the range of sea level which has occurred up to now; over the next 100 years sea level rise may be much larger than has occurred over the past centuries.
- Worldwide change of climate may result in local changes of temperature, precipitation, wind force and wind direction, etc. For the next 10-20 years the changes in the Wadden Sea area are supposed to be small; over the next 100 years important, but until now the changes which may occur are unknown.

The Wadden Sea is also subject to trends in human impacts. In this part of western Europe, *i.e.* a highly developed and densely populated part of the world, the social trends listed below are believed to be important for the Wadden Sea. Lacking better alternatives they have been extrapolated to the next 10-20 years in order to provide scenarios for human impacts. For periods longer than this meaningful scenarios can hardly be made.

- Population density continues to increase slowly. In combination with increases in leisure time and affluence this results in an increased demand for recreation opportunities, among which are active forms of recreation in natural areas.
- Food production becomes more and more concentrated in highly developed cultivation systems. In coastal waters free fisheries are replaced by maricultures requiring less space, but more energy, and producing more wastes. Land reclamation for expansion of agriculture becomes very unlikely (Wolff 1992b).
- International trade continues at least at the same level resulting in the introduction of ever more species from all parts of the world.

• Although some sources of pollution may be increasing, environmental protection becomes more strict. Already the water quality of the rivers debouching into the Wadden Sea shows enormous improvement and it is expected that this trend will continue. The same applies to air pollution. However, whereas "simple" forms of pollution are likely to disappear, or at least greatly decrease, "difficult" forms, such as PCB's and dioxins, are likely to continue to occur, or even increase, because large quantities are still making their way into the marine environment.

• Nature protection changes from a defensive to an offensive strategy.

Scenarios for the management of the future Wadden Sea

At the seventh International Wadden Sea Symposium held on the island of Ameland, the Netherlands, in 1990 (Dankers *et al.*, 1992) four scenarios for the future management of the Wadden Sea were presented:

• multifunctional use of the Wadden Sea, at the same time aiming for a high nature conservation value (Essink 1992);

• a Wadden Sea "free-for-all", without any limits on human exploitation of the area (Muus 1992);

• the Wadden Sea as a pristine nature reserve (Reise 1992); and

• the Wadden Sea in maximum sustainable use (De Vals 1992).

Reise (1992) concluded that a pristine Wadden Sea is impossible to achieve; the maximum possible is probably a number of smaller "zero-use" areas. Similarly, a Wadden Sea free-for-all (Muus 1992) seems very unlikely at present, given the common attitude to protection of natural areas in western Europe. The scenario aiming for maximum sustainable use (De Vals 1992) seems a more likely option at the moment, although it is not in line with the present trends in society as described above. This is the case, however, for the scenario aiming at multifunctional use with a high conservation value (Essink 1992). This scenario extrapolates present ideas and policies, and hence seems the most likely option.

Remarkably enough, global change was hardly incorporated into these four scenarios.

Future projections of biodiversity in the Wadden Sea

Combining elements from these management scenarios with the scenarios for trends in society and environment identified above, we believe that the following developments in the Wadden Sea are possible in the next 10-20 years. In addition, we suggest some possible changes for the next 100 years.

- A very limited impact of climate change is expected for the next 10-20 years. Perhaps a few species of plants or invertebrates from lower latitudes may succeed in establishing themselves temporarily or permanently. However, in view of the species composition of the ecosystems occurring in estuaries further south along the European coasts, we expect only minor ecological changes (see also De Vooys 1990). Over a period of about 100 years more introductions are expected, including new breeding birds and fish species (Costa 1990). Some wintering bird species may disappear, however, because of a shift in their wintering areas to the north, of because of climate induced changes in their northern breeding ranges.
- A very limited impact of increased sea level rise is expected for the next 10-20 years. The Wadden Sea system will largely be able to compensate the expected rise by accretion. Only in some of the saltmarsh areas may present trends in erosion aggravate, resulting in decrease of total marsh area. However, species loss is unlikely in this period. Over 100 years the loss of salt marshes, especially at the barrier islands, may become more serious (Dijkema *et al.* 1990) and some salt marsh plants or invertebrates, already rare at this moment, may disappear from the area altogether. No species losses from sea level rise are expected for tidal flat habitats or shallow water environments.
- Environmental protection in the next 10-20 years is expected to result in a situation in which many noxious substances have decreased very strongly. No further problems are expected from raw sewage, nutrients, heavy metals,

and many pesticides. However, persistent organohalogens are expected to continue to occur in quantities which may have negative effects on biodiversity, especially with regard to the occurrence of marine mammals. The situation over the next 100 years cannot be predicted.

- International trade and transport will continue to introduce new species into the Wadden Sea at a rate of at least 2-3 species per 10 years. In most cases these newcomers will be invertebrates and cryptogamic plants; based on experience gained up to now it is expected that their impact on the existing ecosystem will be slight. Extinctions of indigenous species due to introductions of new ones are expected only as an exception, so there is a net increase of local biodiversity.
- Present trends in society during the next 10-20 years will probably result in a Wadden Sea area in which a considerable part is strictly protected and well guarded, whereas the remainder of the area is open for use by quite a number of human activities. It is expected that areas embanked in earlier centuries will be turned into tidal landscapes again. In the protected areas the present trends of species returning after many years of absence will continue. It is expected that especially the smaller, rapidly reproducing species will profit from this situation. It seems unlikely that large and slowly reproducing species, such as the Dalmation pelican, White-tailed eagle, and Osprey, will return in this situation and within this period. An additional reason is that these species were probably also dependent on the former marshy fringes of the Wadden Sea. Absent species of North Sea fish will not return because the reason for their absence continues to occur outside the Wadden Sea. Anadromous species are likely to return, however, due to the present clean-up and restoration activities in the rivers. It does not make sense to predict the development of society and hence its biodiversity effects over a period of 100 years.

Altogether, we expect for the next 10-20 years a trend of slowly increasing biodiversity in the Wadden Sea. What will happen on a longer time scale is a matter of conjecture. An interesting additional conclusion is that the expected effects of local human activities have an impact which is at least of the same order of magnitude as the effects of climate change.

REFERENCES

Becker PH, Erdelen M (1987) Die Bestandsentwicklung von Brutvoegeln der deutschen Nordseekuests 1950-1079. J. Orn. 126: 29-51

Beeftink WG (1977) The coastal salt marshes of western and northern Europe: an ecological and phytosociological approach. Pages 109-155 in Chapman VJ (ed.). Wet coastal ecosystems. Elsevier, Amsterdam

Beukema JJ, Wolff WJ, Brouns JJWM (eds.) (1990) Expected effects of climatic change on marine coastal ecosystems. Kluwer Academic Publications, Dordrecht

Costa MJ (1990) Expected effects of temperature changes on estuarine fish populations. Pages 99-104 in Beukema JJ, Wolff WJ, Brouns JJWM (eds.) Expected effects of climatic change on marine coastal ecosystems. Kluwer Academic Publications, Dordrecht

Dankers N, Kühl H, Wolff WJ (1981) Invertebrates of the Wadden Sea. Balkema, Rotterdam

Dankers N, Smit CJ, Scholl M (eds.) (1992) Proceedings of the 7th International Wadden Sea Symposium, Ameland 1990. Neth. Inst. Sea Res. Publ. Ser. 20

Dekker R (1989) The macrozoobenthos of the subtidal western Dutch Wadden Sea. I. Biomass and species richness. Neth. J. Sea Res. 23: 57-68

Dijkema KS (1987) Changes in slat marsh area in the Netherlands Wadden Sea after 1600. Pages 42-49 in Huiskes AHL, Blom CWPM, Rozema J (eds.) Vegetation between land and sea. Junk, Dordrecht

Dijkema KS, Bossinade JH, Bouwsema P, de Glopper RJ (1990) Saltmarshes in the Netherlands Wadden Sea: rising high-tide levels and accretion enhancement. Pages 173-188 in Beukema JJ, Wolff WJ, Brouns JJWM (eds.) Expected effects of climatic change on marine coastal ecosystems. Kluwer Academic Publications, Dordrecht

Essink K (1984) Discharge of organic waste into the Wadden Sea: local effects. Neth. Inst. Sea Res. Publ. Ser. 10

Essink K (1985) On the occurrence of the American jack-knife clam *Ensis directus* (Conrad, 1843) (Bivalva, Cutellidae) in the Wadden Sea. Basteria 49: 73-80

Essink K (1992) Multifunctional use of the Wadden Sea aiming for high natural value. Pages 35-43 in Dankers N, Smit CJ, Scholl M (eds.) Proceedings of the 7th International Wadden Sea Symposium, Ameland 1990. Neth. Inst. Sea Res. Publ. Ser. 20

Essink K, Kleef HL (1988) *Marenzelleria viridis* (Verrill, 1873) (Polychaeta; Spionidae); a new record from the Ems estuary (the Netherlands/Federal Republic of Germany). Zoöl. Bijdr. (Leiden) 38: 1-13

Hafton JL van (1974) Zeehonden langs de Nederlandse kust. Wet. Meded. Kon. Ned. Natuurhist. Ver. 109: 1-36

Hartog C den (1959) The epilithic algal communities occurring along the coast of the Netherlands. Wentia 1: 1-241

Hartog C den, van der Velde G (1987) Invasions of plants and animals into coastal brackish and fresh water of the Netherlands. Proc. Kon. Ned. Akad. Wet. C 90: 31-37

Havinga B (1954) Vissen. Pages 253-267 in de Beaufort LF, (ed.). Veranderingen in de flora en fauna van de Zuiderzee (thans Ijsselmeer) na de afsluiting in 1932. Ned. Dierk. Ver., Den Helder

Hoek C van den, Admiraal W, Colijn F, de Jonge VN (1979) The role of algae and seagrasses in the ecosystem of the Wadden Sea: a review. Pages 9-118 in Wolff WJ (ed.) Flora and vegetation of the Wadden Sea. Balkema, Rotterdam

Horst R (1813-84) Verslag omternt onderzoekingen op de oester en de oestercultuur betrekking hebbende, uitgebracht door de Commissie van het Zoölogisch Station. T. Ned. Dierk. Ver. Suppl. 1: 1-673

Koeman JH (1971) Het voorkomen en de toxicologische betekenis van enkele chloorkoolwaterstoffen aan de Nederlandse kust in de periode van 1965 tot 1970. Thesis, Utrecht

Leege O (1905) Die Vögel der ostfriesischen Inslen nebst vergleichender Uebersichtder im südlichen Nordseegebeit vorkommenden Arten. Haynel, Emden

Mitchell ED, Mead JG (1977) History of the grey whale in the Atlantic Ocean. Second Conf. Biol. Marine Mammals, San Diego

Moebius K (1877) Die Auster und die Austernwirthschaft. Wiengandt, Hempel & Parey, Berlin

Muus B (1992) The Wadden Sea "free-for-all", without any limits on human exploitation of the area. Pages 45-48 in Dankers N, Smit CJ, Scholl M (eds.) Proceedings of the 7th International Wadden Sea Symposium, Ameland 1990. Neth. Inst. Sea Res. Publ. Ser. 20

Reijnders PJH (1976) The harbour seal population in the Dutch Wadden Sea: size and composition. Neth. J. Sea Res. 10: 223-235

Reijnders PJH (1978) De GRijze zeehond in het Waddengebied. Waddenbulletin 13: 500-502

Reijnders PJH (1986) Reproductive failure in common seals feeding on fish from polluted coastal waters. Nature 324: 456-457

Reijnders PJH (1992) Harbour porpoises in the North Sea: numerical responses to changes in environmental conditions. Neth. J. Aq. Ecol. 26: 75-85

Reijnders PJH, van Dijk J, Kuiper D (1992) Recolonization of the Dutch Wadden Sea by the grey seal *Halichoerus grypus*. ICES CM 1992/n: 7. ICES, Copenhagen

Reise K (1982) Long-term changes in the macrobenthic invertebrate fauna of the Wadden Sea: are polychaetes about to take over? Neth. J. Sea Res. 16: 29-36

Reise K (1992) The Wadden Sea as a pristine nature reserve. Pages 49-53 in Dankers N, Smit CJ, Scholl M (eds.) Proceedings of the 7th International Wadden Sea Symposium, Ameland 1990. Neth. Inst. Sea Res. Publ. Ser. 20

Reise K, Herre E, Sturm M (1989) Historical changes in the benthos of the Wadden Sea around the island of Sylt in the North Sea. Helgolaender Meeresunters 43: 417-433

Smet WMA de (1976) Evidence of whaling in the North Sea and the English Channel during the Middle Ages. Scientific Consultation on Marine Mammals, Bergen, Norway, 31 August-9 Sepetember 1976

Smit CJ, Wolff WJ (eds.) (1981) Birds of the Wadden Sea. Balkema, Rotterdam

Swennen C (1982) De vogels langs onze kust. Pages 78-100 in Wolff WJ (ed.) Wadden, Duinen, Delta. Pudoc, Wageningen

Verwey J, Wolff WJ (1981) The bottlenose dolphin (*Tursiops truncatus*). Pages 59-64 in Reijnders PJH, Wolff WJ (eds.) Marine mammals of the Wadden Sea. Balkema, Rotterdam

Vlas J de (1992) The Wadden Sea in maximum sustainable use. Pages 55-66 in Dankers N, Smit CJ, Scholl M (eds.) Proceedings of the 7th International Wadden Sea Symposium, Ameland 1990. Neth. Inst. Sea Res. Publ. Ser. 20

Voous KH (1960) Atlas van de Europese Vogels. Elsevier, Amsterdam

Vooys CGN de (1990) Expected biological effects of long-term changes in temperatures on benthic ecosystems in coastal waters around the Netherlands. Pages 77-82 in Beukema JJ, Wolff WJ, Brouns JJWM (eds.) Expected effects of climatic change on marine coastal ecosystems. Kluwer Academic Publications, Dordrecht

Wolff WJ (1973) The estuary as a habitat. An analysis of data on the soft-bottom macrofauna of the estuarine area of the rivers Rhine, Meuse, and Scheldt. Zoöl. Verh. (Leiden) 126: 1-242

Wolff WJ (1983) Ecology of the Wadden Sea. Balkema, Rotterdam

Wolff WJ (1992a) Ecological developments in the Wadden Sea until 1900. Pages 23-32 in Dankers N, Smit CJ, Scholl M (eds.) Proceedings of the 7th International Wadden Sea Symposium, Ameland 1990. Neth. Inst. Sea Res. Publ. Ser. 20

Wolff WJ (1992b) The end of a tradition. 1,000 years of embankment and reclamation of wetlands in the Netherlands. Ambio 21: 287-291

Wolff WJ (in press) Changes in biodiversity of the Wadden Sea in the Netherlands, Germany, and Denmark. in Szaro R (ed.) Proceedings of the Biodiversity Symposium, Sacramento, USA, 1992

Chapter 18: Interactions between tree species and their effects on nitrogen and phosphorus transformations in the forest floor

B. L. Williams
Macaulay Land Use Research Institute
Craigiebuckler
Aberdeen AB9 2QJ
UK

INTRODUCTION

Natural woodlands and forests contain a wide diversity of plant species and the potential for interactions between the different components is great. Mixed conifer plantations are less complex, but provide an opportunity to study interactions that may occur between tree species. Plantations of conifers in Britain have been mainly of single species unless there have been good reasons for planting mixtures. An example of this includes the increasing practice of planting mixed coniferous species on nutrient poor soils. Species mixtures have been planted on podzolized heathland soils and deep peats to establish crops of Sitka spruce (*Picea sitchensis* (Bong.) Carr.) during the past 60 years or more. On heathlands dominated by *Calluna*, the objective was to suppress the growth of heather by planting pine, usually Scots pine (*Pinus sylvestris* L.), in mixture with the spruce (Weatherall, 1957). On the west coast of Scotland, mixed species have been used to establish self-thinning crops on deep peats to surmount the problem of windthrow and to avoid uneconomic thinning. In this situation, larch (*Larix* sp.) was also used as the nurse species (O'Carroll, 1978). The practice of planting mixed species gathered momentum when it was realised that spruce planted in mixture could be established on nutrient poor soils without repeated additions of nitrogen fertilizers and herbicides (Mackintosh, 1983). In a report on the nutritional requirements of Sitka spruce on different site types in Britain, Taylor and Tabbush (1990) suggested that the mixtures effect involved more than just suppression of the

NATO ASI Series, Vol. I 20
Biodiversity, Temperate Ecosystems, and Global Change
Edited by T. J. B. Boyle and C. E. B. Boyle

vegetation. The overall effect appears as an increased supply of nitrogen to the spruce planted in mixture which it does not obtain in pure stands. Similar conclusions had been reached by Richards (1962) in a study of the effects of different *Pinus* species on Hoop pine (*Araucaria cunninghamii* Aik) and by Fisher and Stone (1969) investigating the effects of larch and pine on the growth of herbaceous vegetation.

The mechanism of the mixtures effect is far from clear although many hypotheses have been proposed as to how the spruce acquires a supply of nitrogen in mixture with pine or larch which is not available in monoculture. There are several parts of the soil-plant system where interactions between trees can occur to increase nutrient availablity and this paper reviews the current state of knowledge about these interactions and their effects on soil nutrient transformations and availability. Interest in mixtures has been mainly in, but not solely confined to, the establishment phase of a planted stand and interactions are discussed which may continue beyond canopy closure to different stages of a stand's development.

NUTRIENT CYCLING STUDIES

On nutrient poor soils, the planting of spruce in mixture with either pine or larch has resulted in enhanced growth of the spruce. Planting configurations have ranged from alternate single trees to triplets of each species planted in rows. After establishment, growth of the spruce exceeds that of the nurse species and the latter is usually shaded out and dies back. On more fertile soils, the benefits of mixed planting are not so apparent and the resulting competition between the species does not always favour the spruce (Yanai, 1992).

In a study of Sitka spruce planted in pure stands and in mixture with Scots pine on a heathland site at Culloden Forest in Scotland, the growth rate of spruce began to diverge when the trees were around 8 or 9 years old (Miller *et al.*, 1986). By the time the trees were 15 years old, the pure crop of Sitka spruce contained 67 kg N ha^{-1} compared with 200 kg N ha^{-1} in spruce-pine (Table 1). The difference of 133 kg N between the pure and mixed species was equivalent to a supply of nitrogen of about 22 kg N/ha/yr during the years after growth rates had started to diverge. In spruce-

Table 1. Total contents (kg ha^{-1}) of nitrogen and phosphorus in the 15-year-old standing crop, humus (L+F+H) and soil to 10 cm depth beneath Sitka spruce (SS) and mixed Sitka spruce-Scots pine (SS+SP) at Culloden forest. (Miller *et al.*, 1986)

	TOTAL N			TOTAL P		
	SS	SS+SP	se	SS	SS+SP	se
TREES	67	200**	14.9	11.0	27.4*	2.1
HUMUS	229	407*	36.9	19.2	30.0ns	3.8
SOIL	366	327ns	119.9	52.7	39.3ns	23.0
TOTAL	662	934		82.9	96.7	

*,** significantly different from SS at $P < 0.05$ and 0.01, respectively, ns = not significant

larch mixtures, the spruce foliage exhibited signs of nitrogen deficiency for several years before the beneficial effects became apparent (Carlyle and Malcolm, 1986a).

There have been several suggestions about the origin of the additional supply of nitrogen available to the spruce. Carlyle and Malcolm (1986a) studied Sitka spruce planted in mixture with larch (*Larix eurolepis* Henry) on deep peat. They disproved the hypothesis that the deciduous litterfall from larch supplied nitrogen to the spruce by showing that the quantity of nitrogen reaching the forest floor was insufficient to account for the additional supply in the spruce trees. Where the nurse species was lodgepole pine (*Pinus contorta*) planted on a deep peat, the gradual drying out of the peat which occurs beneath pine could explain the improved growth of the spruce, but other coniferous species such as Scots pine are less able to tolerate the waterlogged conditions and, so far there is no single hypothesis to explain all observations.

Analysis of the forest floor at Culloden showed an increase in the total nitrogen capital of the crop, litter and humus in spruce-pine compared with the ground vegetation, trees and humus in pure spruce (Table 1), indicating that a transfer of nitrogen, presumably from the underlying soil, had occurred (Williams, 1992). An increase in nitrogen capital was not observed beneath spruce-larch planted on deep peat though material from the surface 0-3 cm layer of the forest floor had a greater total nitrogen concentration under spruce-larch than beneath pure spruce (Carlyle and Malcolm, 1986b).

Measurements of rates of soil nitrogen mineralization have been carried out *in situ* beneath spruce-larch (Carlyle and Malcolm, 1986b) and spruce-pine mixtures (Williams, 1992). In the surface 0 - 3 cm on a deep peat, rates of net mineralization, measured over a 4 month period, increased from 22 in pure spruce to 37 kg N /ha in larch - spruce plots (Carlyle and Malcolm, 1986b). In the humus beneath Scots pine - Sitka spruce at Culloden, net mineralization rates increased significantly from 7 kg N /ha under spruce to 15.3 kg N/ha in spruce - pine plots. Net mineralization of nitrogen was greater in the underlying soil than in the humus layer, but differences between the species were not statistically significant. In both studies, the improvement in nitrogen mineralization beneath the mixed crop was confined to the surface horizon and consistent with the increased nitrogen contents. At Culloden, the crop was 15 years old when the samples were taken and already showed very positive signs that nitrogen was more actively cycled in the mixed crop compared with the pure spruce which showed symptoms of nitrogen defficiency. It could not be established from these results that the increased rates of net mineralization were responsible for the enhanced growth and not just an effect of the trees on the forest floor. For unequivocal results that an increased rate of net mineralization of soil nitrogen is the cause of the mixtures effect, further investigations on younger crops are required.

The areas where mixed coniferous tree crops are planted are usually inherently low in plant available phosphorus and applications of fertilizer phosphorus at planting are normally essential (Taylor and Tabbush, 1990). The extent to which mixed planting can relieve phosphorus deficiency has not been established and also its not certain that the mixtures effect will occur if the crop is phosphorus or potassium deficient. The influence of phosphorus fertilization during establishment on the effectiveness of the nurse species will depend on the mechanisms of the interactions. At later stages of the crop's development, any enhanced biological activity beneath mixed crops may be sufficient to sustain the supply of phosphorus for uptake. At Culloden, the phosphorus content of the tree crop above ground was significantly ($P < 0.05$) increased in spruce - pine compared with pure spruce, but there was little evidence for an overall increase in the phosphorus capital of the forest floor and soil as there was for nitrogen (Table 1). Carlyle and Malcolm (1986b) also reported no increases in phosphorus concentrations of the forest floor beneath larch - spruce compared with pure spruce planted on deep peat.

Mixtures of conifers and broadleaved species planted on peaty gley soils have been studied at an experimental site in Gisburn Forest in the north of England. After 26 years, height growth of Norway spruce (*Picea abies*), alder (*Alnus glutinosa*) and oak (*Quercus petraea*) all tended to be greater when planted in mixture with pine than when planted as the pure species (Brown, 1988). The soils at this site were not nitrogen deficient and the benefits of planting mixed species were not nesessarily confined to the establishment phase.

TRANSFERS IN RAINWATER

The enrichment of rainwater in dissolved nutrients and organic matter as it passes through the canopy of forest vegetation has been recognised (McDowell and Likens, 1988), and has prompted the hypothesis that nitrogen is transferred from the nurse species to the forest floor in canopy throughfall and stemflow. Carlyle and Malcolm (1986a) reported small leakage and leaching of nitrogen from senescing larch litter and ammonium concentrations were less in throughfall than in rainfall. Concentrations of phosphorus in throughfall increased relative to those in rainwater, but were very similar beneath both larch and spruce. In the spruce-pine plots at Culloden, fluxes of nitrogen as ammonium and nitrate in throughfall and stemflow were small, 2-4 kg N ha^{-1} yr^{-1}, and were not significantly different beneath the two tree species (Miller *et al.*, 1986). Evidently, nitrogen transfer did not occur in throughfall beneath the spruce-larch and sprue-pine mixtures, but the effect cannot be discounted in other mixed stands.

ROOT INVESTIGATIONS

Root excavations at Culloden showed that the roots of pine and spruce did not intermingle, but that the fine roots of spruce were found mainly in the surface humus and those of pine were at deeper depths (Miller *et al.*, 1986). In detailed quantitative studies of the fine root biomass at the same site, the pattern of rooting with depth was different beneath the two types of stands (McKay and Malcolm, 1988). Beneath the mixed species, the distribution of fine roots was more uniform than beneath the pure spruce where the root biomass was greater at the surface. Furthermore, the biomass of fine roots was greater in pure spruce plots than beneath mixed species. There was no evidence of a direct interaction between the root systems of the two species that would account for the additional amounts of nitrogen in the mixed plots.

MYCORRHIZAL RELATIONSHIPS

Studies on the mycorrhizal associations of tree species planted in mixture have dispelled the idea that fungi, normally associated with pine, simply colonized the root systems of spruce. Ryan and Alexander (1993) showed that beneath mixed Scots pine-Sitka spruce several fungi colonized the roots of both tree species, but only *Suillus sp.* was confined to the roots of Scots pine.

Stribley and Read (1980) showed that mycorrhizal fungi in culture were able to utilize low molecular weight organic compounds such simple amino acids and this was considered to be an advantage for plants growing on soils where rates of net mineralization of soil organic nitrogen were low. Abuzinadah and Read (1986) reported that mycorrhizal fungi associated with native plant species exhibited enhanced proteolytic enzyme activity and could utilize peptides and proteins which they considered to be a part of the adaptation to growth on nutrient poor soils. Enzyme assays with mycorrhizal fungi isolated from the roots of Sitka spruce and Scots pine at Culloden demonstrated that the *Suillus varigatus* had greater proteolytic activity than cultures of *Lactarius rufus* and *Tylospora fibrillosa*, found generally on the roots of both spruce and pine (Ryan and Alexander, 1992). Moreover, *Lactarius rufus* and *Tylospora fibrillosa* were able to utilize protein nitrogen in bovine serum albumin that had been pretreated by incubation with *Suillus* fungus. Therefore, a current hypothesis to explain the enhanced supply of nitrogen states that the mycorrhizal fungi on the roots of spruce are able to utilize organic nitrogen degraded by the action of the pine fungi which are adapted to nutrient poor soils.

The bovine serum albumin used in these studies for the assay for proteolytic activity does not seem a good substitute for soil organic nitrogen. Nevertheless, this approach is beginning to show how mycorrhizal fungi may interact to increase the supply of available nitrogen. The roots of *Calluna* are also infected by ectomycorrhizas having proteolytic activity, but the growth of spruce is not enhanced by heathland vegetation. Indeed, *Calluna* roots produce substances that are toxic to fungi and cause the condition 'heather check' in spruce trees planted on heathlands (Robinson, 1972).

FOLIAR LITTER INTERACTIONS

The litter layer is an obvious site for potential interactions between tree species which are probably less important during the establishment phase, but could occur throughout the development of the crop. Interactions between litters have been studied in litter bag experiments (Blair *et al.*, 1990), in microcosms (Ineson and MacTiernan, 1992) and in leaching columns (Williams and Alexander, 1991). Ineson and MacTiernan (1992) measured CO_2 evolved from 21 combinations of mixed litters and observed positive interactions in 10 of these during 28 weeks incubation. There were no negative effects on CO_2 evolution and no single litter type was responsible for enhanced respiration. Exactly how these interactions are caused is not clear. Litter composition alone may induce changes in the activity of both primary and secondary decomposer organisms affecting both gross mineralization and immobilization processes.

One of the problems of studying these interactions is the marked temporal variations in the dynamics of the processes involved and the need for long as well as short term experiments to monitor the dynamics of the system. Blair *et al*, (1990) reported on the effects of mixing deciduous litters of *Acer rubrum*, *Quercus prinus* and *Cornus florida* in litter bag experiments for up to 378 days. The results showed increased fluxes of nitrogen and significant increases in the numbers of bacterial-grazing nematodes in the mixed litters compared with predicted values from single tree species. In experiments with coniferous litters of Sitka spruce and Scots pine, enchytraeid worms present on the spruce litter proliferated 13 weeks after the needles were mixed (Williams and Alexander, 1989). The rates of net mineralization of nitrogen (Table 2) and release of carbon dioxide in the mixed litters were greater than the values calculated from the pure litters for a 1:1 non-interactive mixture. The increased activities were ascribed to the effects of the worms grazing the fungal biomass which was greater on the initial pine litter than on the spruce needles. Anderson *et al.*, (1983) suggested that faunal grazing of the predominantly fungal biomass in litter layers increases the contribution of bacteria, present in the gut of

Table 2. Cumulative amounts (mg kg^{-1} dry matter) of different nitrogen forms leached with water during 16 weeks at 20°C and then extracted with 1 M KCl from litters of Sitka spruce (SS), Scots pine (SP) and a 1:1 mixture (SS+SP). Values in parentheses calculated for a 1:1 non-interactive mixture.

	SS	SS+SP	SP	se (n=3)
NH_4 -N Leached + extracted	1720a *	1280b	480c	161
NO_3 -N Leached + extracted	2570a	1320b	20c	139
Total mineral - N	4290	2600 (2395)	500	
Organic -N leached	1290a	900b (770)	250c	39

* Values in rows designated different letters, significantly different at $P < 0.05$

fauna, to the litter decomposition processes. Experiments with added enchytraeid worms to Sitka spruce litter showed that the worms accelerated the leaching of all of the major nutrient ions (NH_4, NO_3, PO_4, Ca, Mg, and K) (Williams and Griffiths, 1989).

The first experiment carried out by Williams and Alexander (1989) included pine and spruce litters from separate sites and the enchytraeid worms were imported on spruce litter which turned out to be biologically more active than that of pine. The spruce was growing on a brown forest soil and the pine on an acid soil derived from granitic parent material which probably contributed to differences between the litters. In a second experiment, litters of spruce and pine, sampled from mixed plots of Sitka spruce + Scots pine from Culloden forest, were separated by hand and then leached in columns of individual species and as a 1:1 mixture. Worms were absent from this experiment and the quantities of mineral nitrogen extracted and inorganic phosphorus leached were again significantly greater ($P < 0.05$) in the spruce than in the pine litter (Table 3). Amounts of extracted inorganic phosphate were enhanced and the amounts of extracted mineral and ammonium nitrogen were depressed by mixing the litters compared with the values calculated for a non-interactive 1:1 mixture from the single species columns. The lowering of the mineral - N content on mixing the litters was ascribed to increased immobilization and was consistent with the greater flush of nitrogen from the mixed litters following fumigation with chloroform (an indicator of microbial biomass nitrogen) (Table 4). In contrast to nitrogen, the release of phosphorus following fumigation was not increased by mixing the litters, possibly because the forest floor was not phosphorus limited, having been fertilized during the establishment phase of the crop. In both of these experiments, spruce litter had

Table 3. Accumulated amounts (mg/kg dry litter) of NH_4-N, NO_3-N and PO_4-P leached with distilled H_2O and extracted with 1 M KCl from litters of Sitka spruce (SS), Scots pine (SS) and a 1:1 mixture (SS&SP) during 21 weeks continuous leaching at 20° C. Litters sampled from SS + SP plots at Culloden forest and separated by hand.

	Leached				Extracted			
	SS	SS+SP (n=4)	SP	se	SS	SS+SP (n=3)	SP	se†
NH_4 - N	30	25	9	10.9	2040a	426b	953c	142.2
NO_3 - N	11	12	12	0.4	8	5	4	0.7
Min-N (NH_4+NO_3)	41	37	22	10.8	2048a	431b	954c	142.8
PO_4 - P	79a	49b	18c	2.9	553	650	511	42.2

*values designated different letters significantly different, $P < 0.05$
†se based on the pooled standard error for single columns

Table 4. Flush of nitrogen and phosphorus (mg N and P /kg dry litter) from litters of Sitka spruce (SS) and Scots pine (SP) and a 1:1 mixture (SS & SP) fumigated with chloroform and extracted with 1M KCl after continuous leaching with distilled H_2O for 21 weeks at 20° C.

	N-FLUSH	se† (n=3)	P-FLUSH	se (n=3)
SS	1383	469	1090	116.3
SS+SP	1953 (1082)*	278.3	736 (759)*	102.8
SP	781	266	428	60.7

†se standard error of the mean calculated from pooled standard errors of subsamples from single columns.
* Values calculated for a non-interactive 1:1 mixture

greater biological activity than the pine which is in direct contrast to the effects on tree growth prior to canopy closure. However, it is unlikely that the effects of mixing litters has a significant effect on the nutrient cycle during the establishment phase of a plantation and is more important at later stages.

The chemical composition of litter, particularly C:N:P ratios, has a strong influence on the dynamics of the nutrients during decomposition. For example, mineral nitrogen production is limited if the microbial biomass has a supply of substrates rich in carbon with the result that phases of nitrogen immobilization precede net mineralization (Williams and Alexander, 1989). Changes in the fixation of carbon from atmospheric CO_2 by forests could have important consequences for the dynamics of nutrients in the forest floor if additional amounts of carbon are returned as litter, fine roots and root exudates. However, while mineral forms of nitrogen and phosphorus are controlled by availability of carbon substrates, there is evidence that soluble organic matter containing nitrogen is not regulated in the same way. Nutrients in soluble organic forms are also leached from litters during decomposition and this process is significantly ($P < 0.05$) increased by the interaction between the litters of spruce and pine (Table 2). Soluble organic nitrogen leached from the litter horizon is either mineralized or adsorbed at lower depths in the soil profile and only a small amount may reach surface waters (Williams and Edwards, 1993).

Immobilization of nutrients in the microbial biomass is considered to be a significant process controlling mineral nitrogen release from decomposing plant residues with the activity of secondary decomposers influencing the turnover of microbial material in the soil and litter. The microbial biomass in decomposing litters, as in soil, is a reserve of potentially available nitrogen and phosphorus, but precise measurements of this pool of nutrients is difficult. Advances in the methodology to measure the activity and size of different components of the microbial biomass would increase our understanding of the interactions in the mixed litters.

SOIL BIOLOGICAL ACTIVITY

Decomposition studies in the forest floor beneath mixed tree species have been carried out mainly at Gisburn in the north of England using the cotton strip method (Howson, 1991). At this site, mature stands of mixed tree species included combinations of conifers and broadleaved species. In general, loss of tensile strength of the cotton strip, which is used as an indicator of cellulose decomposition, correlated with the height growth of the trees. Decomposition was greater in soil from beneath Scots pine and alder than beneath stands of Norway spruce or oak (Howson, 1991). In mixed plots containing either alder or spruce, the greatest rates of decomposition were measured in stands containing Scots pine Measurements of earthworm biomass beneath Norway spruce at this experimental site also showed increases when the spruce was in a mixture with either alder or pine confirming that soil biological activity can be greater under some combinations of mixed tree species than others or when planted in pure stands (Brown and Harrison, 1982).

Killham and Rashid (1986) developed an assay for the activity of deaminase enzyme activity in soils and tested the method on the mineral soil horizons beneath spruce and spruce + pine at Culloden forest. The activity of this enzyme, which is responsible for the ammonification of amino nitrogen and is a reflection of gross mineralization of nitrogen, was greater in mixed than in the pure plots. This assay showed differences between the tree species in the underlying mineral soil, in contrast to *in situ* incubation studies, in which significant differences were confined to the humus (L + F + H) layer (Williams, 1992).

CONCLUSIONS

Decomposition sudies have shown that the forest floor floors beneath some combinations mixed tree species are biologically more active than those under pure stands. The exact mechanisms of the processes which bring this about are not clear. Interactions between mycorrhizal fungi associated with different tree species could have an important role to play during the establishment phase of a tree crop on nutrient poor soils. At later stages of a forest's development, interactions between some combinations of foliar litters, which have been observed and reported in laboratory experiments, very probably enhance biological activity in the forest floor and the cycling of nutrients. Moreover, these processes controlling the breakdown and decomposition of litters and plant material are brought about by the activity of a wide range of organisms including primary and secondary decomposers. Predicted increases in carbon inputs to ecosystems, as foliar litters, roots and root exudates, from increased carbon dioxide levels in the atmosphere could very well influence nutrient dynamics in the forest floor and restrict the availability of nitrogen and phosphorus. Under these circumstances, interactions between tree species could become even more significant.

Acknowledgements

The work carried out at MLURI for this project was partly financed by the Scottish Office Agriculture and Fisheries Department and by the European Economic Community. Field experiments at Culloden and Inchnacardoch forests sampled for the studies were designed and managed bythe Research Branch of the Forest Authority.

REFERENCES

Abuzinadah RA, Read DJ (1986) The role of proteins in the nitrogen nutrition of ectomycorrhizal plants. III Protein utilization by *Betula*, *Picea* and *Pinus* in mycorrhizal association with *Heboloma crustiliniforme*. New Phytologist 103: 507-514

Anderson JM, Ineson P, Huish SA (1983) Nitrogen and cation mobilization by soil fauna feeding on leaf litter and soil organic matter from deciduous woodland. Soil Biology and Biochemistry 15: 463-467

Blair JM, Parmelee RW, Beare MH (1990) Decay rates, nitrogen fluxes, and decomposer communities of single and mixed-species foliar litter. Ecology 71: 1976-1985

Brown AHF (1988) Discrimination between the effects on soils of 4 tree species in pure and mixed stands using cotton strip assay. Pages 80-83 *in* Harrison AF, Latter PM, Walton DWH (eds.) Cotton Strip Assay: an Index of Decomposition in Soils. Institute of Terrestrial Ecology Symposium No. 24. NERC, Grange-over Sands

Brown AHF, Harrison AF (1982) Effects of tree mixtures on earthworm populations and nitrogen and phosphorus status in Norway spruce (*Picea abies*) stands. Pages 101-108 *in* Lebrun Ph, Andre HM, De Medts A, Gregoire-Wibo H, Wauthy G (eds.) Proceedings VIII International Colloquium of Soil Zoology. Dieu-Brichart, Ottignes-Louvain-la-Neuve, Belgium

Carlyle JC, Malcolm DC (1986a) Larch litter and nitrogen availability in mixed larch-spruce stands. I Nutrient withdrawal, redistribution and leaching loss from larch foliage at sensecence. Canadian Journal of Forest Research 16: 321-326

Carlyle JC, Malcolm DC (1986b) Nitrogen availability beneath pure spruce and mixed larch + spruce stands growing on a deep peat. I. Net N mineralization measured by field and laboratory incubations. Plant and Soil 93: 95-113

Fisher RF, Stone EL (1969) Increased availability of nitrogen and phosphorus in the root zone of conifers. Proceedings of the Soil Science Society of America 33: 955-961

Howson G (1991) The cotton strip assay: field applications and global comparisons. Pages 217-228 *in* Wilson WS (ed.) Advances in Soil Organic Matter Research: The Impact on Agriculture and the Environment. The Royal Society of Chemistry, Cambridge

Ineson P, MacTiernan K (1992) Decomposition of foliar litter mixtures: a microcosm experiment. Pages 703-706 *in* Teller A, Mathy P, Jeffers JNR (eds.) Response of Forest Ecosystems to Environmental Changes Elsevier Applied Science, London

Killham K, Rashid MA (1986) Assay of activity of a soil deaminase. Plant and Soil 92: 15-21

McDowell WH, Likens GE (1988) Origin, composition, and flux of dissolved organic carbon in the Hubbard brook valley. Ecological Monographs 58: 177-195

McIntosh R (1983) Nitrogen deficiency in establishment phase Sitka spruce in upland Britain. Scottish Forestry 37: 185-193

McKay HM, Malcolm DC (1988) A comparison of the fine root component of a pure and a mixed coniferous stand. Canadian Journal of Forest Research 18: 1416-1426

Miller HG Alexander CE, Cooper J, Keenleyside J, McKay H, Miller JD, Williams BL (1986). Maintenance and enhancement of forest productivity through manipulation of the nitrogen cycle. Unpublished Report to the EC. Contract No. BOS-093 UK

O'Carroll N (1978) The nursing of Sitka spruce. 1. Japanese larch. Irish Forestry 35: 60-65

Richards BN (1962) Increased supply of soil nitrogen brought about by *Pinus*. Ecology 43: 538-541

Robinson RK (1972) The production by roots of *Calluna vulgaris* of a factor inhibitory to growth of some mycorrhizal fungi. Journal of Ecology 60: 219-224

Ryan EA, Alexander IJ (1993) Mycorrhizal aspects of improved growth of spruce when grown in mixed stands on heathlands. Pages 237-245 *in* Read DJ, Lewis DH, Fitter AH, Alexander IJ (eds.) Mycorrhizas in Ecosystems. C.A.B. International, Wallingford UK

Stribley DP, Read DJ (1980) The biology of mycorrhiza in the Ericaceae. VII The relationship between mycorrhizal infection and the capacity to utilize simple and complex organic nitrogen source. New Phytologist 86: 365-371

Taylor CMA, Tabbush PM (1990) Nitrogen deficiency in Sitka spruce plantations. Forestry Commission Bulletin 89, HMSO, London

Weatherell J (1957) The use of nurse species in the afforestation of upland heaths. Quarterly Journal of Forestry 51: 298-304

Williams BL (1992) Nitrogen dynamics in humus and soil beneath Sitka spruce *(Picea sitchensis* (Bong.) Carr.) planted in pure stands and in mixture with Scots pine (*Pinus sylvestris* L.). Plant and Soil 144: 77-84.

Williams BL, Alexander CE (1991) Interactions on mixing litters from beneath Sitka spruce and Scots pine and the effects on microbial activity and N-mineralization. Soil Biology and Biochemistry 23: 71-75

Williams BL, Edwards AC (1993) Processes influencing dissolved organic nitrogen, phosphorus and sulphur in soils. Chemistry and Ecology (In Press)

Williams BL, Griffiths BS (1989) Enhanced nutrient mineralization and leaching from decomposing Sitka spruce litter by enchytraeid worms. Soil Biology and Biochemistry 21: 183-188

Yanai RD (1992) Comparative interactions between Norway spruce and Scots pine at Gisburn Forest, N.W. England. Forestry 65: 435-451

Chapter 19: A Research Agenda for Global Change in Finland: Research Priorities for the Conservation and Sustainable Use of Biodiversity

Rauno Väisänen and Jukka-Pekka Jäppinen
National Board of Waters and the Environment Nature Conservation Research Unit
P.O. Box 250
FIN-00101 Helsinki
Finland

Global environmental problems in connection with the concept of global change were until recently understood mainly as environmental pollution, depletion of ozone layer and global warming of the earth's atmosphere (*e.g.* Liukkonen and Tennberg 1990). The United Nations Conference on Environment and Development (UNCED) resulted in the Convention on Biological Diversity and familiarized the public with the concept of biodiversity. It brought biodiversity unquestionably onto the agenda at the international political level, too (*e.g.* Miller and Barber 1992, Tennberg 1993). More than 150 countries signed the Convention, which emphasizes the need to develop national plans or programmes for the conservation and sustainable use of biodiversity. Such plans are to be based on comprehensive inventories and monitoring of components of biodiversity and of processes having an adverse impact both within and outside protected areas. The Convention also commits signatories to take action in favour of restoration of degraded ecosystems and recovery of threatened species. As a complement to in situ conservation, development of ex situ mechanisms is necessary in certain extreme cases. Furthermore, the Convention emphasizes powerfully the promotion of research and public awareness as well as the integration of biodiversity concerns into national decision-making.

NATO ASI Series, Vol. I 20
Biodiversity, Temperate Ecosystems, and Global Change
Edited by T. J. B. Boyle and C. E. B. Boyle

RESEARCH PRIORITIES

The biodiversity issues are a great challenge to the scientific community (WCMC 1992). Scientists should provide reliable information on the intensity and areal extent of different threats to biodiversity and set national and international research priorities accordingly. These should not include only direct threats such as destruction and fragmentation of habitats, over-exploitation of natural resources, pollution, climate change and introduction of alien species, but also latent causes of biodiversity loss such as population growth, increase in consumption and unequal income distribution (*e.g.* McNeely 1988). It is thus self-evident that interdisciplinary action is necessary. The interaction between ecological and economic sciences is crucial, since environmental decision-making attempts to find (alternative) optimal combinations of ecological and economic objectives. The problems of such an approach include the lack of operational definition of the concept of biodiversity and the lack of objective knowledge for evaluating decision alternatives with respect to components of biodiversity.

Central areas of future research are associated with both the functional aspects of biodiversity (and the importance of habitats and ecosystems) and the increasing need of taxonomic and inventory work (including genetic studies). Research on the present status, minimum viable populations and minimum critical areas of threatened and keystone species is urgently needed. Landscape ecology in general, linking and investigating phenomena in different spatial and temporal scales, and fragmentation of habitats in particular, are undoubtedly providing routes to understand the loss of biodiversity and to reinforce the scientific basis of networks of protected areas, (*cf.* Götmark and Nilsson 1992). The role of biodiversity in the functioning of ecosystems is still poorly understood (di Castri and Younes 1990, Pimm 1991). Between-species interactions, impacts of disturbances and the significance of environmental heterogeneity for biodiversity are all subthemes of this main topic. Such information can be used as the scientific basis for sustainable use of natural resources. The only way to achieve these research goals, even partially, is to develop and improve methods for measurement and indirect assessment of biodiversity. This means *e.g.* the use of remote sensing data, geographical information systems (GIS)-based data processing and modelling of biodiversity.

THE FINNISH NATIONAL BIODIVERSITY PROGRAMME

At present, there is no comprehensive scientific picture of the biological diversity of Finnish nature. However, the threat to the biodiversity is increasing (see Wahlström *et al.* 1993). The National Biodiversity Research Programme (LUMO), 1992—1996, focuses on all biological diversity including the species of organisms and their genetic diversity as well as Finnish ecosystems. The research programme is coordinated by the Nature Conservation Research Unit of the National Board of Waters and the Environment. Detailed knowledge of biodiversity will be acquired by means of long-term research and monitoring, but in the short term the LUMO aims at the understanding of biodiversity by monitoring keystone and indicator species describing environmental changes. In the studies, the emphasis will be on forest ecosystems and traditional rural landscapes as well as on the conservation and sustainable use of their biodiversity and on the specification of their faunistic and floristic knowledge. The central objective of the programme is to gain sufficient knowledge to prevent the impoverishment of the environment and to restore the biodiversity. As a first step of the programme, a national biodiversity status report (*cf.* Sandlund 1992) is under preparation and will be available for the public in early 1994.

The research programme will specify the mechanisms and processes affecting Finnish biodiversity. Processes include land-use and climate changes, the chemicalization of the environment, the excessive and ecologically unsustainable exploitation of biodiversity, environmental changes or destruction and the release of exotic or genetically modified species into nature.

Monitoring of natural and detrimental changes in species and ecosystems valuable for conservation will be developed. Nationally compatible and commensurable data banks and registers for different users will be advanced. In Finland, it may be possible to attain sustainable development by improving especially forestry so that it would be based on ecologically sustainable timber production and conservation of biodiversity.

The general objectives of the research programme can be summarized as follows:

1. To fulfil the obligations of the Convention on Biological Diversity,
2. To find out the dimensions and rate of the possible impoverishment of nature and to predict its future development,
3. To produce information about biodiversity conservation methods,
4. To produce information needed in decision-making as a basis of ecologically sustainable utilization of natural resources and conservation of nature,
5. To develop and evaluate the methods for measuring of biodiversity, which can function as a basis for the improvement of conservation methods and the estimation of the results of the conservation measures taken,
6. To produce the knowledge needed in combining nature conservation and socio-economic activities of society,
7. To expand natural scientific and environmental data banks and registers, and
8. To develop a constant monitoring programme for biodiversity.

The biodiversity programme will be initiated with five scientifically emphasized subprogrammes. Their results will be utilized in the subsequent socio-economic related subprogrammes, connected with administration, legislation, community planning, education and the obligations of international agreements and cooperation. The feasibility and applicability of the results for the needs of environmental administration and its decision-making is to be emphasized. The LUMO will also produce material for the Environment Data Centre (EDC) of the National Board of Waters and the Environment, for the creation of a national and commensurable Environmental Data System (EDS). At the final stage, the results will be published as a report series which will be composed of the summaries of research reports and scientific papers connected with the research programme. The results should also be published as popular articles. A national biodiversity action plan will be also produced during the research programme.

Principal Research and its Objectives

The biodiversity research programme is being realized in accordance with a plan approved by the LUMO steering committee. It is based on eight subprogrammes, which are given in the Appendix.

1. General Environmental Changes and Biodiversity

To avoid unnoticeable loss or disappearance of species or valuable habitats, biodiversity monitoring is needed. The aim of the first subprogramme is to gather elementary knowledge about Finnish biodiversity and to develop a consistent and permanent biodiversity monitoring system for terrestrial and aquatic ecosystems. For example, the Nature Conservation Research Unit started continuous monitoring of nocturnal macrolepidoptera using a network of 112 light traps throughout Finland (Väisäinen 1993). Thus, it is a modification of the famous British Rothamsted Insect Survey (Taylor *et al.* 1978).

Noticing the quantitative or qualitative changes in biological diversity requires an extensive ecological monitoring system and, as part of it, an environmental specimen bank composed of a long-term series of environmental specimens. The development and integration of the biodiversity monitoring system can be started with the help of existing monitoring systems of the environmental state, and long-term environmental investigations.

The aim is also to investigate the relationship between natural changes of biodiversity and environmental changes caused by man (*e.g.* hazardous chemicals, climate change), and to evaluate the effect of direct and indirect environmental changes caused by man. This goal is connected with ecological impact studies and long-term environmental monitoring, including the Integrated Monitoring Programme (IMP) (Nihlgård and Pylvänäinen 1992). In the beginning, the main emphasis of field work will be laid on different soil organisms (plants, animals, terrestrial microbes). It will also attempt to combine the knowledge and experience gained in the national acidification (HAPRO) and climate change (SILMU) research programmes with biodiversity research (see Kaupi *et al.* 1990, Kanninen and Antilla 1992). Within

LUMO it is possible to continue the investigations which were overlooked or received insufficient attention in the above research programmes.

2. Measuring and Monitoring Methods for Biodiversity

The tight schedule imposed by the rapid destruction of the environment and increased social need for scientific research requires environmental research which is more cost-effective and productive than before. Thus, new and more effective sampling and processing methods are needed in conservation ecology. The aim of the second sub-programme is to study, evaluate, apply, and evolve remote sensing techniques, GIS applications, mathematical-statistical methods, and modelling in the analysis of biodiversity. When possible, traditional strenuous field inventories are to be supplemented and substituted by more sophisticated and cost-effective methods (*e.g.* Margules and Austin 1991). The aim is to develop effective and reliable biodiversity measuring methods and to investigate the possibilities of using remote sensing methods in measuring biodiversity. The work combines map interpretation, satellite image interpretation, and fieldwork.

The objective is to survey the total area and compartment size of different habitats, especially forests and peatlands, and the effect of possible changes in them on animal and plant populations. The purpose is to investigate long-term changes and to anticipate their effects on biodiversity by modelling studies. The work covers both protected areas and commercial forests, and provides basic information for the management and planning of nature reserves.

The habitats and distribution of species can be predicted to same degree on the basis of different environmental variables. For example, the distribution of forest insects or fungi could be based on the structural variables of forest stands such as age of trees or tree species composition of a stand. Predicting or modelling can be based also on the variables of environmental data produced by remote sensing (*e.g.* Lauver and Whistler 1993), for example on the timber volume of forest stands, the distribution of habitats and tree species etc. The aim is to combine long-term field inventories with more economic sampling and mathematic modelling. The ultimate aim is a national, scientifically reliable and practical method for measuring biodiversity for conservation and sustainable use.

Prior to modelling biodiversity, several environmental variables have to be examined simultaneously in relation to the distribution of organisms and habitats. However, this requires not only new information but also the development of the structure and composition of existing geographical information systems used in Finland (integration and utilization of GIS in the measuring of biodiversity).

The purpose of the second subprogramme is also to define the relationship of the measured components of biological diversity which are essential to conservation in different scales. Parameters and processes important at one scale are frequently not important or predictive at another scale. Ecological problems often require the extrapolation of fine-scale measurements for the analysis of board-scale phenomena (Turner *et al.* 1989). Since it is uncertain whether it is possible to reach a maximal conservation result in a larger scale by protecting the greatest biodiversity in a smaller scale, some attention is paid to the relationship of species or habitat richness and rarity or the proportion of threatened species or habitats in different spatial scales. This is connected with the planning and decision-making on conservation which can take place at international, national, provincial or municipal levels.

3. Changes in Species Composition and Endangered Species

According to the report of the Committee for the Monitoring of Threatened Animals and Plants in Finland (Rassi *et al.* 1992, *cf.* Rassi and Väisänen 1987), the number of threatened species amounts to 1,692 species. The environmental changes have been so thorough and sudden that especially the most specialized species have not been able to adapt to them. Consequently, these species have become endangered.

There are also significant deficiencies in the taxonomical knowledge of species. The above mentioned Committee based its work on only about 42 percent of the known species in Finland, while the vulnerability of 58 percent of the species could not even be estimated. More information for the basis of vulnerability evaluation of species is needed. By means of more accurate information the unnoticed disappearance of species could be prevented. This could give the protection measures the right order of priority and improve their efficiency. Unfortunately, the economic support for taxonomic studies have been very limited.

Keystone species are species on which several other species in ecosystems are dependent. For example, a keystone species can be a broad-leaved hardwood tree, which provides habitat to several species of insects and fungi. Species which are very sensitive to certain environmental changes are called indicator species. For example, these changes include the changes caused by chemicalization, agriculture or forestry. By their existence or absence or abundance the indicator species can reflect the changes taking place in the environment. The aim of the LUMO is to obtain a detailed picture of the habitat level distribution of the ecologically important keystone and indicator species and their present existence and development in the future. The research tries to find out whether some species or species groups can be used as indicators of biodiversity in certain habitats.

The conservation research must be increased on the genetic level, too. More information is needed also on intraspecific genetic variation in small populations and the relationship of population ecology and genetics (*e.g.* Soulé and Simberloff 1986). The purpose of the LUMO is to investigate how differentiated the Finnish populations of threatened and rare species are and, on the other hand, investigate the geographic pattern of genetic variation in common species (*e.g.* forest trees, fish, game birds and mammals). The purpose is to detect the genomes or genes, which have most conservational, economic or social value. The general aim is to maintain and protect the genetic variation of Finnish species.

The population dynamics of species in a patchy environment will be studied. The purpose is to investigate the fundamental factors which threaten species in changing habitats. More common species which are ecologically and/or taxonomically close to endangered and rare species can be used as reference material. The study involves the investigation of the factors that affect the existence of small populations, such as inbreeding, random variation of birth and death rates and isolation of populations. Much attention will be paid on migration and dispersal barriers between populations and other aspects of metapopulation dynamics (Gilpin and Hanski 1991).

The aim is to produce information on the population ecological effects of habitat changes, especially in small populations, for planning conservation measures.

Of the 1,692 species classified as threatened about 500 belonged to the categories endangered and vulnerable requiring both concrete and immediate conservation measures (Rassi *et al.* 1992). The aim of the LUMO is to coordinate conservation, management and monitoring measures of endangered and vulnerable species, to prepare species conservation plans (specified in Finnish law), to coordinate their implementation and to develop the research of such species.

The purpose is to obtain information on the habitat requirements of threatened species and to find out which factors expose species to disappearance or local extinction. The effects of traditional and new conservation methods is to be compared. The aim is to find out the most cost-effective conservation measures by which the threatened species can be saved. Apart from the scientific forum, the results will be published also as reports directed to the environmental administration and general public.

The threatened species (plants and animals) have approximately 30,000 habitats in Finland and the observations linked with them can exceed 100,000. The gathering of this detailed information to a register requires nationally centralized recording and registration. The Nature Conservation Research Unit gathers detailed information on organisms valuable for conservation and threatened species to a data base (UHEX register) which is a part of the national Environmental Information System (EIS). The register is for the use of environmental administration and, with some restrictions, of other users of information.

The aim is to increase the information of the UHEX register and to improve its usefulness in regional administration and in environmental impact assessment (EIA). The final result should be a nationally compatible species data bank, which is composed at least of the UHEX register, the register of vascular plants and the register of the stock of fish. The data bank could be utilized in land use planning and also in the management and protection of animal and plant species.

4. Endangered Ecosystems and Development of the Network of Protected Areas

In order to attain the highest possible effectiveness and economy, environmental administration needs ecological information on areas with high conservation priorities and on the effectiveness of nature conservation based on conservation areas (see *e.g.* Margules and Austin 1991). This will include investigations of the quality, quantity, management and biogeographical location of the protected areas as well as research on their minimum possible size. When the habitats of species are split into smaller pieces as a consequence of different land-use forms, the species can become isolated in these patches (*e.g.* Niemelä *et al.* 1988, Lord and Norton 1990, Hansson 1992). The aim is to find out the changes in the area and unit size of habitats and the ecological effects of these changes (fragmentation of habitats) and to obtain necessary information for habitat monitoring and the development of conservation area networks. The purpose is also to investigate conservation, management and restoration methods in different biotopes, *e.g.* on grass-herb forest types, peatlands and traditional rural landscapes.

The aim of this subprogramme is to produce information for the establishment of a network of protection areas based predominantly on scientific criteria. The purpose is to identify habitats and ecosystems valuable for conservation, which contain *e.g.* great biodiversity, endemic species, rare and/or threatened species, and conservational, social, scientific, economic or cultural value. The aim is to obtain basic information on significant habitats *e.g.* traditional rural landscapes, primeval forests, fens, rocky outcrops etc. This forms a basis for the development of the network of conservation areas with high conservation priorities and further protection programmes. The information should also form a basis for further ecological research in Finland. One goal is a nationally compatible information system of valuable ecosystems, a biotope data bank (a part of the EIS), which would include *e.g.* the wilderness area data bank, the habitat register of birds and the register of eskers.

The success or sufficiency of conservation measures has not been evaluated in Finland. However, to get the measures as justified as possible, of the right dimensions, effective and economic, the efficiency of protection measures should be measured and optimized. The aim is to find out whether it is possible to develop methods and systems of the measurement of protection measures by which one can produce indicators of conservation evaluation for the use of authorities who are planning and realizing the management of conservation areas.

5. Ecologically Sustainable Use of Nature and Natural Resources

The Biodiversity Convention emphasizes the preservation of biological diversity also in the areas of economic use. Along with the conservation of biodiversity the Convention tries to promote the sustainable use of biodiversity. The utilization of nature is sustainable when it does not exceed the regeneration capacity of natural resources.Effective forestry and land use are the most important reasons for the endangerment of Finnish species, habitats and ecosystems. Forestry is threatening primarily almost half (41 %) of the threatened species (Rassi and Väisänen 1987, Rassi *et al.* 1992, Väisänen *et al.* 1993). In the future, forestry and forestry policies in Finland should aim at both sustainable wood production and ecological sustainability. This means that the forest ecosystem must be treated as an entity, and biodiversity also has to be preserved as much as possible in managed areas. This requires integration of the conservation and other uses of nature (see Kangas and Kuusipalo 1992).

The aim of the LUMO is to direct the use of biotopes so that the needs of forestry, conservation and other forms of land-use are taken into account in a balanced way. The comprehensive project includes nature conservation in forestry, agriculture, the fishing industry, the use of game and reindeer management, soil excavation and construction, biotechnology, recreation and in community planning. One aim is to produce applied information to the environmental administration on ecological and sustainable urban and regional planning. The project will complete the environmental impact assessment process (EIA) with the results gained from the LUMO. But before that we need to measure the impacts of the intensity, frequency and distribution of different uses of nature on populations and the duration and reversibility of these effects. The research should produce models of the effects of the use of biodiversity. The aim is to define tentatively the critical or highest acceptable levels of decrease in biodiversity in Finnish nature.

The aim is to evaluate the economic effects caused by conservation (*e.g.* the purchase of conservation areas, employment considerations, the positive economic consequences of the preserving of biological diversity, developing nature tourism etc.). The aim is also to develop a value theoretical framework, which can be applied in the conservation of biodiversity and in environmental decision-making. The ethical-philosophical grounds of conservation and the valuation of non-commodities of nature are emphasized in relation to the utilitarian point of view. The result will be an evaluation of the material and immaterial benefits of nature conservation in relation to the costs caused by conservation.

6. Biodiversity in Relation to Legislation, Administration, Community Planning and Education

The causes of biodiversity decline are largely socio-economic, and solutions will require interdisciplinary approaches (McNeely 1988, WCMC 1992). The conservation of biodiversity should be included as a part of socio-economic research and social activity on every level, so that ecologically sustainable development should be possible.Among the aims are the production of necessary information for the development of legislation and administration, improved research and decision-making methodologies, and the collecting of ecological knowledge for working out a target programme for nature conservation. An important part of this is the transferring of conservation knowledge and other information on biodiversity to education at different levels (day nursery, school, vocational school, university and official education). Especially, the development and strengthening of academic teacher training programmes is emphasized. The implications of the Convention on Biological Diversity on Finnish nature protection legislation are also to be investigated. At a later stage the focus will be on the relationship of ecological information and decision-making.

7. International Aspects of Biodiversity

Although the possible impoverishment of Finnish nature can be less dramatic than in the tropics, every country has her own national and international responsibility for the preservation of biodiversity. Since the resources of developing countries to fulfil their national obligations in the research and conservation of their biodiversity are often limited, the LUMO tries to apply the results gained in the biodiversity research programme to Finnish development cooperation and international research cooperation. The investigations will probably be biodiversity investigations in one or more developing countries required by the Biodiversity Convention (*e.g.* the preparation of national action plans, which aim at the protection of the biodiversity of developing countries). The subprogramme will probably be realized in cooperation with the Finnish International Development Agency (FINNIDA) of the Finnish Ministry for Foreign Affairs and the Academy of Finland.

8. The Summary of Results and Conclusions

The results of the LUMO will be published to decision-makers, administration, the scientific community, mass media and public through reports, scientific publications, news transmissions and a multi-volume report series. The results will also be published in separate progress reports before the final stage, as fast as possible and in a form such that both administrative and private decision-makers and planners could put them in practice as soon as possible.

On the basis of these results a national biodiversity action plan will be compiled. The action plan is a partial answer to research and conservation needs, which have been specified in the Biodiversity Convention and which are binding upon the signatory countries. The action plan will include basic knowledge of Finland"s biodiversity, its protection and ecologically sustainable use. The plan should provide concrete action propositions and a necessary conservation strategy to preserve biodiversity in conservation areas and in areas of economic use. The strategy will be created at least from the following elements:

- improvement of the knowledge of biodiversity in Finland
- development of more cost-effective biodiversity evaluation
- strengthening of conservation activities
- sustainable use of biodiversity

- integration of biodiversity protection knowledge in different areas and activities
- organizing a constant biodiversity monitoring system
- organizing the necessary national cooperation, division of work and costs

The implementation of the action plan should also include a training programme for the experts and other personnel of future biodiversity monitoring systems (ecologists, taxonomists, technical and supporting personnel and amateur botanists, ornithologists and entomologists). In matters concerning conservation, sustainable use and monitoring of biodiversity, the training should also be directed to the personnel of the environment, agriculture and forestry administration, forestry professionals and private forest owners (see also Council of State 1991).

REFERENCES

Council of State (1991) Sustainable development and Finland. Council of State Report to Parliament on Sustainable Development in 1990. The Government Printing Centre, Helsinki

di Castri F, Younès T (1990) Ecosystem function of biological diversity - Biology International, Special Issue

Gilpin M, Hanski I (eds.) (1991) Metapopulation dynamics: empirical and theoretical investigations. Academic Press, London

Götmark F, Nilsson C (1992) Criteria used for protection of natural areas in Sweden 1909—1986. Conservation Biology 6: 220—231

Hansson L (1992) Landscape ecology of boreal forest. Trends in Ecology and Evolution 7: 299—302

Kangas J, Kuusipalo J (1992) Estimation and use of biodiversity in forest management planning. A paper published at the SSFE-MFU Valuing Biodiversity Seminar, 8-9. October 1992, Korpilampi, Espoo, Finland. The Finnish Forest Research Institute, Joensuu Research Station

Kanninen M, Anttila P (eds.) (1992) The Finnish Research Programme on Climate Change. Progress Report. Publications of the Academy of Finland 3/92. VAPK -Publishing, Helsinki

Kauppi P, Anttila P, Kenttämies K (eds.) (1990) Acidification in Finland. Springer Verlag, Berlin, Heidelberg, New York

Lauver CL, Whistler JL (1993) A hierarchical classification of Landsat TM imagery to identify natural grassland areas and rare species habitat. Photogrammetric Engineering and Remote Sensing 59: 627—634

Liukkonen E, Tennberg M (eds.) (1990) Global change and Finland. Seminar report. The Finnish Institute of International Affairs, Research Reports 7/90, Helsinki

Lord JM, Norton DA (1990) Scale and spatial concept of fragmentation. Conservation Biology 4: 594—598

Margules CR, Austin MP (eds.) (1991) Nature conservation: cost effective biological surveys and data analysis. CSIRO, Australia

McNeely JA (1988) Economics and biological diversity: developing and using economic incentives to conserve biological resources. IUCN, Gland, Switzerland

Miller K, Barber C (1992) Biodiversity after UNCED: next steps. Species 18: 31—32

Niemelä J, Haila Y, Halme E, Lahti T, Pajunen T, Punttila P (1988) The distribution of carabid beetles in fragments of old coniferous taiga and adjacent managed forest. Annales Zoologici Fennici 25: 107—119

Nihlgård B, Pylvänäinen M (eds.) (1992) Evaluation of integrated monitoring in terrestrial reference areas of Europe and North America. The pilot programme 1989—1991. Environment Data Centre, Helsinki

Pimm SL (1991) The balance of nature? Ecological issues in the conservation of species and communities. The University of Chicago Press, Chicago

Rassi P, Väisänen R (eds.) (1987) Threatened animals and plants in Finland. English summary of the report of the Committee for the Conservation of Threatened Animals and Plants in Finland. Ministry of the Environment, Helsinki

Rassi P, Kaipiainen H, Mannerkoski I, Ståhls G (1992) The report of the Committee for the Monitoring of Threatened Animals and Plants in Finland (in Finnish). Ministry of the Environment. Komiteanmietintö (1991) 30, Helsinki

Sandlund OT (1992) Biological diversity in Norway. A country study. Direktoratet for Naturforvaltning. DN-report 1992-5b, Trondheim

Soulé ME, Simberloff D (1986) What do genetics and ecology tell us about the design of nature reserves? Biol. Conserv. 35: 19—40

Taylor LR, French RA, Woiwood IP (1978) The Rothamsted Insect Survey and the urbanization of land in Great Britain. Pages 31-65 *in* Frankie GW, Koehler CS (eds.) Perspectives in urban entomology. Academic Press, New York

Tennberg M (1993) Global environmental problems and their challenges to Finland (in Finnish). The Finnish Institute of International Affairs, Challenges to Foreign Policy 1/1992, Helsinki

Turner MG, O"Neill RV, Gardner RH, Milne BT (1989) Effects of changing spatial scale on the analysis of landscape pattern. Landscape Ecology 3: 153—162

Väisänen R (1993) Valtakunnallinen yöperhosseuranta (Abstract: Moth monitoring network of light traps established in Finland). Baptria 18(1): 9—11

Väisänen R, Biström O, Heliövaara K (1993) Sub-cortical Coleoptera in dead pines and spruces: is primeval species composition maintained in managed forests? Biodiversity and Conservation 2: 95—113

Wahlström E, Hallanaro E-L, Reinikainen T (1993) The state of the Finnish environment. National Board of Waters and the Environment, Environment Data Centre and the Ministry of the Environment, Helsinki

WCMC (World Conservation Monitoring Centre) (1992) Global biodiversity. Status of Earth"s living resources. Chapman and Hall, London

APPENDIX The realization stages of the Finnish National Biodiversity Programme (LUMO), 1992—1996, and the principal research issues in its eight subprogrammes. These issues consist of the central fields of research. Their mutual importance as regards the conservation of biodiversity in Finland is indicated by an increasing number of asterisks (from 1 to 3) for the first stage. The priorization is based both on the present state of knowledge in each field and the proportional strength of threats.

FIRST STAGE OF REALIZATION (Mainly Nature Research)

---

1. General Environmental Changes and Biodiversity

Basic Knowledge about Biodiversity and Monitoring of Biodiversity in Terrestrial Ecosystems * * *
Basic Knowledge About Biodiversity and Monitoring of Biodiversity in Aquatic Ecosystems * * *
The Effect of Pollution and Climate Change on Biodiversity *

2. Measuring and Monitoring Methods of Biodiversity

The Development of Measuring and Monitoring Methods of Biodiversity * * *
Modelling Biodiversity * * *
Integration and Utilization of GIS in the Measuring of Biodiversity * * *
The Relationship of the Components of Biological Diversity * * *

3. Changes in Species Composition and Endangered Species

The Improvement of Vulnerability Evaluation ***
The Distribution and Number of Keystone and Indicator Species *
Intraspecific Genetic Variation ***
Effects of Ecosystem Changes on Population Ecology ***
Threatened Species and their Ecological Characteristics ***
Species Data Banks ***

4. Endangered Ecosystems and Development of the Network of Protected Areas

Habitats with High Conservation Priorities ***
Fragmentation of Ecosystems ***
Habitat Management and Restoration **
Evaluation of the Success of Conservation Measures ***
Biotope Data Information Systems *

5. Ecologically Sustainable Use of Nature and Natural Resources

Measurement of the Impacts of Nature Use **
Economic Evaluation of the Benefits and Costs of Nature Conservation **
Biodiversity in Community Planning *
The Integration of Use and Protection of Nature ***

SECOND STAGE OF REALIZATION (Socio-Economic Research)

---

6. Biodiversity in Relation to Legislation, Administration, Community Planning and Education

7. International Aspects of Biodiversity

Ecological Research Required by International Agreements
Biodiversity and Development Cooperation

8. The Summary of Results and Conclusions

Reporting and Publishing the Results
National Biodiversity Strategy - Integrating the Conservation Knowledge to Everyday Use

---

Chapter 20: Biodiversity, Global Change and forest management: a Québec perspective

Agathe Cimon
Ministry of Forests
Government of Québec
930 ch Sainte-Foy, 6th Floor
Québec City (Québec)
Canada

INTRODUCTION

This chapter is not concerned with research, but is rather the observations of a member of the professional staff of the Direction de l'environnement at the ministère des Forêts, responsible for monitoring the environmental impact of forest management activities. They therefore represent the standpoint of a forest manager.

The ministère des Forêts needs knowledge and tools in order to carry out its role effectively, that is to protect forest ecosystems, develop them correctly, and ensure the sound management of resources. Research plays an essential part in acquiring this knowledge and designing these tools. The object of this presentation is thus to explain what Québec's forest managers hope to learn from research in light of the two major challenges of the 1990s: climatic change and the conservation of biodiversity.

The paper is divided into three sections. In the first, I will give an overview of Québec's forests, emphasizing their particularities from the biophysical, socio-economic and managerial perspectives. This background information is helpful in relating the issues of global change and biodiversity to the specific Québec context. In the second section, I will explain the approach taken by the ministère des Forêts with regard to the two themes of this NATO Advanced Research Workshop. I will also present the research priorities that have been identified, not in the form of an

NATO ASI Series, Vol. I 20
Biodiversity, Temperate Ecosystems, and Global Change
Edited by T. J. B. Boyle and C. E. B. Boyle

exhaustive list of projects, but rather as a description of the research areas that should be concentrated on in the short term. Lastly, I will conclude with remarks on the importance, in forest management, of both research and monitoring.

A PORTRAIT OF QUÉBEC'S FORESTS

Biophysical and socio-economic aspects

From north to south, Québec is split into three vegetation zones: tundra, taiga and forest. The forest zone is subdivided into three sub-zones: the boreal coniferous forest, the mixed forest and the hardwood forest. Québec possesses the most extensive woodlands in Canada (757,900 km2), which cover 46% of its territory. Almost 80% of Québec forests (in the "intensive inventory zone") are productive and accessible; they contain 75% softwoods, in reverse proportion to the planetary situation as a whole.

Natural disturbances have played and continue to play an important role in the shaping of the landscape and the regeneration of forest stands. The most striking examples in Québec are the forest fire cycle in spruce stands and the recurrent spruce budworm epidemics in fir stands. It is estimated that over a 17-year period of infestation, the spruce budworm has destroyed 235 million cubic metres of timber, roughly the equivalent of ten years' harvest, and fire destroys on average 42,000 hectares of forest each year.

Québec, like the rest of Canada, is characterized by a low population density (4.6 inhabitants per km^2), for the most part concentrated in the southern portion of its territory. The majority of Québec's population lives along the St. Lawrence valley, a zone with strong agricultural, urban and industrial potential, in which forest ecosystems have undergone major changes because of human activity. Further north, the coniferous forest, where forestry is mostly extensive, has not undergone as much pressure from the population.

The forest industry holds first place in Québec's manufacturing sector for the value of both output and exports. Wood products account for 16% of the value of deliveries, or $12.5 billion annually. Overall, the forest industry provides nearly

190,000 direct and indirect jobs, and the survival of more than one hundred towns and villages depends on it completely.

The wealth of Québec's plant and animal life and its forest landscapes also attract over a million vacationers and tourists every year. The leisure and tourism industry generates annual revenues of over $2.3 billion.

Forest management

In Québec, in contrast to several European countries and to many American states, 90% of the forest territory is state-owned, and the ministère des Forêts is the main forest manager. Québec's forests are thus a collective asset that is managed by means of acts and regulations defining the standards to be applied for managing, developing and conserving the forest environment, as well as for harvesting wood.

The Forest Act, passed in 1986, altered decades of forestry practices. It is based on two guiding principles: sustained yield and the multiple use of resources.

To reach its objectives, Québec has three basic tools:

(a) The public land-use plan. This plan is a legal instrument that fixes the intended socio-economic use of each portion of the public territory in order to protect, conserve and develop the resources it contains.

(b) The Regulation respecting standards of forest management for forests in the public domain. This regulation defines the sylvicultural practices, as well as protection and conservation measures to be adopted according to the intended use of each territorial unit. The standards are decided jointly with the two other government departments involved in forest resource management, the ministère de l'Environnement and the ministère du Loisir, de la Chasse et de la Pêche (that is the department in charge of leasure, fish and game).

(c) Timber supply and forest management agreements (TSFMAs). A TSFMA is an agreement between the government and a mill owner. The ministère des Forêts fixes production objectives (harvesting objectives) and gives the TSFMA holder responsibility for the harvesting, management and protection of resources. The contract covers a defined, but not exclusive, territory.

Long-term planning is one of the requirements imposed on TSFMA holders. They must present to the Government, for approval, a general management plan covering the contract's 25-year span, a five-year management plan and an annual intervention plan. These plans allow harvesting activities, the sylvicultural work indicated in the contract to obtain a sustained yield, and the measures needed to provide for overall protection of resources to be planned effectively. These plans are available to the public. Control and monitoring activities are also provided for to ensure both that sylvicultural prescriptions are respected and that they are improved, or that their impact is lessened, where necessary. Control and monitoring are also important elements for guiding applied research.

This new approach to forest management and development has, in recent years, given rise to two major projects : the integrated resource management project, and the proposed forest protection strategy. The latter, which went through public hearings in 1991, reflects the wish of the ministère des Forêts to use preventive rather than curative measures in the field of forest protection, while respecting the ecology of each forest environment.

QUÉBEC'S APPROACH TO DEALING WITH BIODIVERSITY AND GLOBAL CHANGE

Global climatic change and the conservation of biodiversity are among the new challenges facing the managers of Québec's forests. The Québec government has undertaken, by way of government orders, to abide by the agreements reached in Rio de Janeiro regarding these two points. We shall now see how the ministère des Forêts intends to deal with them from the management perspective and the principal areas in which research is needed.

Global climatic change

Given that the production of the manmade gases, especially CO_2, CFC and methane, which contribute to the greenhouse effect is increasing steadily, generalized global warming seems inevitable from the point of view of several world experts. It has been predicted that temperate ecosystems will be among the hardest hit and that the repercussions on the boreal forest will be considerable. However, much uncertainty remains as to the characteristics and amplitude of the changes anticipated, as well as the reliability of the simulation models used and the effects on the various ecosystems. In light of Québec's geographical situation and the importance of the forest sector in its economy, the ministère des Forêts follows the global climatic situation with particular interest.

Québec has recently acquired the tools necessary to collect information on environmental conditions and the evolution of natural forest ecosystems. The ministère des Forêts operates, alone or in collaboration with the ministère de l'Environnement, two forest monitoring networks: first, the forest ecosystem control network (French acronym RESEF), and secondly, the atmospheric pollution measurement network in the Québec forest and agricultural environments (French acronyms REMPAFAQ and RAMPAFAQ) (Figure 1).

The **RESEF** is made up of 27 stations found in protected zones, at a distance from local sources of pollution, throughout the Québec forest territory. The main objectives underlying the setting up of the network are : the long-term monitoring of the evolution of forest ecosystems, the acquisition of reference data for the realization of comparative studies, and, lastly, the evaluation of the role of various stress factors, of natural or human origin, on the forest environment. The data collected at these stations have, until now, mainly concerned the composition and evolution of the forest cover, the state of health of the forest, soil characteristics, and the history of disturbances.

The **REMPAFAQ and RAMPAFAQ** is made up of 18 stations located in rural and forest areas. Several stations have been intentionally placed near RESEF plots in order to better understand the characteristics of the local situation. All stations are equipped to measure the ozone concentration and the principal meteorogical variables, while some of them are also equipped to measure nitrogen oxides.

A considerable amount of data from these two networks are presently available, although much work remains to be done before their analysis is completed.

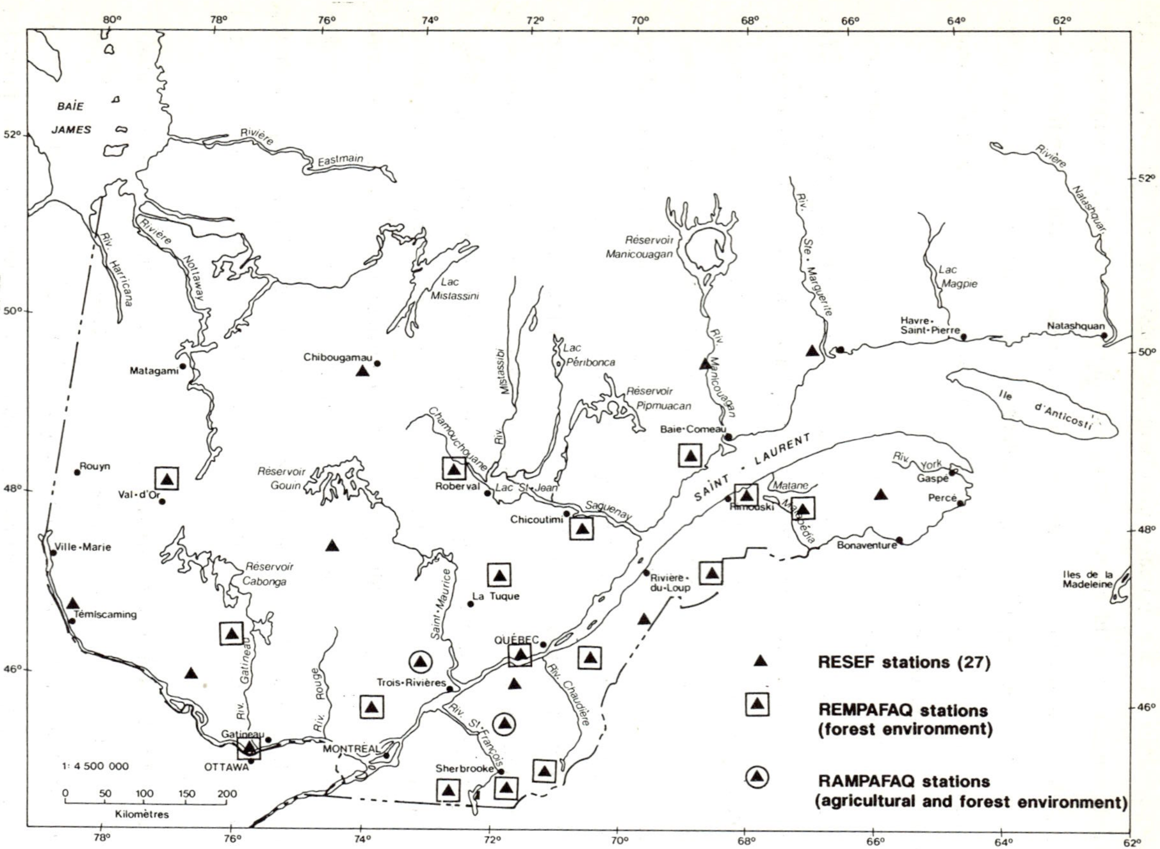

Figure 1. Forest monitoring networks in Québec

The data could eventually be made available to researchers, and the ministère des Forêts is open to joint research projects in areas over which it has competency.

The ministère des Forêts is currently defining its own agenda of research into the relations between Québec forest ecosystems and anticipated climatic changes. In view of present uncertainty about the carbon cycle and the way it is affected by forestry practices, it is expected that research work will revolve around the three following themes:

- inventories of the biomass contained in forest ecosystems, in the different Québec forest zones;
- the role played by the forest as source, sink and pool in the carbon cycle;
- the impact of present-day forestry policies and practices on the carbon cycle in our ecosystems.

The directions emerging from this workshop will also be taken into account when making the final selection of our activities, in order to unite our efforts with those of the international community.

It is to be hoped that, in parallel with this work, we will become more knowledgeable about forecasting the evolution of greenhouse gas concentrations; modeling the resulting climates at an adequate scale; and foreseeing the evolution of soils and forest ecosystems in the anticipated conditions. The alignment of our management practices will be based on an integration of all such areas of knowledge.

Conservation of biodiversity

Conserving biodiversity is the second challenge facing the ministère des Forêts. As a forest manager, the department must ensure that all forest management activities respect biodiversity. This responsibility is all the more important since we will eventually have to face climatic changes and, where possible, adopt practices which will improve the capacity of ecosystems to resist and adapt to such changes.

The nature of our ecosystems, the current way in which our resources are used and the legal framework of acts and regulations all contribute to making the preservation of the forest environment biodiversity less problematic in Québec than elsewhere in the world. The situation, however, varies from one area of the province to another. Thus, the majority of the vascular plant species liable to be designated threatened or vulnerable are found in the vicinity of the St. Lawrence river, while the environment in public-owned forests, generally found at some distance from highly urbanized zones, is presently little affected by the problem of endangered species.

Nevertheless, the possibility of human activity in the forest having an impact on biodiversity cannot be ruled out. The ministère des Forêts has therefore begun work on a report on the state of biodiversity in the forest environment. This study will allow us to evaluate, and eventually modify, the current management framework in view of the new concerns.

Biodiversity can be defined as "*the variety and variability among living organisms and the ecological complexes of which they occur*". Therefore, we will have to improve our knowledge of the composition, structure and function of ecosystems, at all levels of biological organization and on all spatial scales. Every manager of a natural environment concerned with conserving biodiversity must also translate theoretical considerations into concrete action and measurable objectives. In our view, the three main elements which should guide an approach to biodiversity conservation are:

- the protection of rare, vulnerable or threatened species;
- the setting aside of certain portions of the territory in a network of protected zones; and
- the adoption of management practices that encourage the maintenance of biodiversity in the forests under management.

FUTURE WORK

An upcoming report to be prepared by the ministère des Forêts will allow areas in which research is needed in relation to the conservation of biodiversity to be specified. At this stage, we foresee the broad areas covered as being the three following ones :

A) Population dynamics

It is important to improve our understanding of the mechanisms governing population dynamics, so as to better evaluate the situation of the species found in Québec's forests and their ability to adapt to various disturbances, natural and manmade, such as climatic changes, the fragmentation of habitat, etc.

A further step will be to ascertain how such knowledge can be translated into our management practices. Prominent among the questions with which we are currently concerned are:

- How should we take certain concepts such as metapopulation, source population, sink population, and viable population into account in our management decisions?
- How should the notion of genetic diversity be integrated into our management decisions?

B) The measurement of the effectiveness of the protected zones network

We need to have a better understanding of the criteria which will allow the effectiveness of the protected zones network to be evaluated. In our view, such evaluation must involve not only the question of the percentage of territory protected, but also a set of variables including the representativeness, configuration and degree of interconnection of protected natural spaces, as well as their relationships with adjacent territories. This approach will enable us to make a more effective choice of sites for the establishment of parks and reserves, and to better coordinate the development of adjacent territories.

C) Management practices that encourage the maintenance of biodiversity

It is important to ensure that semi-natural ecosystems are developed in a way that allows their fundamental ecological characteristics to be perpetuated. Therefore, the impact of natural disturbances must be better understood and compared with that of forest management activities. Analysis must be carried out both at the level of forest stands (alpha-diversity) and landscape (beta-diversity). It is particularly important to broaden our knowledge of the way in which ecosystems behave on the latter scale since many forest management decisions are concerned with landscapes.

CONCLUSION

It is clear that much more needs to be known about global climatic change, biodiversity and forest management. Since it will take considerable time to obtain results in several cases, research must be directed, at once, towards the resolution of concrete problems. However, the importance of monitoring results as a complement to fundamental research must not be neglected if we are to improve our knowledge of the effects of global change and forestry practices on forest ecosystems. The ministère des Forêts places great emphasis on monitoring to validate and improve its management practices, an approach which requires the use of adequate change indicators. In a context of adaptive management, attention will have to be paid in the short term to the development of such indicators.

Chapter 21: Research Needs and Opportunities: the Response of Forest Biodiversity to Global Change

Robert C. Szaro
USDA Forest Service
Research - 1NW
P.O. Box 96090
Washington, DC 20090-6090
U.S.A.

Biodiversity has become a major global environmental issue over the last 20 years, during a period of growing recognition that forests and climate are inextricably linked at global, regional, and local scales and play a vital role in providing environmental and economic benefits to the world's people (World Conservation Union *et al.* 1991, Keystone Center 1991, Lubchenco *et al.* 1991, Peters and Lovejoy 1992, Reid *et al.* 1992). Global environmental change, biodiversity, and forest ecosystems are inextricably linked. As a consequence, sustaining global forest productivity, health, and biodiversity was a central theme in many of the discussions at the United Nations Conference on the Environment and Development held in 1992. The challenge for nations, government agencies, organizations, and individuals is to protect and enhance biodiversity in an unstable environment while at the same time continuing to meet people's needs for natural resources. The diversity of life constitutes the foundation for all human resources, including the future richness of life for our children for future generations.

We are currently uncertain even of the magnitude of that foundation resource. Estimates of the number of species of plants, animals and microorganisms ranges from 5 to 80 million, with a growing consensus that the actual value is around 5 million (May 1992). To date, only a small fraction (approximately 1.4 million species) has

NATO ASI Series, Vol. I 20
Biodiversity, Temperate Ecosystems, and Global Change
Edited by T. J. B. Boyle and C. E. B. Boyle

been scientifically described and catalogued. Species appear to be disappearing at an unnatural rate with estimates of the current global extinction rate as high as 17,500 species per year (Wilson 1989), compared to the natural "background" extinction rate over the past 600 million years of approximately one species per year (Raup and Sepkoski, 1984). However, unlike previous periods of massive extinctions caused by natural catastrophe, this time the extinctions are the result of undesired human influences. Most species extinctions are due to the degradation, destruction or alteration of habitats, ecological communities and even whole ecosystems. The U.S. Council on Environmental Quality (1991) noted that ignoring the long-term health of ecosystems will prove disastrous, since, beyond the value of any individual species they contain, healthy ecosystems collectively provide life support services for the planet.

The loss of biodiversity is a problem not just because of what we know about it, but also because of what we do not know (Wilson and Peter 1988). It is mainly the lesser-known (or unknown) species, remote habitats, and unstudied ecosystems that are being eliminated and with their extinction the potential of these species is lost before it is ever investigated. Moreover, the scientific community has just begun to realize the diversity of life in the oceans (Grassle *et al.* 1991). For example, in the 1970's scientists described new marine life forms in hydrothermal vent communities, including 20 new families, 50 new genera, and 100 new species. There may be as many as a million undescribed species inhabiting the deep sea. With each new discovery and investigation there exists the potential for significant advancement in basic and applied sciences, including food and nutrition, biotechnology, medicines, raw materials, and understanding global climate.

We will never be able to demonstrate an immediate utilitarian reason for preserving every species and habitat and thus should proceed with extreme caution (Nations, 1988). Who can tell us which level of extinction will seriously disrupt the web of life that we depend on as human beings? Or, as Norton (1988) put it, we do not know the "disaster point" to which biodiversity can be reduced without irreversible harm to the biosphere, or perhaps more precisely, to ourselves.

Land-managers around the globe are faced with unprecedented uncertainty in assessing future resource conditions and with unprecedented risk of catastrophic change. However, before rational management decisions can be made we need additional information on how the biodiversity of forest ecosystems will respond to global change with particular focus on answers that will allow for the long-term productivity, sustainability, resilience, and diversity of our environment. Although we have an extensive understanding of biophysics, hydrology, climate, and ecology, our knowledge base leaves considerable uncertainty regarding the magnitude and timing of integrated ecological effects.

RESEARCH NEEDS AND OPPORTUNITIES

Palæoecological evidence clearly links changes in biodiversity with changes in global climate. Current evidence clearly links changes in biodiversity with other global changes, such as deforestation, which feed-back into global climate change through alteration of the carbon cycle. We simply cannot study biodiversity in isolation from global change, nor can we study global change while ignoring biodiversity impacts. Research on forest biodiversity and global change should be designed to provide a sound scientific basis for making regional, national, and international management and policy decisions regarding forest ecosystems in the context of global change challenges (USDA Forest Service 1992). Predicted changes in ecosystems include redistribution of plant and animal species, altered hydrologic regimes, and changes in soil formation processes. Climate change could have drastic effects on forests as well as change the boundaries of the world's biomes (McNeeley *et al.* 1990). These changes may alter ecosystem processes, which in turn, may alter evapotranspiration and albedo, hence influencing climate (Committee on Earth and Environmental Sciences 1990). How these complex positive and negative feedback mechanisms might affect global climate and biodiversity and what roles human-influenced processes might play are not well characterized. Understanding these feedback mechanisms is crucial in determining national and international policy responses to global change (Committee on Earth and Environmental Sciences 1990).

Fundamental Research Questions

The USDA Forest Service's Global Change Research Program recommended that research should provide the scientific basis to address three broad questions concerning global change and forest ecosystems (USDA Forest Service 1992):

1. What processes in forest ecosystems are sensitive to physical and chemical changes in the atmosphere?
2. How will future physical and chemical climate changes influence the structure, function, and productivity of forest and related ecosystems, and to what extent will forest ecosystems change in response to atmospheric changes?
3. What are the implications for forest management and how must forest management activities be altered to sustain forest productivity, health, and diversity?

These three questions can be restated in policy relevant terms: 1) Is there are a problem? 2) How serious is the problem? and 3) What can be done about the problem?

Generic Research Problems

There are a number of biological, physical, social, and other concepts that should be recognized in any scientific endeavour examining the interrelationship between biodiversity and global change. These might be termed generic or fundamental research problems, several of which are particularly important to consider in setting research priorities (Lugo, personal communication):

Complexity: the extreme biodiversity, species richness, or heterogeneity that characterizes forests in general, particularly those in the tropics.

Resiliency: forests, particularly tropical ecosystems, are often characterized as fragile, but they have remarkable ability to recover from some types of disturbance.

Periodicity: of climate, disturbance, and forest response to these.

Productivity: total site productivity, not just of wood products or other market-valued outputs.

Soil fertility: especially its variability, and with emphasis on nutrient cycling.

Sustainability: implicit in the term is the assumption that the forest is a renewable resource but it remains a vague but highly prized characteristic that is highly dependent on the scale at which it is assessed.

Invadability: how resistant or vulnerable to intrusion or disruption by exotic species are forests whether managed or otherwise?

Valuation: how to estimate the worth or contribution of forests, especially of non-market benefits.

Extinction: what is the relationship between percentage of land cleared and percentage of species lost, and how does high complexity change this relationship?

Inventory

Implementing biodiversity conservation goals and assessing the impacts of global change will require resources and knowledge. The role of science in the conservation of biodiversity during a period of rapid environmental change is critical. More research to improve methodologies, distributional and status information, and strategies based on sound information will ultimately provide the basis for all sound policy and management decisions. Current scientific understanding of ecological processes is far from perfect. Better inventories and assessments are needed of current conditions, abundances, distributions, and management direction for genetic resources, species populations, biological communities, and ecological systems. Even in the United States, with the intensity of scientific effort focused on natural resources, it is extremely difficult to assess the overall status of biodiversity. Rarely have studies been done examining the all vascular plants and vertebrates and their relationships in any given ecosystem, let alone the thousands of other species found within it. Particularly as Maini (1992) pointed out that "*forests are a rich repository of planet earths' genetic heritage*". Forests are usually delineated by the presence of a few dominant species but this barely touches the surface of their species richness. For example, in two pine systems in the southeastern United States, tree species make up less than 10% of the plant and vertebrate animal species. And this does not take into account all the thousands of other species likely to be found. May (1992) states "*If we speak of total number of species, then to a good approximation everything is a terrestrial insect*". In fact, if one uses an estimate of 3-5 million as total species on earth then vascular plants (5.4 - 13.3 %) and vertebrates (0.9 - 1.5%) only represent 6.3 to 14.8%. This illustrates the extent of the problem as the traditional approaches to forest management have not incorporated a consideration of the vast majority of species (Erwin 1988). Even in an area the size and geographic scope of a typical National Forest in the United States, native biodiversity can easily encompass thousands of species of plants and animals, dozens to hundreds of identifiable biological communities, and an incomprehensible number of pathways, processes, and cycles through which all that life is interconnected. Obviously, it is not possible to address each and every aspect of this complexity. Therefore, identification is of specific aspects of diversity, such as distinct species, biological communities, or ecological processes that warrant special consideration (Salwasser 1990).

Biodiversity at the species, community and ecosystem levels can be assessed by surveys (analyses of the geographic distribution of the biota) and inventories (catalogues of the elements of biodiversity present at a site or in a region). Surveys and inventories are critical in helping determine which areas are in greatest need of preservation or management, and in identifying threatened species and habitats. A wide range of such activities needs to be supported, including:

- Rapid Assessment Programmes, which give a "snapshot" of species richness in an area;
- Surveys, to map out the distribution of the Earth's ecosystems. Such analyses would be greatly aided by the development of a common and consistent scheme to classify communities and ecosystems. There is currently no widely accepted vegetation/ecological classification system;
- Broad-based inventory efforts, focused on poorly known habitats (soil communities, tropical and marine systems) that take advantage of existing protected sites, such as UNESCO-MAB's Biosphere Reserves and analogous national networks, wherever possible. However, degraded and multiple-use habitats should also be included, as well as better-understood taxa, such as

vertebrates and vascular plants, that can be used as "benchmarks" for habitat quality; and

- Intensive inventories, which attempt to determine all of the species present, from microbes to vertebrates, should be undertaken at two or three sites around the world.

Such Analyses will require intense effort and a long time frame (a decade or more), but they are key to answering such questions as how biodiversity at the species level affects diversity at the ecosystem and community levels) (Solbrig 1991).

It is a challenging task to formulate an integrated, concise and relevant approach to inventory. However, there are some recommended broad principles outlined in the Keystone Biodiversity Report (1991) on the components of an inventory programme. These include:

- The inventory should be hierarchical and "top-down" in the sense that landscape level assessments such as the U.S. Fish and Wildlife Service's "gap analysis" (Scott *et al.* 1993) are used to identify priorities for inventory at the local level, and local assessments are used to identify priorities at the site level.
- The inventory should make maximum use of existing data management systems.
- The inventory should be landscape based in the sense that abundance and distribution of plant and animal species are correlated with soils, vegetation, plant and animal community characteristics, and landscape features.
- The inventory at a minimum should include natural vegetation, all vertebrate and vascular plant species and at least some indicator species of non-vascular plants and invertebrates, and some indicators of other elements of biological diversity, such as sensitive communities or human-influenced processes and elements of structural diversity.
- Provision should be made for systematic inventories of all candidate, threatened, endangered, and sensitive species and for all other elements that are imperiled due to human activities or natural events.

- Inventories should be guided by a master plan that coordinates acquisition of aerial photography, soil survey, vegetation survey, and vertebrate inventory that ensures compatibility of data within and among agencies.
- The inventory should be compatible with, and feed information directly into, development and implementation of Geographic Information System (GIS) methodology, monitoring programmes, and research activities.
- The inventory should provide the basis for determining species (including genetic level assessment), species groups, population guilds, habitats, landscapes or processes that require more intensive studies.
- The inventory process should identify levels or intensities of inventory that are appropriate for each level of planning, type of management activity or impact, type of land classification or degree of rarity or sensitivity of the element being inventoried.
- The inventory should have a strong element of quality control and assurance, including setting specific standards of accuracy and precision, timing the inventory to encompass the life-cycles of the target elements, standardizing methods and databases to the extent possible, and using trained personnel to conduct the inventories.

The need for more specific data and more efficient ways for collecting and managing data will lead to significant changes in inventory processes including the use of methods and technology that will: (1) provide resource estimates for specific geographic units and evaluate the reliability of such estimates; (2) display estimates and units spatially; (3) make maximum use of existing information and new technology, such as remote sensing and geographic information systems; (4) provide a baseline for monitoring changes in the extent and condition of the resource; (5) eliminate redundant data collection, develop common terminology, and promote data sharing through corporate data bases; (6) utilize information management systems to provide maximum flexibility for data integration, manipulation, sharing, and responding to routine and special requests; and (7) provide up-to-date data bases using modelling techniques, accounting procedures, and re-inventories.

Monitoring

Present concepts of monitoring vary depending on who is expressing them, their background, and the objectives of the monitoring being considered. There is a need for greater coordination, with considerable direction and standardization set at both national and regional levels. Monitoring means different things to different people. Just what it is depends solely on monitoring objectives. This should emphasize the need to exercise great care in defining the objectives for monitoring. Monitoring should provide sufficient information about the abundance of animals or plants targeted for monitoring to assure that current management practices are not threatening the long-term viability of their populations (Verner 1986). But viability concerns have added an additional layer of complexity to monitoring programmes particularly when we try to derive a number for minimum viable population size. This theoretical concept espoused by Michael Soulé (1987) is useful from the standpoint that there probably is some minimum size population threshold that when crossed will lead to the demise of a species' population. The level of this threshold clearly varies with the species concerned, for passenger pigeons it probably was in the hundreds of thousands while for desert pupfish the number may be as low as 200. But as useful a concept as this may seem, in the real world it is close to impossible to determine minimum viable population size with any degree of certainty. There are too many variables involved; demographic variables such as immigration, emigration, recruitment, birth rates, survivorship, dispersal mechanisms, etc., catastrophic events, habitat loss and changing climatic conditions. It is difficult to envision that we can determine minimum viable population size for every species. Perhaps a more subjective approach, *i.e.* that larger numbers of a species are obviously better than fewer numbers of the same species, should be considered. Maybe we should try to maintain species with as large a population size as practical.

Monitoring efforts are often severely hampered by the lack of prior planning and thought given to the desired results from any given monitoring effort. It is not enough to select a management indicator species, guild, or other monitoring target with the idea that this will allow us to assess the impact of any given management activity (Szaro and Balda 1982, Szaro 1986). Ideally, the results from monitoring should feed-back into the system to correct or fine tune management activities.

Effort should not be wasted on monitoring systems that fail to give the level of confidence needed to deduce the most likely effects of management activities on wildlife, fish or sensitive plant resources. In an era when mankind's activities are the dominant force influencing biological communities, proper management requires understanding of pattern and process in biological systems and the development of assessment and evaluation procedures that assure protection of biological resources (Goldsmith 1991, Karr 1987). It is essential that appraisals of these resources give us the ability to forecast the consequences of human-induced environmental changes accurately. But we have a long way to go in this process.

Monitoring should also have a strong element of quality control and assurance, including setting specific levels of accuracy and precision, timing the inventory to encompass the life-cycles of the target species, as much as possible, and standardizing methods and databases for all Forest Service organizational units, especially when monitoring the same species. However, whatever is done must be as cost-effective as possible. Some possibilities are risk analysis, increasing the scope of monitoring efforts, and determining the needs of monitoring objectives. We might want to limit direct monitoring only on high risk species, on a priority basis, while relying on habitat relationships for most other species. Monitoring efforts should be spread over as large a geographical base as possible (and feasible) to reduce the cost per forest and to increase the scope of applicability. Whenever possible, monitoring should only be asked to detect **declining trends** because of the potential cost savings (almost 90%) (Verner 1986).

Along with this is the need to develop a quality control and assurance programme that ensures: (1) objectives are measurable (and thus capable of being monitored); (2) appropriate measurement techniques and procedures are being used; and (3) management thresholds are clearly identified and incorporated into the planning process so that, if crossed, they automatically trigger a reanalysis of the planned activities.

The advent of geographic information systems holds future promise for the development of a comprehensive biological information system (Davis *et al.* 1990). Much work and coordination will be involved to bring this to pass including building in mechanisms to check and update continually the information in the database. It is not enough to set up the system and then use it without regard to the dynamics of ecological systems.

Ecosystem Disturbance Research

Ecosystems by their very nature are in a constant state of flux, always shifting and changing from one condition to the next, and a true steady state probably never exists. Reproductive strategies and other ecological characteristics of many ecosystems are strikingly adjusted to disturbance regimes resulting from catastrophes such as fire, insect outbreaks, and flooding. Global climate projections suggest that drought cycles, precipitation patterns, temperature extremes, strong winds, and intense storms may change the future. Changes include redefining regional vegetation patterns that are affected by such factors as global carbon dioxide build-up, regional effects such as El Niño and monsoon systems, and local effects, often involving fire management (McNeeley *et al.* 1990). Large-scale disturbances such as the Yellowstone fires of 1988 and the gypsy moth epidemic in northeastern North America may become even more prevalent with changes in climatic factors driving both the occurrence and severity of fire, insects, and disease episodes. This in turn may drive dramatic changes in ecosystem composition, structure, and function as differing mixes of species respond to altered climatic regimes. We should expect variable responses by resident plant and animal species and that these responses are likely to be highly unpredictable (McNeeley *et al.* 1990).

Earth is a planet that is affected dramatically by fire. Mankind has both feared and used fire since the earliest days of civilization. Forest and savannah ecosystems have been modified by fire worldwide. Today forest and savannah fires are used as a principle resource management tool by many indigenous people. Yet forest and savannah fires also cause economic and environmental resource loss. Wildfires destroy people's housing and wildlife habitats. They are a principle factor in deforestation, land conversion, and greenhouse gas emissions. Further research is needed on the role of forest fires as an integral component of many forest ecosystems needed to sustain those ecosystems. Results need to reduce the potential for destructive wildfires and practices that contribute to excessive biomass burning in order to lessen economic impacts and greenhouse gas emissions. Fire research needs include:

- Determining the changes in frequency and severity of fire weather resulting from global atmospheric changes;
- Modelling the processes by which fire affects the current equilibrium between climate and forest ecosystems;
- Predicting the net result of new fire regimes on the production of trace gases and particulate matter in the atmosphere;
- Assessing the contributions of biomass burning in tropical, temperate and boreal forests to production of greenhouse gases;
- Assessing the historic relationships between fire and land use practices and how those historic precedents suggest future scenarios;
- Determining how human activities influence global biogeochemical cycles; and
- Developing management practices that emulate natural disturbances in terms of frequency, severity, and spatial distribution and determining their relationships to biodiversity, ecosystem health, and sustained productivity.

Other major contributors to a decline in forest biodiversity are pests, especially new and/or introduced pests. One needs only to look at the effects of past epidemics in North America such as chestnut blight, dutch elm disease, larch case bearer, and white pine blister rust, and follow the progress of current epidemics of gypsy moth, beech bark disease, butternut canker, and dogwood anthracnose to see that the character and biodiversity in native North America forests have and are continuing to change in response to these pests. Nor are the effects of these pests limited to the trees and shrubs they attack, for many of these trees provide food and shelter to the areas' wildlife. Examples include the dependency of black bear fecundity and cub survival in northeastern North America on beech mast which has been significantly reduced by beech bark disease and the impact of introduced weed pests which threaten to swamp out most of the native birds, trees, and other plants of many native Hawaiian forests.

These pests were not introduced by forest managers, but by and large, are the result of introductions through increased travel and commerce in the world today. Climate change has the potential to greatly magnify the impact and alter the distribution of many of these pests. Insect and disease research needs include:

- Determining how climate change influences the severity and frequency of insect and disease outbreaks;
- Determining the importance of such outbreaks on ecosystem stability and change;
- Predicting how host species stressed by climate and air pollution are further impacted by insects and diseases;
- Determining the resistance or vulnerability to intrusion, disruption or invasion by exotic species of forests, rangelands, and associated aquatic ecosystems whether managed or not;
- Learning to control exotic pests by developing methods that do not disrupt the normal structure and function of ecosystems;
- Developing methods to assess and predict the impacts of insects, disease, and environmental stresses on individual trees, stands, and landscape patterns;
- Understanding the role of native insects and diseases as keystone species that regulate the structure and function of forest ecosystems.

Ecosystem Function and Dynamics Research

The dynamics of many forest plant communities present a complex set of interacting variables affected by both time and environment. Superimposed on the equilibrium fluctuating between the physical environment and plant species is the animal species component, which utilizes these habitats as breeding and wintering habitat, migratory corridors, and as areas for brief stopovers. Change in climate that affects forest ecosystems and their biodiversity has been the norm and not the exception but the effects of that change on forest biodiversity differs with the size and extent of the remaining remnant forests (Botkin and Nisbet 1992). The issue is not the change *per se*, but the rate of change combined with the occurrence of these changes on an increasingly fragmented landscape that prevents migration and dispersal in response to these changes (Botkin and Nisbet 1992). Many species will not be able to redistribute themselves fast enough to keep up with the projected changes, potentially resulting in considerable alteration in ecosystem function and structure (Reid *et al.* 1992). Those first and most greatly affected will probably be those with highly specialized habitat requirements of limited geographic distribution rather than those species with broader habitat tolerances and with the potential to migrate (Botkin and Nisbet 1992). Where species occur at their distributional limits, in transition zones, small changes in climate are likely to promote disproportionately large responses in plant species (Melillo *et al.* 1990). Those biological processes that are most sensitive to climatic change and, further, that regulate biogeochemical cycling, such as production and decomposition, must be considered as key indicators for monitoring and predicting the impacts of global change on species and ecosystem distribution and survival (Mooney 1991). Ecological understanding and knowledge are urgently needed to plan for the management of sustainable ecological systems, to determine the influence of both natural and human-induced perturbations on ecosystem sustainability, to evaluate consequences of a wide range of human activities, and to detect and monitor environmental changes.

Research that determines the structure and function of ecosystems and their key components must be accelerated. In terrestrial ecosystems, efforts should be expanded to link the behaviour of below-ground components and processes to above-ground data. This is especially important for understanding how carbon allocation to plant shoots and roots is altered by environmental conditions. The focus should be on identifying ecosystem controls and defining the mechanisms and feedback loops that constrain and stabilize an ecosystem. Currently, there is considerable uncertainty in many ecosystems as to which parameters are relevant indicators of ecosystem condition. Ecosystem function and dynamics research should focus on:

- Understanding basic ecosystem functions and processes, such as elemental cycling and responses to a changing physical and chemical environment and the cumulative effects of management actions on the ecology and viability of populations and communities;
- Determination of plant and animal species composition and the critical characteristics that ensure viable functioning of each component in the whole system;
- Understanding the role of soil and hydrologic processes and how they affect forest ecosystems;
- Determination of patterns and indicators of ecological systems responding to stress;
- Understanding the role of biodiversity in controlling ecological processes and the complex suite of ecological processes that shape patterns of diversity;
- Improving our understanding of the ecologic, social, and economic implications of the landscape patterns created by past and present resource management - especially forest fragmentation, edge effects, and linkages among landscape ecosystems;
- Development, demonstration, and enhancement of forest based technologies that are economically and environmentally sound while reducing further growth in emissions of greenhouse gases, and increasing sinks for these gases;
- Understanding the effects of both spatial and temporal aspects of habitat patterns, including fragmentation, on species interactions, community structure, and ecosystem processes;
- Understanding the response of species, ecological communities and natural and managed ecosystems to elevated levels of carbon dioxide, climate change, and physical/ chemical stresses; and
- Evaluating the sustainability of management systems and land and resource use in temperate and tropical ecosystems, considering social needs and ecosystem limitations.

CONCLUSIONS

Forests are an integral part of the Earth's life support systems by playing a crucial role in regulating the atmosphere and climate (World Conservation Union *et al.* 1991). Consequently, the need for scientific information as a foundation for resource management decisions continues to increase dramatically. Even a decision not to manage is essentially a management decision. The need to assemble and format new and existing research results into packages that can be used by managers is great. As such, efforts to synthesize science, and in so doing identify information voids and set priorities, should be increased. Long-term observations of biological systems are invaluable in providing information on their response to change and in providing the understanding needed for predictions regarding the future (Mooney 1991). Research on computer-based decision support models and expert systems hold promise of providing an effective interface between management and research. Ideally we want to be able to predict what kinds of organisms will be present under a given climatic scenario and how these organisms will interact with their new environment (Mooney 1991).

Policy decisions need to balance the needs of society with the best scientific knowledge. To facilitate this, the interface between social and ecological models must be improved. An essential task is to improve the ability to quantify social demands for both consumptive and non-consumptive goods. These demands must then be weighed against the need to maintain ecosystems and their attributes. To provide the scientific basis for ecosystem management requires a combination of (1) basic long-term scientific inquiry; (2) issue-focused, problem solving interdisciplinary research and technology development; (3) scientific support to the ongoing management of lands and resources, such as through ecological classifications, inventories, ecological analyses, and monitoring and evaluation; and (4) integrating science into the policy-making process.

REFERENCES

Botkin DB, Nisbet RA (1992) Projecting the effects of climate change on biological diversity in forests. Pages 277-293 *in* Peters RL, Lovejoy TE (ed.) Global warming and biological diversity. Yale University Press, New Haven

Committee on Earth and Environmental Sciences (1990) Our changing planet: the FY 1991 research plan of the U.S. global change research program. Washington DC

Davis FW, Stoms DM, Estes JE, Scepan J, Scott JM (1990) An Information Systems Approach to the Preservation of Biological Diversity. Intntl. J. Geogr. Information Systems 4: 55-78

Erwin TL (1988) The Tropical Forest Canopy: the Heart of Biotic Diversity. Pages 123-129 *in* Wilson EO, Peter FM (eds) Biodiversity. National Academy Press, Washington, DC

Goldsmith FB (ed) (1991) Monitoring for Conservation and Ecology. Chapman and Hall, New York

Grassle JF, Lasserre P, McIntyre AD, Ray G (1991) Marine Biodiversity and ecosystem function. Biol. Int. Special Issue 23: i-iv, 1

Karr JR (1987) Biological Monitoring and Environmental Assessment: A Conceptual Framework. Environmental Management 11: 249-256

Keystone Center (1991) Final Consensus Report of the Keystone Policy Dialogue on Biological Diversity on Federal Lands. The Keystone Center, Keystone, CO

Lubchenco J, Olson AM, Brubaker LB, Carpenter SR, Holland MM, Hubbell SP, Levin SA, Macmahon JA, Matson PA, Melillo JM, Mooney HA, Peterson CH, Pulliam HR, Real LA, Regal PJ, Risser PG (1991) The sustainable biosphere initiative: An ecological research agenda. Ecology 72: 371-412

Maini JS (1992) Sustainable Development of Forests. Unasylva 169, Vol. 43, 3-8

May RM (1992) Past Efforts and Future Prospects Towards Understanding How Many Species There Are. *in* Solbrig OT, van Emden HM, van Oordt PGWJ (eds) Biodiversity and Global Change, International Union of Biol. Sci., Paris, Monograph No. 8

McNeely JA, Miller KR, Reid WV, Mittermeier RA, Werner TB (1990) Conserving the world's biological diversity. IUCN, Gland, Switzerland; WRI, CI, WWF-US, and the World Bank, Washington, DC

Melillo JM, Callaghan TV, Woodward FI, Salati E, Sinha SK (1990) Effects on ecosystems. *in* Houghton JT, Jenkins GJ, Ephraums JJ (eds) Climate change: the IPCC Scientific Assessment, Cambridge University Press, Cambridge

Mooney HA (1991) Biological response to climate change: an agenda for research. Ecological Applications 1: 112-117

Nations JD (1988) Deep ecology meets the developing world. Pages 79-82 *in* Wilson EO, Peter FM (eds) Biodiversity. National Academy Press, Washington, DC

Norton B (1988) Commodity, amenity, and morality: The limits of quantification in valuing biodiversity. Pages 200-205 *in* Wilson EO, Peter FM (eds) Biodiversity. National Academy Press, Washington, DC

Peters RL, Lovejoy TE (eds) (1992) Global warming and biological diversity. Yale University Press, New Haven

Raup DM, Sepkoski JJ (1984) Periodicity of extinction in the geologic past. Proc. Natl. Acad. Sci. USA 91: 801-805

Reid W, Barber C, Miller K (1992) Global Biodiversity Strategy: Guidelines for action to save, study, and use Earth's biotic wealth sustainably and equitably. World Resources Institute, Washington, D.C.; The World Conservation Union (IUCN), Gland, Switzerland; United Nations Environment Programme (UNEP), Nairobi, Kenya

Salwasser H (1990) Conserving Biological Diversity: A Perspective on Scope and Approaches. Forest Ecology and Management 35: 79-90

Scott JM, Davis F, Csuti B, Noss R, Butterfield B, Groves C, Anderson H, Caicco S, D'Erchia F, Edwards TC jr, Ulliman, J, Wright RG (1993) Gap Analysis: protecting biodiversity using geographic information systems. Wildlife Monograph 123: 1-141

Solbrig OT (ed) (1991) From Genes to Ecosystems: A Research Agenda for Biodiversity. International Union of Biological Scientists, Cambridge, MA

Soule ME (1987) Viable Populations for Conservation, Cambridge University Press, NY

Szaro RC (1986) Guild Management: An Evaluation of Avian Guilds as a Predictive Tool. Environmental Management 10: 681-688

Szaro RC, Balda RP (1982) Selection and Monitoring of Avian Indicator Species: An example from a Ponerosa Pine Forest in the Southwest. USDA Forest Service General Technical Report RM-89, Rocky Mountain Forest and Range Experiment Station, Fort Collins, CO

United States Council on Environmental Quality (1991) Environmental Quality: Annual Report: 135-150

USDA Forest Service (1992) Forest Service Global Change Research Program. USDA Forest Service Washington DC

Verner J (1986) Future Trends in Management of Nongame Wildlife: A Researchers Viewpoint. Pages 149-171 *in* Hale JB, Best LB, Clawson RL (eds) Management of Nongame Birds in the Midwest-- A Developing Art, North Central Section, Wildlife Society, Madison, WI

Wilson EO (1989) Threats to biodiversity. Scientific American 261: 108-116.

Wilson EO, Peter EM (eds) (1988) Biodiversity. National Academy Press, Washington, DC

World Conservation Union, United Nations Environment Programme, World-Wide Fund for Nature (1991) Caring for the Earth. A Strategy for Sustainable Living. The World Conservation Union (IUCN), Gland, Switzerland; United Nations Environment Programme (UNEP); World Wide Fund for Nature (WWF)

Chapter 22: Biodiversity and Climate Change: Identifying an Economic Research Agenda

Donald G. Hodges and James L. Regens[1]
Department of Forestry
Mississippi State University
Box 9681
Mississippi State, Mississippi 39762-9681
USA

INTRODUCTION

During the past two decades, environmental problems have expanded in terms of both their geographical scope as well as complexity. As a result, scientific research and policy deliberations have shifted from an emphasis on predominately localized, single-resource problems such as ambient air or water quality within relatively discrete spatial domains to consideration of a myriad of interrelated problems at a regional or even global scale. This raises challenges for developing integrated environmental research programmes and, ultimately, linking insights from scientific findings to policy- making.

Climatic change and biological diversity are typical of such issues. Successfully addressing these problems requires not only increasing our knowledge about biological and physical processes but also involves learning how to value remedial actions and, more importantly, how to make inherently difficult policy decisions under conditions of substantial biophysical uncertainty. Developing effective policies to address climate change and biodiversity will depend on providing decisions makers with some idea of the value of biodiversity. This value could be in monetary terms, or could be provided

---

[1] Center for Bio-environmental Research, Tulane University (SL-1), 1430 Tulane Avenue, New Orleans, Louisiana 70112, USA

NATO ASI Series, Vol. I 20
Biodiversity, Temperate Ecosystems, and Global Change
Edited by T. J. B. Boyle and C. E. B. Boyle

just as effectively in terms of social costs and benefits. We identify some key elements of an economic research agenda for valuing biodiversity and the impact due to climate change based on a review of existing economic knowledge pertaining to climate change and biodiversity.

Biological diversity has become a salient concern since 1987 for several reasons. Massive deforestation combined with ecologically disruptive agricultural practices in developing countries have triggered international concern about diminished biodiversity in tropical areas. Similarly, human activities associated with industrial activity, timber production, housing development, and energy generation have fostered concern about endangered species and critical habitats in temperate ecosystems.

Despite an increase in the amount of scientific research over a relatively short time horizon, the current biophysical knowledge base vis-à-vis biodiversity has a large degree of uncertainty in a number of key areas. These gaps can be readily identified through a quick survey of the literature. For example, baseline inventories are incomplete in terms of the number of species or species characterization, especially on a landscape ecology or larger spatial scales. Wilson (1988), for instance, noted that estimates of the current number of living species ranges from 5 to 30 million, with only 1.4 million species actually being described. And, although most biological scientists agree that the magnitude of species extinction has increased dramatically as the level of human activity has increased, estimates of the aggregate extinction rate vary from 10,000 to 20,000 per year and are likely to fluctuate spatially. Finally, time series data that might permit truly longitudinal rather than static analysis or modelling are extremely rare, particularly for entire ecosystems as opposed to single species with specific habitats. Such limitations constrain the degree to which research findings from the biological or physical sciences can be generalized.

Moreover, as noted above, biophysical stresses are not the only environmental factor adversely affecting biodiversity. It is worth noting the extent to which anthropogenic factors emerge as potentially key causal agents. Flint (1992) identified four categories of anthropogenic factors that have contributed to increased rate of biodiversity loss in the past two decades: Development Pressure; Market Failure; Intervention Failure; and Immediate Causes.

Development pressures include not only the impact of the rising global population, but it also entails the intensification of agriculture in response to the population growth. The intensification has negatively affected diversity by emphasizing a small number of plant varieties for food production. In fact, three species - wheat, corn, and rice - account for 54 percent of the developing countries caloric consumption (CGIAR 1987 in Flint 1992). This increased reliance on fewer species enhances the likelihood of insect and disease outbreaks or impacts from climatic events.

Market failure has contributed to biodiversity losses because of the problems associated with any non-market goods, which include externalities, market imperfections, and uncertainty. The primary market failure, however, is that many of the goods and services provided by biodiversity are public goods in which consumption is non-rival and non-exclusive. As a consequence, little incentive exists for conserving the public good.

Intervention failure involves both ineffective positive intervention, such as failing to effectively implement conservation policies, and unintentional negative intervention which consists of the unintended consequences of policies designed to address other public problems.

Immediate causes include such factors as human-induced habitat loss and fragmentation, pollution, and climate change. While the impact on biodiversity of the first two are readily demonstrated, identifying the actual biophysical impact of a changing climate within a temporal horizon which permits policy actions to attenuate such an impact is much more problematic. Effects on plant and animal species of the projected temperature increase of 1 to 3 degrees Celsius over the next century are likely to vary substantially, and will depend on numerous variables. These include the ability and speed at which species can adapt or migrate, the change in other climatic variables such as precipitation, and the impact of future development on habitat and resource availability. As a result, the effects of climate change as well as the adaptation of individual species inevitably will vary substantially, especially as one moves to considering biodiversity linkages to climate change on a global scale.

OVERVIEW OF ECONOMIC RESEARCH ON CLIMATE CHANGE

Much of the research attempting to assess the economic implications of climate change has focused on impacts to agriculture and forestry. This can be attributed to two factors. First, agriculture and forestry are two of the primary, natural resource-based industries in much of the world and by their nature could be affected significantly by any change in environmental conditions. Moreover, because most of the products from agriculture and forestry are market goods, they are easily evaluated using existing economic tools.

Evaluating the economic impacts of climate change on agriculture is relatively straightforward, although not devoid of the need to make assumptions whose empirical validity is somewhat debatable. The primary effect of a changing climate on agriculture is an increase or decrease in crop yields as a result of temperature and precipitation variations. Evaluations of the global impact of climate change on agriculture is mixed, with some regions of the world potentially benefiting in terms of agricultural production while others will experience reduced yields and incomes. On balance, projections based on Global Circulation Model scenarios suggest that commodity yields from temperate climates are likely to be adversely affected by climate change. Yields in the European Community, for example, are projected to decline by 20 percent (Kane *et al.* 1989). Conversely, if agriculture displaces boreal forests, northern latitude countries could experience selective increases in agricultural yields depending on actual land allocation patterns. Given existing production patterns, however, overall agricultural yields are projected to decline by approximately 7 percent, resulting in an annual economic loss of $40 billion on a global basis (Cline 1992).

Evaluating the impacts of climate change on forestry is somewhat more problematic for several reasons. First, management practices are not as uniform for timber production as they are for agriculture. Moreover, forests produce a variety of market and non-market goods. As above, since estimating the impacts on market goods is much more straightforward than evaluating non-market goods and services, most of the work to date has focused on timber production impacts rather than on products such as recreation or aesthetics (Regens *et al.* 1989).

Generally, the impacts on forestry have been estimated by first determining the theoretically plausible shift in the range of commercially important timber species such as loblolly pine or Douglas- fir attributable to climatic changes. Once the new ranges have been delineated, changes in stand productivity, management practices and harvesting costs, and mill location must be projected. Based on such data, as well as the resulting changes in employment patterns, the economic impact of climate change can be estimated (Hodges *et al.* 1992).

VALUING BIODIVERSITY

Estimating the value of biological diversity at a fixed point in time and, by extension, the value of the impacts of climate change on biodiversity, is much more difficult analytically than is the intellectual challenge posed in terms of assessing timber production or agricultural commodities. The first problem involves operationalizing what constitutes the economic value of reductions in biological diversity. That is, is it distinct from the value of biological resources *per se*? Flint (1992), for instance, argues that "*The value of biological diversity is the difference between the current or future value of a diverse range of genes/species/ecosystems and the value of a less-diverse range. It is not the gross value of all naturally derived goods and services.*" Such a conceptualization differs considerably from those used in previous estimates of biodiversity's economic value. Nonetheless, Flint's definition more closely resembles the question that policy makers must face when selecting alternative courses of action regarding biodiversity. That is, what are the trade-offs, beyond simply the existence of a given species, between selecting a policy that reduces biodiversity and one that maintains biodiversity?

Unfortunately, from a biophysical standpoint, answering this question in terms of biodiversity has proven to be impossible, given the current level of biological information. As the preceding section demonstrated, empirical data as opposed to qualitative appraisals suitable for delineating biological diversity, let alone the impacts of climate change on biodiversity, are sparse. As the estimates of the number of species and the rate of species loss that were mentioned above indicate, scientists have not been able to answer the most basic questions with any degree of precision. We still do not know how the loss of some species in a community will influence the survival of other species (Randall 1986). Thus, any policy deliberations regarding biodiversity must contend with a large degree of uncertainty, including what the outcomes of various alternatives may be as well as the values of these outcomes (Batie and Shugart 1989).

CURRENT ECONOMIC ANALYSIS TOOLS

Environmental economists have been developing techniques such as hedonic prices or travel cost models for valuing non-market goods and services for several decades. Their application has become more prevalent as outdoor recreational activities such as sport fishing or other non-market goods such as air quality have become important amenities.

For purposes of quantitative analysis, economists traditionally equate value with the satisfaction derived by an individual from a particular good or service. For private exchanges of goods and services (i.e., market transactions), it is possible to assign directly prices so that the price of many commodities can be equated to the satisfaction or benefit that commodity provides. With public or collective goods such as biodiversity or environmental quality, direct measures of price typically are absent so it becomes necessary to develop proxies. As a consequence, economists have been forced to develop alternative means of estimating the satisfaction or value of these non-market goods. Those environmental benefits can be allocated to three categories of values: Use; Option; and Existence.

Use value is defined as the benefit derived from using the particular good or service. This may be in a consumptive use as in the case of most market goods such as an automobile or in a non-consumptive use such as bird-watching. Flint (1992) recognizes use value from biodiversity as arising in three areas: agriculture, medicine, and industry. Many of these benefits are well known and include such items as improved crop varieties and unique chemical compounds used in medicine. In terms of policy deliberations, however, the problem still remains of identifying the increment of the total value that can be attributed to a particular policy alternative.

Option value is defined as the amount that an individual or society would be willing to pay to preserve the option of utilizing a resource in the future. A related concept is the quasi-option value that represents the value of choosing not to take irreversible actions if new information is expected to become available in the future (Arrow and Fisher 1974). Such a value could be particularly important in valuing biodiversity, given the lack of currently available information. In all likelihood, policy deliberations will involve situations in which a primary concern is to maintain as much flexibility as possible for future activities.

The third category of benefits is the existence value, which is defined as the satisfaction derived from the knowledge that a resource exists whether it is ever used or not. For example, one might derive satisfaction simply from knowing that an endangered species exists. Existence values have been used extensively in debates over the value of remote wilderness areas or certain wildlife species.

Option and existence values can be estimated based on revealed preferences. These preferences may be based on past decisions or hypothetical situations. In the case of past decisions, economists rely on the costs incurred in earlier decisions to protect non-market goods. Numerous environmental values have been estimated by such means. Cline (1992) argues that the economic value of the wages and resulting economic activity foregone to protect portions of the U.S. Pacific Northwest could represent the economic value of the Northern Spotted Owl. That is, society determined that the value of protecting the owl outweighed the value of the economic gains.

The assumptions implicit in such an approach have clear limitations. As the Pacific Northwest decision clearly illustrated, defining foregone opportunities is extremely difficult. Several analysts estimated the potential job and economic loss associated with the Spotted Owl. These estimates varied substantially, based on the assumptions on which the analysis was based. Successfully utilizing information on the economic value of past decisions also requires that a similar case exists. In the case of biological diversity, past decisions are almost nonexistent or may have limitations to how much they can be generalized. As a result, accurately predicting current, non-consumptive use values as well as future use **and non-use values** based on prior decisions is somewhat tenuous.

The second approach to estimating option and existence values, the Contingent Valuation Method (CVM), appears to be a more promising technique for valuing biodiversity in the context of public policy deliberations. First, much of the policy debate surrounding biodiversity reflects differences in preferences which to some degree are antonymous from existing scientific understanding of the genetic, environmental, and ecological processes associated with biodiversity. How then, can we obtain quantitative estimates of those preferences? Because the CVM approach to valuing biodiversity directly elicits responses from individuals, it offers a technique for assigning monetary values to the non-consumptive use, option, and existence components of the overall benefits which discrete individuals, and by implication societal aggregates, assign to biodiversity. Individuals are asked how much they are "willing to pay" for preventing or mitigating environmental damages. Willingness to pay serves as a proxy for the benefits value. CVM typically uses either an iterative bidding process or a referendum format (i.e., simple yes/no for a fixed price) to establish how much an individual would pay.

FUTURE RESEARCH NEEDS

Based on the preceding review of the current state of the art for economic valuation of biodiversity and climate change, three topics merit additional research: option and existence values, inter-generational considerations, and basic biodiversity information such as resource inventories. The first primarily requires work by economists to develop methods for more accurately capturing environmental benefits. The second is both empirical and normative in nature since the issue of inter-generational considerations raises philosophical issues, especially in terms of discounting future benefits, and encompasses the humanities as well as social sciences. The third lies within the domain of linking the biological sciences to economic analysis.

Option and Existence Values

Although the Contingent Valuation method offers the most promise of existing tools for valuing biodiversity, several improvements could enhance its effectiveness. Because the starting point can influence how much individuals are asked what they would be "willing to pay," the use of iterative bidding seems inappropriate for valuing biodiversity. Instead, research should focus on the utility of the referendum format. Moreover, since results can be sensitive to framing and other aspects of instrument design, strategies to net out or eliminate such bias need to be tested in a variety of settings. Similarly, CVM estimates can be sensitive to strategic bias as well as possible free-riding. Individuals may inflate or deflate benefits estimates to emulate their policy preferences in an attempt to influence the total value. They also do not pay in reality since the "market" is a hypothetical one and real trade-offs in terms of disposable income are not involved. Finally, while the approach assumes that individuals have adequate information with which to respond, actual information levels for issues as complex and uncertain as biodiversity and climate change are likely to be limited. Research should focus on clarifying the extent to which these problems are tractable.

Inter-generational Considerations

An important issue with any long-term economic analysis is how to incorporate the concerns for future generations. Because we are dealing with a time period of a century or more with climate change, inter-generational considerations are a significant problem. Discount rates typically employed in conventional economic analyses range from 5 to 10 percent. Over the period involved with climate change, however, such discount rates would essentially reduce the benefits received accruing to future generations to zero. A dollar received 200 years from today as a benefit for current biodiversity protection, for example, would be worth only $0.000058 in present terms. This traditional approach poses a significant barrier to considering future generations adequately, and has been criticized by numerous economists in addressing such long term issues (d'Arge *et al.* 1982, Cline 1992, Randall 1986).

Several alternatives have been offered to deal with the question of discount rates for such analyses. Cline (1992) notes that several economists have argued that zero is the appropriate discount rate because, rather than involving an individual's time preference, analyses over long time periods require estimating the time preferences of a large number of individuals. Ultimately, the question is more philosophical than economic and will likely be resolved in the political arena. If a lowered discount rate is warranted, economists have developed several theoretically sound alternatives that could be employed (Cline 1992).

Basic Biodiversity Questions

The final and perhaps most critical research need for valuing biodiversity is obtaining the information required to address some basic questions regarding biodiversity conservation. Two of the more pertinent problems are (1) identifying biodiversity conservation priorities and (2) determining the "optimal" level of biodiversity. Answering these questions would be simplified by utilizing economic principles in formulating the problem. At least as important, however, is obtaining the biological information needed to address the two problems.

Identifying conservation priorities poses the most difficult problem from a social science perspective. While setting priorities involves explicitly identifying species and ecosystems that could be sacrificed, it does focus attention on the more critical resources. Failure to prioritize will still lead to reduced diversity but the sacrifices will not necessarily be those species or ecosystems that are the least important. The obvious problem is identifying the most critical resources. A confounding problem identified by Flint (1992) in dealing with global issues like biodiversity is resolving the different priorities of developed and developing countries. Before answers to the social preference questions can be answered, however, basic biological knowledge on the productive potential and the sustainability of alternative levels of diversity must be developed. Priorities cannot be determined until information regarding the biological response to alternative actions is available.

Similarly, the amount of biological diversity that must be maintained depends heavily on basic biological information. We must have a better understanding of the current level of diversity within individual ecosystems, as well as how the ecosystem and its productive capacity is affected by a reduction in diversity. With such information, economists could develop some realistic boundaries on the value of biodiversity.

CONCLUSIONS

In summary, the role of economics in the biodiversity debate is very limited at the present time. The tools exist to value non-market goods such as those provided by biological diversity, although they must be enhanced to evaluate some of the benefits provided by a diverse environment. Moreover, the problem of accounting for future generations must be resolved before realistic and acceptable values can be determined. The principal obstacle to providing useful information for decisions regarding diversity remains the lack of data on current levels of diversity on an ecosystem basis and the response of ecosystems and species to a marginal reduction in that diversity.

The primary contribution of economics to policy deliberations concerning biological diversity and climate change is providing a means of explicitly identifying the trade-offs of various alternatives. Too often, these trade-offs are only implicitly identified or ignored. Economic analysis should not be viewed by economists or non-economists as providing the final answer on such issues. Like the biological sciences, it represents one source of information that policy makers can utilize in determining the final course of action. The final decision involves many more considerations than the biological or economic implications, and often is primarily based on these other factors - as the current controversies surrounding the forests of the U.S. Pacific Northwest and international whaling agreements demonstrate.

REFERENCES

Arrow KJ, Fisher AC (1974) Environmental preservation, uncertainty, and irreversibility. Quarterly J. Economics 55:313-319

Batie SS, Shugart HH (1989) The biological consequences of climate changes: an ecological and economic assessment. Pages 121-130 *in* Rosenberg NJ, Easterling WE, Crosson PR, Darmstadter J (eds). Greenhouse Warming: Abatement and Adaptation, Resources For the Future, Washington, DC

CGIAR (1987) Priorities and future strategies. TAC Secretariat. FAO, Rome

Cline WR (1992) The Economics of Global Warming. Institute for International Economics, Washington, DC

d'Arge RC, Schulze WD, Brookshire DS (1982) Carbon dioxide and intergenerational choice. American Economic Review 72(2):251-256

Flint M (1992) Biological diversity and developing countries. Pages 437-469 *in* Reader A, Markandya A, Richardson J (eds). Environmental Economics: St. Martin's Press, New York

Hodges DG, Cubbage FW, Regens JL (1992) Regional forest migrations and potential economic effects. Environmental Toxicology and Chemistry 11:1129-1136

Kane S, Reilly J, Buclin R (1989) Implications of the greenhouse effect for world agriculture commodity markets. US Department of Agriculture, Washington, DC

Randall A (1986) Human preferences, economics and the preservation of species. Pages 79-109 *in* Norton BG (ed.). The Preservation of Species: The Value of Biological Diversity. Princeton University Press, Princeton, NJ

Regens JL, Cubbage FW, Hodges DG (1989) Climate change and U.S. forest markets. Pages 303-309 *in* Topping JC (ed.). Coping With Climate Change. Climate Institute, Washington, DC

Wilson, EO (ed) (1988) Biodiversity. National Academy Press, Washington, DC

Chapter 23: Biodiversity Research Needs and Opportunities, the Role of International Scientific Organisations: the *Diversitas* Example

Talal Younès
IUBS
51 Blvd de Montmorency
75016 Paris
France.

INTRODUCTION

Before introducing *Diversitas*, the IUBS-SCOPE-UNESCO Programme on Biological diversity, in particular the role of international scientific organisations in identifying research needs and opportunities, it is important to recall how and why it was made possible for the topic of biological diversity, of vague and diffuse interest among scientists, to become, within less than a decade, a major theme of interest to the general public as well as to the policy and decision makers worldwide.

Indeed, the public perception of biological diversity moved from the simple appreciation of the variety and beauty of life forms present on our food tables or in collections of stamps, to concern for the protection of rare and endangered species. Public awareness vis-a-vis the loss of biological diversity and the need for its conservation is mainly due to the rapid conversion of tropical landscapes, both forests and savannas.

The emergence of many active "SAVE..." parties, such as "SAVE the Dolphins", "SAVE the Elephants", "SAVE the Forests", "SAVE the Earth" were quite realistically followed by a "SAVE the Money" party. This last group, in pointing out an important dimension of our lives, the economic one, brought together both the concerns for "environment" and for "development", and thus provided the central theme of the most important gathering in our recent history, the United Nations

NATO ASI Series, Vol. I 20
Biodiversity, Temperate Ecosystems, and Global Change
Edited by T. J. B. Boyle and C. E. B. Boyle

Conference on Environment and Development (UNCED), better known as the "Earth Summit", held in Rio de Janeiro, in June 1992.

To scientists, particularly biologists, the most relevant of the UNCED documents is undoubtedly the "Convention on Biological Diversity", now signed by 161 states, and which came into force at the end of 1993. The agreement on biodiversity contains provisions intended to ensure effective national action to curb the destruction of living species, habitats and ecosystems. But it also draws attention to the general lack of information and knowledge regarding biodiversity, as illustrated by the many myths that are still in existence, such as:

- We know how much biodiversity exists;
- It is all in the tropics;
- People have always reduced it;
- We know its function;
- Extinctions are unnatural;
- We can (and should) save every species;
- Protected areas are enough;
- A "hands-off" approach is the best protection;
- We know its economic value;
- Biodiversity is antagonistic to development;
- We cannot act until we know more.

Furthermore, the Convention on Biological Diversity underlines the urgent need to develop scientific, technical and institutional capacities to provide the basic understanding upon which to plan and implement appropriate measures. Several points need urgent attention by the scientific community:

1. the need to consider biodiversity as a building block for development;
2. the need to obtain better knowledge of species numbers, their distributions and rates of reduction and extinction;
3. the need to bring the best and most up-to-date information on biodiversity to the attention of policy makers; and
4. the need to step up greatly education and awareness efforts on biodiversity.

This implies that studies on biodiversity need to be not only increased, but more especially reconsidered in the global time and space scales in order to answer the needs of land managers, politicians, teachers, non-governmental action groups and the private sector, etc. Information needs to be collated and processed so that it can be used most efficiently and effectively by these different user-groups.

In order to meet the challenges outlined above, studies on biodiversity need to be more consolidated and coherent, and to follow a given direction. If action is not taken along these lines, scientific work on biodiversity will continue to be dispersed and weak. It is necessary to bring together ongoing and past studies on biodiversity and review and revise them as appropriate in a global context, thus stepping up the time and space scales. *Diversitas*, the IUBS-SCOPE-UNESCO Programme on Biological Diversity has been launched with a great responsibility, that is to pave the "road after Rio".

DIVERSITAS - OBJECTIVES, MAJOR THEMES AND PRIORITY HYPOTHESES

Diversitas, by its very definition, must cover the diversity of all living beings on the planet. Our major objective in *Diversitas* is to identify the scientific questions and to promote collaborative research that require international cooperation, since we all know that good science starts and ends with good questions.

Major themes

The major *Diversitas* themes were first laid down at the IUBS-SCOPE meeting on "*Ecosystem Function of Biodiversity*", in 1989, at the US Academy if Science (NAS), in Washington, DC. They are as follows:

1. Ecosystem Function of Biodiversity;
2. Origins, Maintenance, and Loss of Biodiversity;
3. Biodiversity Inventory and Monitoring; and
4. Conservation of Wild Relatives of Cultivated Plants and Domesticated Animals.

In consequence, it will address everything from whales to microbes, and thereby cover two aspects that have been somewhat neglected up until present: marine biodiversity and microorganism biodiversity.

Scientific Questions and Hypotheses

The workshop "*From Genes to Ecosystems: A Research Agenda for Biodiversity*", held at the Harvard Forest Station, in June 1991, resulted in a broad synthesis and development of the outlines for a biodiversity programme. Four sets of priority hypotheses and recommendations were developed at genetic, species to community, and ecosystem levels, and to deal with the problem of inventory and monitoring species diversity and their changes around the world:-

a) Biodiversity at the intra-specific level (from genes to species)

- The consequences of intra-specific genetic diversity for biodiversity at higher levels (eight hypotheses);
- Spatial fragmentation, genetic bottlenecks and biodiversity (eight hypotheses);
- Speciation: the connection to micro- and macro-evolution (four hypotheses); and
- Biodiversity and life-history (two hypotheses).

b) Research issues for species diversity at the community level

- Functional redundancy, species diversity, and the stability of communities and ecosystems (eight hypotheses);
- Regional biogeography and local species richness (seven hypotheses);
- Disturbance, human impacts, and the structure of communities and ecosystems (five hypotheses); and
- Ecological and evolutionary differences between marine and terrestrial ecosystems.

c) Biodiversity from communities to ecosystems

- Biodiversity and ecosystem attributes (four hypotheses);
- Strategy for testing hypotheses (four hypotheses); and
- Observational requirements.

d) Monitoring and inventory of biodiversity

i) Documenting biodiversity patterns:

- Sampling and analytical approaches;
- Data collection;
- What to sample, how to sample, and where to sample;
- Data analysis and data quality; and
- Exploratory data analysis.

ii) Monitoring biodiversity:

- Rationale for a monitoring system for biodiversity;
- Global versus local monitoring;
- Proposal for setting up a minimum core network of sites for monitoring changes in biodiversity.

THEME 1. ECOSYSTEM FUNCTION OF BIODIVERSITY

As natural ecosystems are being increasingly affected by human activities, resulting in disruption of system interactions, losses of populations, and even species, there is increasing concern about how we are modifying the ecosystem processes that maintain these systems and that benefit humankind. At present there is no answer available to this question. It has been said that "*in the end, it is these ecological questions about the role of diversity that most need to be answered, not the numerical ones*".

The *Diversitas* theme "Ecosystem function of biodiversity" aims to synthesize our knowledge of the functional role of biodiversity and to help design an experimental programme. It has been designed to address two questions:

1. Does biodiversity "count" in system processes (*e.g.* nutrient retention, decomposition, production, *etc.*), including atmospheric feed-backs, over short and long-term time spans, and in the face of global change (climate change, land use change, species invasions)?
2. How is system stability and resistance affected by species diversity and how will global change affect these relationships?

Within the framework of the programme theme "Ecosystem function of biodiversity", a series of activities were planned for the period 1991-1994, culminating in an overall synthesis meeting at Asilomar, California, in 1994.

The first phase was launched at a meeting on background issues, held in October 1991, in Bayreuth, Germany. This meeting brought together ecologists and population biologists, both disciplines being directed toward evaluating the consequences of human-driven disruptions in natural systems. In particular, there was an examination of the degree of redundancy within systems, the ubiquity of keystone species, the tightness of species interactions, the resilience of systems to perturbations, and the interaction of landscape units. The proceedings were edited by Mooney and Schulze, for publication by Springer Verlag.

The second phase consists of a series of meetings focusing on specific biotic regions of the world. Twelve regions (estuaries, lagoons, and mangroves; Mediterranean; islands; boreal forest; tundra; coral reefs; savanna; up-welling systems; tropical forests; lakes and rivers; temperate forests; and arid zones) have been selected, representing particularly critical areas in terms of threats of diversity losses, or are particularly sensitive to global change effects, or are especially amenable to experimentation.

Within each of the above systems, the same questions will be posed in order to get a uniform treatment of the nature of the diversity of that system, to know how that particular system is being modified, and to learn of potentially differential structural/functional relationships among systems. Each regional meeting will address the following issues:

1. Natural diversity of systems: species, populations, functional groups, systems, and landscapes;
2. Impact of change on diversity: climate and atmosphere, land-use and invasions;
3. Assessing diversity's role on ecosystem function: additions (invasion analogue), subtractions (harvesting, disease, *etc.*), fragmentation, and disturbance;
4. Reconstructing and maintaining diverse systems; and

5. Refining our knowledge through explicit experiments and long-term observations.

Also, a number of meetings were organized to address special topics such as "Ecosystem Functional Units", "Ecosystem Reconstruction", and "Managing Biodiversity under Climate Change.

THEME 2. ORIGINS, MAINTENANCE, AND LOSS OF BIODIVERSITY

The second *Diversitas* theme, focusing on the genetic and population levels, will be developed following the conceptual framework outlined at the Harvard Forest workshop, 1991. The specific topics to be addressed in this theme include the following:

1. The origins of biodiversity;
2. The paeleo-records: the history of present biodiversity;
3. The dynamics of biodiversity at the population level;
4. The ecology and genetics of extinctions:
   - problems of measuring current and predicting future extinction rates,
   - genetic erosion, its measurement and consequences,
   - the IUCN established categories and other systems for determining threats to species and other taxa,
   - Minimum Viable Populations (MVPs) and Population Viability Analysis (PVA), and
   - metapopulation.
5. The dynamics of biodiversity at the ecosystem level; and
6. The role of humans in shaping biodiversity:
   - the effects of human management practices and harvesting techniques in the long-term perspective,
   - multiple time dimensions, including origins of agro-biodiversity and domesticates, and
   - introduced species.

A special emphasis will be put on the modelling approach to understand and possibly predict how species and genetic variability can affect the dynamics of species and population extinctions. Also, this theme will address the following particular points:

- the role and effects of parasite/host and mutualistic relationships on biodiversity changes;
- studying and quantifying extinction rates;
- linking biodiversity issues, particularly at the level of genetic variability and population dynamics, with global change concerns (greenhouse effects as well as land-use changes);
- population viability analysis and genetics and dynamics of metapopulations or fragmented populations in a changing environment; and
- the ecology and genetics of rarity: patterns, causes, and consequences.

THEME 3. MONITORING AND INVENTORY OF BIODIVERSITY: THE NETWORK APPROACH

Within the framework of *Diversitas*, the inventory and monitoring theme is quite different, by its nature, from the other ones, since biodiversity inventory and monitoring necessitate a closer interaction with countries, more funding, and a long-term commitment from national authorities and research organizations. Because of this, *Diversitas* is viewed as an enabling mechanism , enhancing existing individual or country efforts by providing standardisation of methodologies, and acting as a catalyst for future efforts. The major issues to be addressed here are:

- the taxonomic basis of inventory;
- the ecological dimensions of monitoring;
- the need for standard sampling techniques; and
- the adoption of a "network" approach.

The term "network" is used loosely here to refer to a group of sites and stations linked by a common understanding of purpose, and using comparable (if not identical) methods, and for which the data generated are subject to global analysis. "Networking" is ensured by the sharing of information and experience, exchanges of personnel, and regional and global meetings of participating scientists and representatives of their institutions. Five biomes were selected for the establishment of the network: tropical rainforests, tropical savannas, temperate rainforests and temperate deciduous forests, temperate grasslands, and tundra.

The use of a scaling approach is also necessary throughout this programme theme. There are several reasons for this, including the new awareness of the global coverage of biodiversity and the global impact of human activities upon it, the appreciation of the hierarchical view of life, from the DNA molecule to the landscape, and the seeming acceleration of history, such that rates of extinction are estimated to be 1,000 times faster than the norm over evolutionary time. Space scaling will consist of combining extensive and intensive biodiversity inventories using a suite of comparable sites (in the same biome), and similar sampling methodologies should provide a holistic view at the global level. The sites for intensive inventory form the controls against which the more extensively treated sites can be calibrated, and these extensive sites are the replicates, covering different spatial areas.

Also, time scaling is essential. As is true with most biological sciences, there is a need to reinstate the role if history and the role of change with time in studies of biodiversity.

MARINE BIODIVERSITY

Concepts related to biodiversity are mostly derived from terrestrial studies where, on a global scale, there is a clear pattern of increasing diversity from the poles to the tropics. In the marine realm this is not the case, and certain groups of organisms may reach maximum diversity in polar regions (*e.g.* mammals), temperate regions (*e.g.* seaweeds), or the tropics (*e.g.* corals, mangroves). For most animal phyla the pattern is simply not known because strictly comparative data are missing.

The focus of conservationists in the oceans is mainly on individual large species such as whales, dolphins, turtles, and, to a lesser extent, some of the larger sharks. In fact, for the marine environment, only now is biodiversity beginning to be described and the implications of its differing facets being appreciated.

For the moment, we are far from being able to understand the importance of biodiversity to the functioning of ecosystems and their processes. An in-depth consideration of this problem is necessary since it could have important implications for the management of marine ecosystems and the way in which they are eventually exploited. Furthermore, the scales of the processes determining the distributions of many marine species are considerably larger than those on land. Conservation of marine ecosystems and processes thus becomes a rather different problem from the conservation of individual marine species.

The recent discovery of very small plankton (picoplankton) has major implications for the primary production and turnover in oceans. These widespread unicellular pico-eukaryotes, cyanobacteria, and pro-chlorophytes appear to play a significant role as a sink for atmospheric carbon dioxide.

Except for certain marine areas where habitat alteration is extensive (coastal wetlands, mangroves, lagoons, for example), species extinction does not appear to be as great a problem as for land areas due to the greater ranges and fecundity of most marine species.

BIODIVERSITY OF MICROORGANISMS

Microorganisms exhibit the greatest breadth of genetic diversity on Earth, and are essential for the survival of all organisms, as basic components of food chains, and as mutualists. Less than 5% of the world's microorganisms have yet been described, and it is not improbable that their real number on earth exceeds even that of the insects.

Microorganisms, *i.e.* algae, bacteria (including the cyanobacteria and mycoplasmas), fungi (including lichens and yeasts), protozoa, viroids, and viruses, are vital to the function and maintenance of the Earth's ecosystems and biosphere. As major contributors in biogeochemical cycles, they perform unique and indispensable activities in the circulation of matter, on which all larger organisms, including humans, depend. They constitute a genetic resource of great potential for contributing to the sustainable development of the planet as well as human, animal, and plant health. Urgent attention to redress our ignorance of many key aspects of scientific understanding of their distribution and functions is required. In collaboration with the International Union of Microbiological Societies (IUMS), a series of activities were undertaken within the framework of *Diversitas* in relation to microorganism diversity.

CONCLUSION

Biodiversity represents the ultimate result of life's evolutionary processes and mechanisms over very large time and space scales. In addressing the very complex and multi-faceted issues of biodiversity conservation and management, we urgently need to fill the many gaps that still exist in our knowledge of biodiversity function, extinction, and distribution.

Although *Diversitas* involves geneticists, physiologists, population biologists and ecologists, the contribution of taxonomists is crucial. The programme must address the question of removing the "taxonomic impediment" *i.e.* the acute shortage of trained taxonomists, the lack of specialists for many important groups of organisms, and the existence of only a small number of reference collections which are, in most cases, far from the place of origin of the organisms being studied.

Moreover, *Diversitas* should try to demonstrate the viability of new types of partnerships, advocated by UNCED, between intergovernmental and non-governmental organizations in taking up this challenge. It should be able to take up the opportunities of new funding sources for biodiversity. The partnership exhibited through *Diversitas* provides a good example of how to draw on the strengths and complementarity of non-governmental and intergovernmental scientific organizations, *e.g.* IUBS, SCOPE, and UNESCO. Also, a partnership has been developed with ICSU and its family, such as the IUMS, together with a number of international organizations, in particular with UNEP, for the preparation of the "*Global Biodiversity Assessment*" (GBA) as a follow-up to the Convention on Biological Diversity.

In many countries, national initiatives are being developed and designed in close collaboration with the *Diversitas* programme, for example in China, France, Japan, Switzerland, and the USA. Furthermore, to engage a most needed dialogue between scientists, key economic actors, policy makers, and media experts, concerned with biodiversity and development, an international conference "*Biodiversity and*

Development: Towards a New Partnership", co-sponsored by the IUBS and UNESCO, will be organized in 1994.

RELEVANT LITERATURE

Di Castri F, Younès T (1990) Ecosystem function of biological diversity. Biology International, Special Issue 22

Di Castri F, Robertson-Verhnes J, Younès T (1992a) Inventorying and monitoring biodiversity. Biology International, Special Issue 27

Di Castri F, Robertson-Verhnes J, Younès T (1992b) The network approach for understanding global biodiversity. Biology International 25: 3-9

Grassle JF, Lasserre P, McIntyre AD, Ray GC (1991) Ecosystem function of marine biodiversity. Biology International, Special Issue 23

Hawksworth DL, Colwell RR (1992) Biodiversity amongst microorganisms and its relevance. Biodiversity and Conservation 1: 219-356

Lugo A (1988) Diversity of tropical species: questions that elude answers. Biology International, Special Issue 19

Maury-Lechon G, Hadley M, Younès T (1984) The significance of species diversity in tropical forest ecosystems. Biology International, Special Issue 6

Mooney H, Schulze ED (in press) Biodiversity and ecosystem function. Springer Verlag, Berlin

Simpson B (1988) Biological diversity in the context of ecosystem structure and function. Biology International 17: 15-17

Solbrig OT (1991) Biodiversity: scientific issues and collaborative research proposals. MAB Digest 9

Solbrig OT, Nicholis G (eds.) (1991) Perspectives on biological complexity. IUBS Monograph Series

Solbrig OT (ed.) (1991) From genes to ecosystems: a research agenda for biodiversity. IUBS Monograph Series

Solbrig OT, van Emden H, van Oordt P (eds.) (in press) Biological diversity and global change. IUBS Monograph Series

The Economist (1991) Species galore; and Counting species. The Economist 7724 (14-20 Sept. 1991): 17, and 99-100

UNESCO (1992) *Diversitas*: IUBS-SCOPE-UNESCO Program on biodiversity. Information leaflet

World Resources Institute, World Conservation Union, United Nations Environment Programme (1992) Global Biodiversity Strategy: guidelines for action to save, study, and use Earth's biotic wealth sustainably and equitably. World Resources Institute, Washington, DC

Chapter 24: Conclusions, and a Research Agenda for Biodiversity, Temperate Ecosystems, and Global Change

T.J.B. Boyle
Canadian Forest Service
351 St. Joseph Blvd.
Hull, Quebec, Canada

On the final day of the NATO Advanced Research Workshop, research priorities were discussed in four working groups, which dealt with:

- The impact of global change on populations and species,
- The impact of global change on ecosystems and landscapes,
- The relationship between biodiversity and sustainability, and
- International cooperation in research.

After discussion in the working groups, a plenary session addressed issues related to linkages across two or more of the groups.

RESULTS

There was agreement in all working groups that although additional fundamental research on the relationship between biodiversity and global change is important, the fact that impacts of global change are already well advanced in temperate ecosystems demands that top priority be given to research on topics that can have a quick or immediate impact. For example, rapid methods of measurement and assessment are essential for quantifying effects of global change on ecosystems, but such methods have not yet been developed or, if developed, are not widely used.

There was agreement that the goal for managing ecosystems under conditions of global change should be the maintenance of ecological functions. Conservation of biodiversity is therefore <u>not</u> a goal in itself, but stable levels of biodiversity are indicators of a "healthy" ecosystem, in which ecological functions are being

NATO ASI Series, Vol. I 20
Biodiversity, Temperate Ecosystems, and Global Change
Edited by T. J. B. Boyle and C. E. B. Boyle

maintained. Stated another way, ecological sustainability is dependent on conserving biodiversity. This generated much discussion on the meaning of ecological sustainability, and the distinction between ecological and economic sustainability. There was agreement that ecological and economic sustainability are not synonymous but that, in the long term, economic sustainability is dependent on ecological sustainability. There was also substantial discussion on the question of whether ecological sustainability is synonymous with ecological stability, the view of most participants being that they are synonymous.

The main conclusions of the four discussion groups are summarised below:

Populations and species

i) The highest priority should be given to the development of methods for rapid assessment. Such methods may involve remote sensing data, or protocols for ground-based data collection. In the latter case, censusing of "indicator species" is a technique that is already utilised in some ecosystems. However, research is required on what species are suitable as indicator species, and the most urgent need is for those ecosystems at greatest risk.

ii) High priority should also be given to understanding what are the vital attributes of species which govern their role in ecosystem functioning and, again, to develop methods for assessment of these attributes. The concept of "keystone species", although not unanimously accepted as a useful concept, is clearly related to this issue. It should be noted that keystone species will not necessarily be, and indeed most likely will not be, indicator species, and *vice versa*.

iii) The effect of influences at higher levels in the biological hierarchy on the attributes of species needs to be better understood. Changes at the ecosystem level, and even at the landscape level, will affect the role of species in the ecosystem. Because it is the usual situation that changes are effected at the landscape or ecosystem level, which subsequently affect species and populations, these interactions must be understood if accurate predictions of the consequences of global change are to be possible.

iv) Although impacts on species and populations are typically initiated at higher levels in the hierarchy, it is the genetic diversity which determines the response. Different species (or populations) will have differing adaptive potentials, depending on their genetic variability and structure. Quantification of adaptive potentials in evolutionary and genetic terms is therefore also needed to predict responses to change.

v) There is an enormous existing store of data, and data collection capability already far exceeds analytical capability. Therefore, synthesis and analysis of existing data is far more useful, in order to make rapid progress, than supporting the collection of new data.

Ecosystems and Landscapes

i) Description and inventory of ecosystems is a high priority, and this requires the development and use, at least within geographic regions, of common protocols, data standards, and classification systems.

ii) Empirical and/or quantitative studies are required on selected ecosystems - those that are of greatest significance (in a parallel concept to "keystone species"). Such case studies can be used to build, and subsequently test theories on the impacts of global change. Suitable empirical/quantitative methods must first be developed or, where already developed, must be applied consistently in order to allow cross-comparisons.

iii) Vulnerability/sensitivity analyses are also required on selected ecosystems - those that are most indicative of change (in a parallel concept to "indicator species"). The material balance type of approach may be a useful methodology.

iv) Risk assessment needs to be more sophisticated and comprehensive. "Environmental impact assessments" are required in many countries for policies and programmes, but such assessments do not usually have an adequate scientific basis. For example, an assessment of the risks associated with exotic species requires predictive models of species-environment interactions, and these do not yet exist, or are rudimentary.

v) Scenarios for trade-offs between conservation and exploitation need to be undertaken, in order to assess the likely consequences of such decisions.

vi) Finally, it needs to be remembered that categorization of biodiversity, especially at the level of ecosystems and landscapes, is to some extent arbitrary. Consequently, the interaction among different levels is important to consider.

Biodiversity and Sustainability

i) Elucidation of the relationship between biodiversity and sustainability was recognized as important. Figure 1 shows several possible relationships between sustainability and diversity. Depending on whether the actual relationship is more like line A or line D will obviously be significant in terms of the impact of reductions in diversity on ecological sustainability.

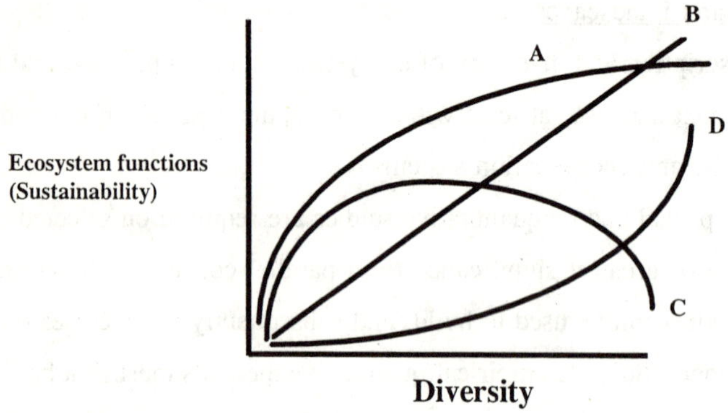

Figure 1. Possible relationships between biodiversity and ecological functions, or sustainability.

ii) In examining the linkage between biodiversity and sustainability, some methods are already available. These need to be assessed in terms of their adequacy, but it is certain that new methods will also need to be developed.

iii) The development of flexible management systems that maintain biodiversity without requiring an understanding of which relationship applies, was identified as a having a very high priority, because the time taken to determine the correct relationship may compromise future efforts to maintain ecological sustainability.

International cooperation

i) Reasons for international cooperation were identified as including the following:

- pooling of expertise and resources facilitates and improves research;
- spatial scale of many ecosystems and disturbances transcends political boundaries; and
- scientific training and capability is variable, so transfer of knowledge, and capacity building is important.

ii) Mechanisms for international cooperation may be by:

- funding of research - through bi-lateral projects, the Global Environment Fund (GEF), UNEP, international NGOs, or regional organizations such as the IAI, CSCE, NATO, etc.; and
- information exchange - for which agencies such as UNEP, UNESCO, IUBS and IUCN, as well as regional organizations, are well equipped.

iii) Subjects for international scientific cooperation include:

- species inventories for different ecosystems, including taxonomy and life history description;
- ecosystem characterization, through comparative studies using common methodologies, and historical land use analysis;
- model development and validation, including both biophysical and socio-economic issues; and
- monitoring, through cooperation on data collection networks and the development of quality control/assurance protocols.

Linkages and Over-riding Issues

A conceptual framework for research on the impacts of global change on biodiversity is multi-dimensional. One dimension is the different levels of biodiversity, another consists of the different biogeographical and/or socio-economic regions, a third consists of the various global change issues, and a fourth is made up of the objectives of research activities, in dealing with impacts, processes, etc. Table 1 depicts, for any biogeographical/socio-economic region, and for any global change issue, the remaining two dimensions. For each level of biodiversity, management goals are identified, together with major indicators related to those goals, impacts or consequences of change, and possible research methods.

As discussed above, it is important to remember that the distinctions among levels of biodiversity are somewhat arbitrary, and as a result, so too are the divisions among goals, indicators, and impacts. One goal applies to all levels of diversity, and this is the goal of <u>maintaining adaptability and elasticity/resilience</u>. Meeting this goal will ensure both that the impacts of global change are kept within tolerable limits, and that adaptation to new conditions is possible. The indicators proposed at each level also include some overlap and duplication, such that the total effort required is less than is apparent from the Table.

Table 1. Scientific requirements for each biogeographic/socio-economic region, and for each global change issue.

Level	Goals	Major Indicators	Major Impacts	Research Methods
Landscape	• Maintain adaptability, elasticity, and resilience • Sustainable productivity and development	• Presence/absence of critical ecosystems • Presence/absence of threatened ecosystems • Ecosystem diversity • Landscape heterogeneity	• Loss of ecosystems • Fragmentation/ degradation • Change of disturbance regime	Landscape variability analysis
Ecosystem/ community	• Maintain adaptability, elasticity, and resilience • Maintain ecosystem health	• Presence/absence of critical species • Presence/absence of threatened species • Community and species diversity • Nutrient/chemical concentrations in steady-state	• Loss of habitat • Soil degradation/ erosion • Destruction of food webs • Decoupling of material cycling • Fragmentation of species ranges • Change in disturbance regime	Material balance
Population/ species	• Maintain adaptability, elasticity, and resilience • Conservation of biodiversity	• Presence/absence of critical species • Presence/absence of threatened species • Genetic diversity • Evolutionary adaptability	• Extinction • Increased isolation • Increased inbreeding and loss of genetic diversity	Population viability analysis

INDEX

E

F

The ASI Series Books Published as a Result of Activities of the Special Programme on Global Environmental Change

---

This book contains the proceedings of a NATO Advanced Research Workshop held within the activities of the NATO Special Programme on Global Environmental Change, which started in 1991 under the auspices of the NATO Science Committee.

The volumes published as a result of the activities of the Special Programme are:

Vol. 1: **Global Environmental Change.** Edited by R. W. Corell and P. A. Anderson. 1991.
Vol. 2: **The Last Deglaciation: Absolute and Radiocarbon Chronologies.** Edited by E. Bard and W. S. Broecker. 1992.
Vol. 3: **Start of a Glacial.** Edited by G. J. Kukla and E. Went. 1992.
Vol. 4: **Interactions of C, N, P and S Biogeochemical Cycles and Global Change.** Edited by R. Wollast, F. T. Mackenzie and L. Chou. 1993.
Vol. 5: **Energy and Water Cycles in the Climate System.** Edited by E. Raschke and D. Jacob. 1993.
Vol. 6: **Prediction of Interannual Climate Variations.** Edited by J. Shukla. 1993.
Vol. 7: **The Tropospheric Chemistry of Ozone in the Polar Regions.** Edited by H. Niki and K. H. Becker. 1993.
Vol. 8: **The Role of the Stratosphere in Global Change.** Edited by M.-L. Chanin. 1993.
Vol. 9: **High Spectral Resolution Infrared Remote Sensing for Earth's Weather and Climate Studies.** Edited by A. Chedin, M.T. Chahine and N.A. Scott. 1993.
Vol. 10: **Towards a Model of Ocean Biogeochemical Processes.** Edited by G. T. Evans and M. J. R. Fasham. 1993.
Vol. 11: **Modelling Oceanic Climate Interactions.** Edited by J. Willebrand and D. L. T. Anderson. 1993.
Vol. 12: **Ice in the Climate System.** Edited by W. Richard Peltier. 1993.
Vol. 13: **Atmospheric Methane: Sources, Sinks, and Role in Global Change.** Edited by M. A. K. Khalil. 1993.
Vol. 14: **The Role of Regional Organizations in the Context of Climate Change.** Edited by M. H. Glantz. 1993.
Vol. 15: **The Global Carbon Cycle.** Edited by M. Heimann. 1993.
Vol. 16: **Interacting Stresses on Plants in a Changing Climate.** Edited by M. B. Jackson and C. R. Black. 1993.
Vol. 17: **Carbon Cycling in the Glacial Ocean: Constraints on the Ocean's Role in Global Change.** Edited by R. Zahn, T. F. Pedersen, M. A. Kaminski and L. Labeyrie. 1994.

Vol. 18: **Stratospheric Ozone Depletion/UV-B Radiation in the Biosphere.** Edited by R. H. Biggs and M. E. B. Joyner. 1994.

Vol. 19: **Data Assimilation: Tools for Modelling the Ocean in a Global Change Perspective.** Edited by P. O. Brasseur and J. Nihoul. 1994.

Vol. 20: **Biodiversity, Temperate Ecosystems, and Global Change.** Edited by T. J. B. Boyle and C. E. B. Boyle. 1994.

Vol. 21: **Low-Temperature Chemistry of the Atmosphere.** Edited by G. K. Moortgat, A. J. Barnes, G. Le Bras and J. R. Sodeau. 1994.

Vol. 22: **Long-Term Climatic Variations – Data and Modelling** Edited by J.-C. Duplessy and M.-T. Spyridakis. 1994.